HANDBOOK OF ELECTRONIC INSTRUMENTATION, TESTING, AND TROUBLESHOOTING

HANDBOOK OF ELECTRONIC INSTRUMENTATION, TESTING, AND TROUBLESHOOTING

Vester Robinson

RESTON PUBLISHING COMPANY, INC., Reston, Virginia 22090

A Prentice-Hall Company

Library of Congress Cataloging in Publication Data

Robinson, Vester.
Handbook of electronic instrumentation, testing, and troubleshooting.

Bibliography: p
1. Electronic instruments. 2. Electronic measurements. 3. Electronic apparatus and appliances–Maintenance and repair. I. Title.
TK7878.4.R6 621.381'028 74-9657
ISBN 0-87909-327-7

THIS BOOK IS DEDICATED
to MY WIFE and HAPPINESS
FLORENCE BELLE

A Prentice-Hall Company

Box 547

Reston, Virginia 22090

10 9 8 7 6 5 4 3 2 1

Printed in the United States of America.

PREFACE

This book is intended to meet four basic needs of electronic technicians, engineers, and experimenters. These are:

1. To serve as a guide in choosing measuring instruments to fit the specific work situation.

2. To describe the available instruments as to purpose, function, and principles of operation.

3. To develop procedures and test circuits showing practical applications of the instruments described.

4. To provide a condensed, logical, timesaving method of troubleshooting with techniques that have proved consistently reliable.

The text is divided into three parts. Part I, Electronic Instrumentation, covers a variety of electronic instruments grouped according to their principal purpose. Many sample instruments are described in considerable detail along with photographs and functional circuits. It is, of course, impossible to include all types of electronic instruments. The coverage is therefore concentrated on the instruments most likely to be needed in laboratory work, electronic repair shops, and system maintenance. Where there is a distinct difference in general purpose and laboratory equipment, this is pointed out with sample instruments for both applications.

Part II, Testing Practices, utilizes the instruments described in Part I to explain many practical applications. A variety of test circuits are developed, with instructions as to what instruments to use, how to make the connections, and what indications to expect. Many of the tests and measurements are common to nearly all practical electronic situations, but others are of a more specialized nature. The test circuits and instrument applications range from measuring current in a basic dc circuit to analyzing the spectrum of a pulse-modulated carrier wave.

Part III, Techniques for Maintenance, Troubleshooting, and Malfunction Analysis, describes procedures of interest to all persons engaged in electronic activities of any nature. The techniques of maintaining electronic systems are covered in a condensed but very practical fashion. Any engineer or technician is likely to become involved with a system at some point in his career. The principles covered here will enable him to grasp the job quicker and to perform in a more professional manner. A six-step, time-honored troubleshooting procedure is developed in detail. Following these steps will improve the efficiency of the troubleshooter regardless of his range of experience. These techniques are logical, and they are designed to locate all types of trouble. They work consistently, and in the long run they save many frustrating hours that are likely to occur with a less scientific approach. Maintenance is covered from the standpoint of both preventive and corrective activities, and detailed malfunction analysis is conducted on several common circuits. In this analysis the circuit is covered from all angles—first assuming a possible trouble symptom, and then analyzing the circuit for all malfunctions that could possibly cause this symptom.

Additional support information is provided in the appendices in the form of conversion tables, color codes, frequency allocations, modulation index, and transistor data. More than 300 illustrations are used to clarify the text, including many revealing photographs. A detailed, cross-referenced index provides quick access to any desired information. A bibliography is also included.

A great many people have contributed to the contents of this book. My instructors, students, and associates over the past 30 years have unconsciously made many valuable contributions. Unfortunately there is no way to thank each one individually, but their efforts are appreciated nonetheless. Several manufacturers of commercial electronic instruments have been very generous both with literature and illustrations. An expression of gratitude is extended to all these companies and especially to the following:

American Electronic Laboratories
General Radio
Hewlett Packard
Honeywell
Radio Corporation of America
Sencore
Simpson Electric
Superior Manufacturing and Instrument
Tektronix

A special thanks to Matt Fox who encouraged the writing of this book and rendered much valuable assistance and encouragement along the way.

Vester Robinson

CONTENTS

Part III

TECHNIQUES FOR MAINTENANCE, TROUBLESHOOTING, AND MALFUNCTION ANALYSIS, 301

part one

ELECTRONIC INSTRUMENTATION

This part, consisting of Chapters 1–8, illustrates and describes the functions of basic test circuits and how these circuits are combined into complex instruments. It contains diagrams and functional descriptions of typical instruments. Operating techniques to avoid erroneous measurements are discussed. Safety precautions for the protection of persons and equipment are explained.

chapter one

BASIC INSTRUMENTS

This chapter treats such common devices as ammeters, voltmeters, ohmmeters, and multimeters. Both dc and ac versions of these meters are covered. The most common meter movements are discussed along with typical meter circuits. In general terms, these meters are grouped as either electric instruments or electronic instruments. The electric instruments are composed of all passive components such as resistors, inductors, and capacitors. The electronic instruments also use passive components, but in addition they incorporate one or more active components in the form of either electron tubes or solid-state devices. Both types are discussed in this chapter.

METER MOVEMENTS

A meter is composed of meter movements, circuits, a dial, and a case with suitable terminals. The movement consists of the mechanism that enables the pointer to move across the scale when it is activated. The three most popular movements are D'Arsonval, taut band, and dynamometer.

D'Arsonval Movement

The D'Arsonval movement is based on the principle of a moving electromagnet pivoted in a uniform air gap between the poles of a fixed, permanent magnet. This principle is illustrated in Figure 1–1. With polarities as shown, there is a repelling force between like poles that applies torque to the pivoted coil. The torque is proportional to the magnitude of the current being measured. The pointer is secured to the electromagnet and swings across the dial as the coil turns.

Figure 1–2 provides a better picture of the meter structure. Notice that the movable coil is arc-contoured and axially pivoted. Each end of the shaft is mounted in jeweled bearings. The activating current enters through one spring, passes through the coil windings, and exits via the other spring. Repelling force between the poles of the permanent

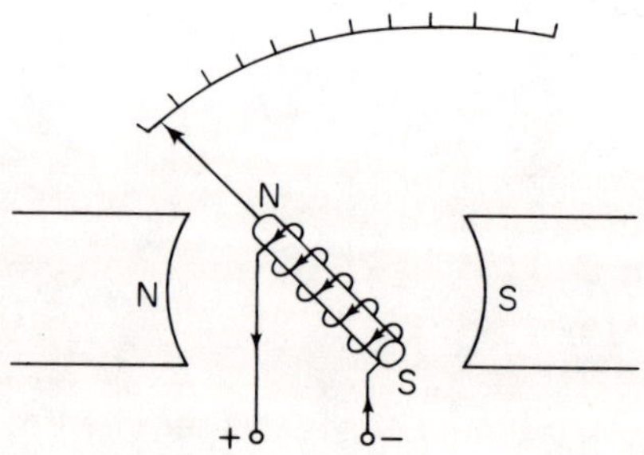

Figure 1–1 D'Arsonval Principle

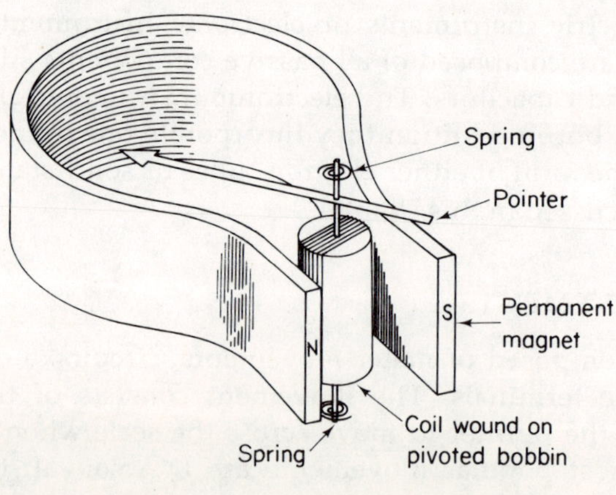

Figure 1–2 D'Arsonval Movement

magnet and those of the electromagnet causes the coil, bobbin, and pointer to rotate. The springs slow the action and return the pointer to zero when current ceases.

The *pole pieces* (not shown) are contoured pieces of soft iron alloy used to fill the spaces between the bobbin and the poles of the permanent magnet. The pole pieces extend the poles and create a small, uniform air gap around the bobbin.

The D'Arsonval movement has been the favorite of manufacturers for a number of years, but it is gradually being replaced by the *taut band* movement.

Taut Band Movement

The taut band movement utilizes the same principles as the D'Arsonval —movable coil and fixed magnet. The primary difference between the two is the method of mounting the movable coil. See Figure 1–3. The taut band meter has a solenoid mounted in a cradle and surrounded by a ring-bar magnet. The cradle is secured to a support bracket, which is in turn suspended between two *taut* bands. Current being measured energizes the solenoid. The interaction of the magnetic fields rocks the cradle to the side and moves the pointer along the scale.

Movement of the cradle exerts a twisting force on the steel bands. These twisted bands supply the torque to return the pointer to zero when the current is removed. This meter movement is rugged, accurate, and reliable. It has no bearings to wear out and features a constant level of sensitivity throughout the range of movement.

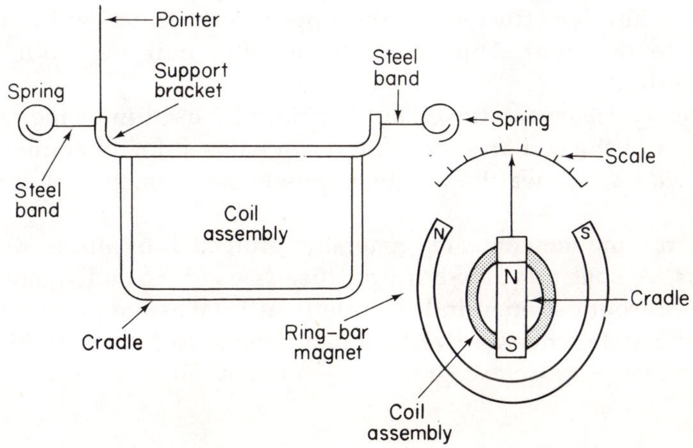

Figure 1–3 Taut Band Movement

Dynamometer Movement

The dynamometer movement uses a pivoted coil between two fixed coils. Measurement of the current, or some part of it, energizes all three coils. The resultant magnetic fields exert a torque on the movable coil. A series arrangement of the three coils is shown in Figure 1–4.

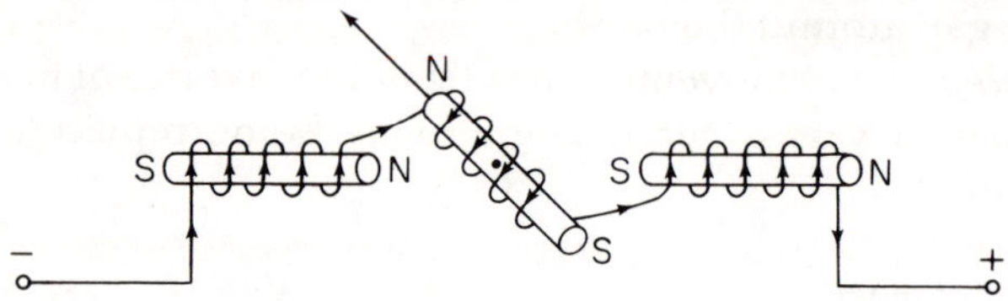

Figure 1–4 Dynamometer Movement

Applications

All three movements are primarily current measuring devices. The precise relationship of current, voltage, resistance, and power enables direct conversion from current to other desired units. The D'Arsonval and taut band movements are widely used in ammeters, voltmeters, and ohmmeters. When used with ac, the input is rectified in order to keep current in a constant direction through the coil.

The *galvanometer* is a highly sensitive meter usually found in the laboratory. It is constructed by a slight modification of either the D'Arsonval or taut band movements. The center of its scale is zero, and the pointer can swing in either direction from this point. Refer back to Figure 1–3. Assuming that the pointer position is showing zero current, current through the coil in one direction will deflect the pointer to the left. Current in the opposite direction will deflect the pointer to the right. Thus the galvanometer indicates both polarity and magnitude.

The *dynamometer* movement is primarily used in ac meters. With variations in the coil arrangement, it can be used for an ammeter, voltmeter, wattmeter, watthour meter, power factor meter, or frequency meter.

Meter movements are generally protected by fuses or circuit breakers. A special quick-burning fuse is used. For adequate protection, the fuse must open in less time than it takes the pointer to move across the scale when an overload is applied. No fuses should ever be used for this purpose except those recommended by the manufacturer.

DC METER CIRCUITS

A test instrument is formed by combining a meter movement with appropriate circuits, scales, and case. The basic instruments measure current, voltage, and resistance.

Ammeter

Current-measuring instruments are named according to the range of current they measure. These instruments include microammeters, milliammeters, and *ammeters*. Since they differ only in the quantities they measure, such fine distinctions will not be made here. For this discussion, all current-measuring instruments will be called ammeters. See Appendix A for unit prefixes and conversion process. Figure 1–5 illustrates a typical dc ammeter circuit and a few ammeter scales.

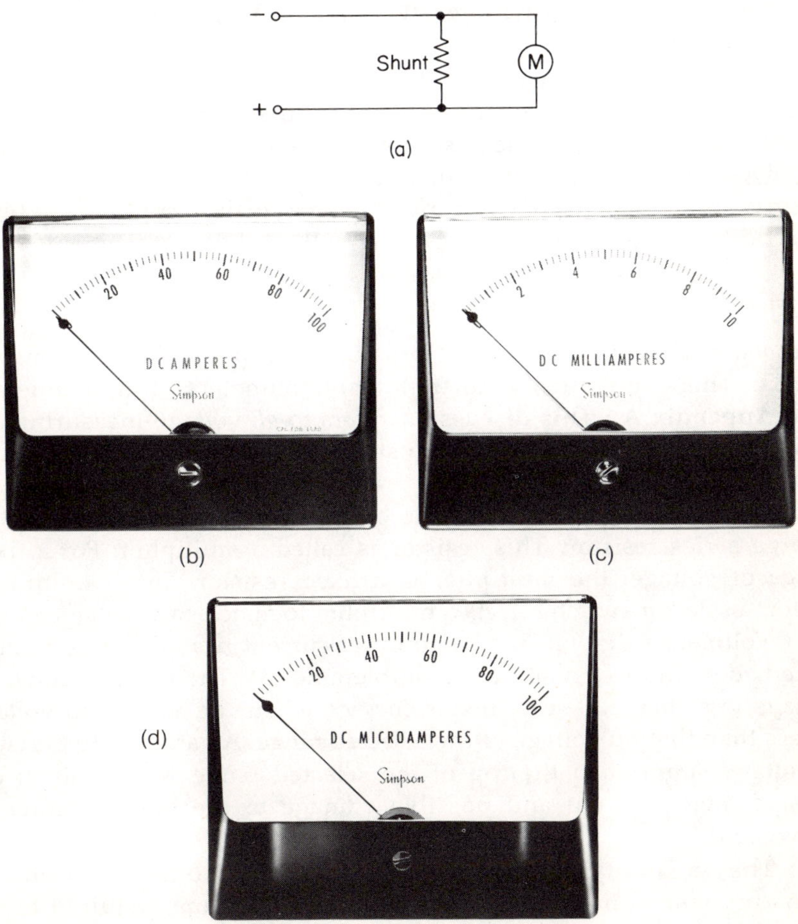

Figure 1–5 dc Ammeter Circuit and Scales (Courtesy Simpson Electric Company)

Most meter movements give full-scale deflection when 200 μA (or less) of current is passing through the windings of the movable coil. When measuring any value of current above the specified full-scale current, a shunt is necessary. A meter with a fixed range, such as the samples in Figure 1–5, has fixed internal shunts. Multirange meters have several shunts that are switched in to match the selected range. Some meters have external shunts that can be changed in a mechanical fashion.

The shunt in any case is a precise, minute resistance. It carries a prescribed percentage of the current being measured. For example, when measuring 10 A (amperes) on a 200 μA movement, the shunt will carry 9,999,800 μA of current and will allow only 200 μA to pass through the meter movement. This will give a full-scale deflection that is calibrated as 10 A.

This type of meter is placed in series with the circuit under test, and polarities must be observed; that is, the negative lead must be connected to the point nearest the negative terminal of the power supply, and the positive terminal connected to the point nearest the positive terminal of the power supply. Even a small current in the wrong direction can damage either the pointer or the delicate movement.

Voltmeter

The *voltmeter* measures electric pressure: the potential between two points. There are microvoltmeters, millivoltmeters, and voltmeters. (See Appendix A.) This discussion refers to *all* voltage-measuring devices as voltmeters. Figure 1–6 illustrates a sample circuit and a scale for a dc voltmeter.

Current through the movable coil of the voltmeter is limited by a large series resistor. This resistor is called a multiplier. For a fixed range of voltage, the multiplier is a fixed resistor. For a multirange meter, a selector switch selects a multiplier to match each voltage range.

Voltmeters draw a small amount of current from the circuit being tested, but this is usually a negligible amount. The current drawn is always less than full-scale meter current when the measured voltage is less than the top voltage of the selected range. An attempt to measure a voltage larger than the top of the selected range will result in excessive meter current and possible damage to the meter pointer or movement.

The sensitivity of a voltmeter is expressed in ohms per volt. The ohms-per-volt rating represents the number of ohms required to extend the meter range by one volt. A meter movement that has a 1 mA full-scale current has a sensitivity of 1000 Ω/V. If the full-scale current

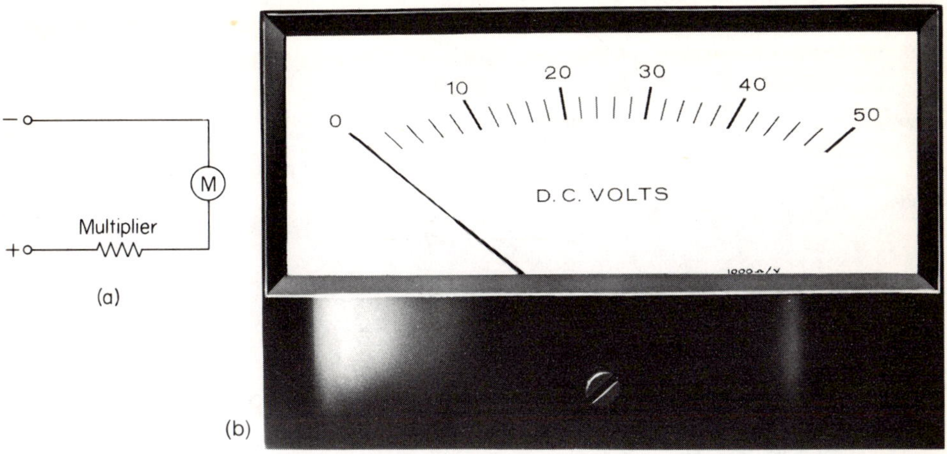

Figure 1–6 Voltmeter Circuit and Scale (Courtesy Simpson Electric Company)

is 100 μA, the sensitivity is designated as 10,000 Ω/V. The higher the ohms-per-volt rating, the more sensitive the meter and the smaller the loading effect on the circuit. A good voltmeter has a sensitivity of 20,000 to 100,000 Ω/V.

Series Ohmmeter

The series ohmmeter is the one most frequently encountered. It contains a series circuit consisting of a battery, a meter movement, a fixed resistor, and a variable resistor. The circuit is complete when the external leads are shorted together or connected across a conducting component.

The series ohmmeter is an inverse scale meter; a full-scale deflection is zero ohms. The circuit and scale for this meter are illustrated in Figure 1–7. R_1 is a dropping resistor and R_2 is the zero-adjust. Before each resistance measurement, the leads should be shorted together and R_2 adjusted for full-scale deflection of the pointer. If the pointer won't adjust to zero, the battery needs replacing. When the leads are placed across an external resistor, meter current is reduced. A half-scale resistance reading is obtained when the external resistance exactly matches the total internal resistance.

Most ohmmeters have multiple ranges. These meters have a range selector switch marked RX1, RX10, RX100, etc. This switch selects different values for the dropping resistor R_1. On range RX1 the scale is read directly; on RX10 the scale is multiplied by 10; on RX100 the scale is multiplied by 100.

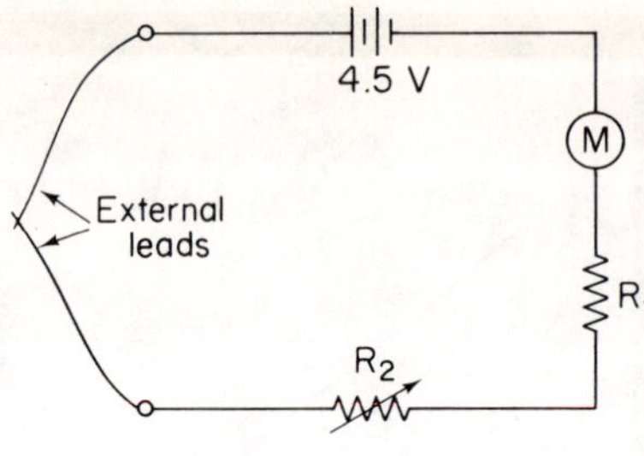

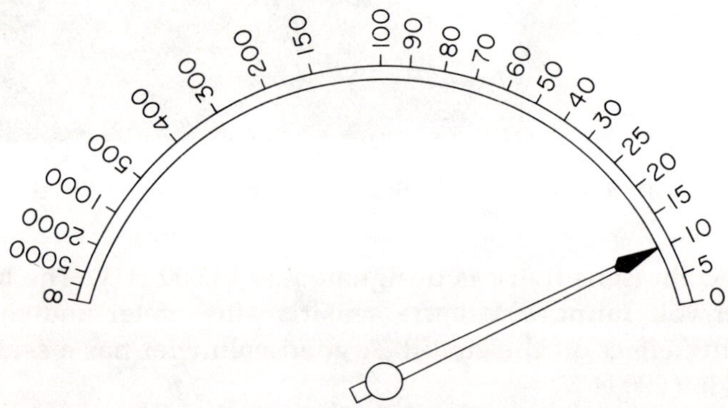

Figure 1–7 Series Ohmmeter: Circuit and Scale

Shunt Ohmmeter

The shunt ohmmeter is so named because the resistance being measured is in parallel with the meter movement. The scale is a direct scale; zero is on the left. Figure 1–8 illustrates the basic circuit and the scale. The switch opens the internal circuit when the meter is not in use to protect the battery. When the switch is closed, the circuit is completed. Shorting the external leads together allows all current to bypass the meter movement and provides a zero reading.

With the switch closed and the external leads open, R_1 is adjusted for full-scale deflection of the pointer. Now, a resistor placed between the leads is in parallel with the meter movement. Current divides in inverse proportion to the resistance of the two branches. A resistor larger than the meter resistance causes an indication between half- and full-scale. A resistor less than the meter resistance causes an indication between zero and half-scale. A half-scale indication is obtained when the measured resistance is exactly *equal* to the meter resistance.

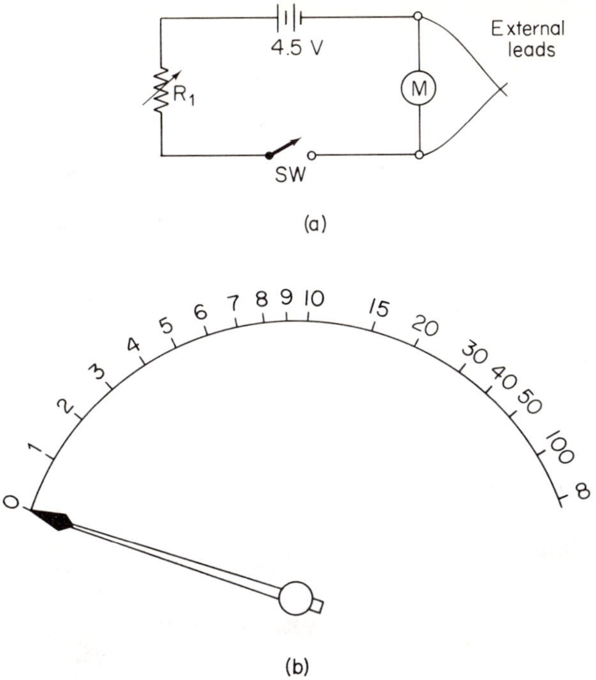

Figure 1–8 Shunt Ohmmeter: Circuit and Scale

The shunt ohmmeter is especially valuable for measuring very small precision resistors. Accurate measurements can be made of resistance values less than one milliohm.

Multirange shunt ohmmeters have a range selector switch that selects various values of resistance in series with the meter movement. This enables fairly accurate measurements of large, as well as small, resistance values. For instance; when 1000 ohms is in series with the meter movement, a 1000-ohm external resistor would cause a half-scale deflection of the pointer.

Multimeter

The *multimeter* incorporates several instruments into one case, and they all utilize the same meter movement. A multimeter circuit is illustrated in Figure 1–9, and Figure 1–10 is a photograph of a modern multimeter.

The circuit in Figure 1–9 is greatly simplified. SW_1 is a function switch. It selects the type of measurement—current voltage or resistance. SW_2 is a range switch. It is used when SW_1 is in the ohms position,

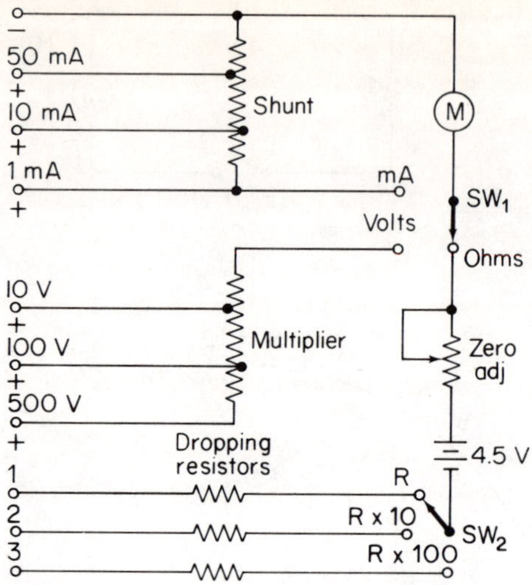

Figure 1–9 Multimeter Circuit

Figure 1–10 Modern Multimeter (Courtesy RCA)

and it selects different ranges for the ohmmeter. This meter has ten jacks for plugging in the external leads. The black lead is plugged into the "−" jack. The (red) positive lead is moved to any of the remaining jacks. For current measurements, the choice of jacks is 1 mA, 10 mA, or 50 mA. For voltage measurements, the choice is 10 V, 100 V, or 500 V. When measuring resistance, the red lead is inserted into jack 1, 2, or 3 to match the position of SW_2.

The multimeter pictured in Figure 1–10, the RCA WV-520A, is a modern instrument for use in laboratory, shop, and field. It features seven ranges of current, eight dc voltage ranges of either positive or negative voltage, six ac voltage ranges, and four resistance ranges. The scales include resistance, dc, ac (0–1.5, 0–5, and 15 up), and dB. The heavy black line is a mirror to prevent parallax error when reading the indication. The dB reading is in respect to 1 mW at 600 ohms. The meter sensitivity is 10,000 ohms per volt ac and 100,000 ohms per volt dc.

AC METERS

Basic Meter

The basic ac meter circuit is identical to the basic dc meter with one exception: the ac meter has a rectifier in its input circuit. The rectifier may be either half-wave or full-wave and is generally constructed with solid-state diodes. The reason for its presence is simple: the meter movement is for dc only. The rectifier converts the ac to a pulsating dc, which keeps current in a constant direction through the meter coil. Figure 1–11 illustrates an ac meter circuit with a full-wave, bridge rectifier.

The solid arrows represent the current path during the half cycles when point *A* is negative and point *B* is positive. The broken arrows trace the current path during alternate half cycles when the polarities are reversed.

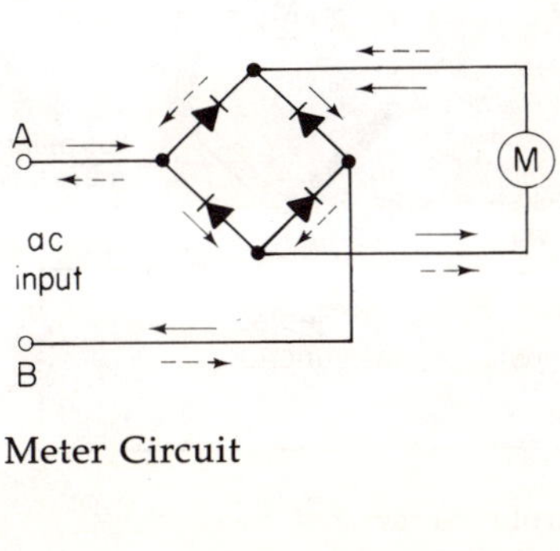

Figure 1–11 Basic ac Meter Circuit

ac Values

When dealing with ac there are several values to choose from: average, effective, peak, and peak-to-peak. Average value of a pure sine wave is zero; therefore this value is of little use. The peak value is the amplitude of the sine wave at either point of maximum excursion. The peak-to-peak value is measured between positive and negative peaks; it is twice the peak value.

The value most often considered is the effective value, also known as the root-mean-square (rms) value. This value represents the square root of the average of the squares of all instantaneous values in a sine wave. This value is 0.707 of the peak value. In terms of heating effect, a given effective ac will produce the same heating effect as that same value of dc. The effective value is used so frequently that all ac values are assumed to be effective values unless otherwise stated.

Some meters have scales for reading either (or both) effective and peak-to-peak values. When using meters without this convenience, it is sometimes necessary to perform a mental conversion.

$$\text{Effective} = 0.707 \times \text{peak}$$

$$\text{Peak} = 1.414 \times \text{effective}$$

$$\text{Peak-to-peak} = 2 \times \text{peak}$$

These values and relationships apply to both alternating current and voltage.

Thermocouple Meter

The *thermocouple* meter is usually classed as an ac meter, but it is also effective in dc measurements. In ac circuits it can be used with any frequencies up to a high RF. A sample circuit is shown in Figure 1–12.

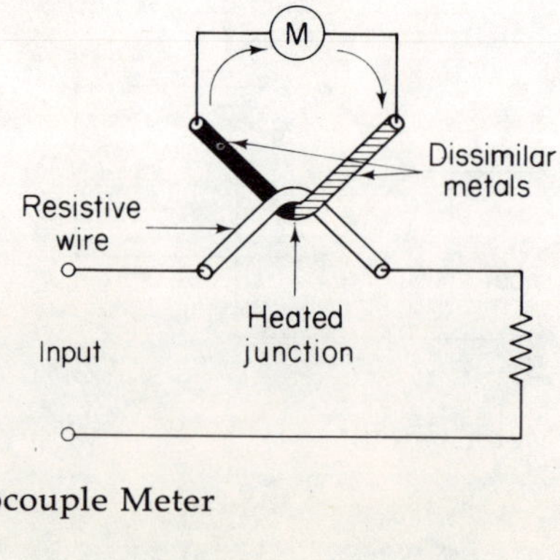

Figure 1–12 Thermocouple Meter

Heat at the junction of the two dissimilar metals generates a difference in potential. This causes current through the meter movement. Regardless of the type of current in the test circuit, the meter current is always dc. The meter current is directly proportional to the heat at the junction; the heat, in turn, is directly proportional to the square of the current being tested. As a result the meter scale is a square-law scale.

Clip-on Meters and Probes

Since alternating currents set up a constantly changing magnetic field about a conductor, it is possible to measure this current by transformer action. The principle of operation and a clip-on probe are shown in Figure 1–13. The arrangement shown here can be used for inputs to any ac measuring device. Some manufacturers combine the probe with a self-contained meter for direct readings.

Dynamometer Ammeter

The *dynamometer* movement can be used in either ac or dc meters without the need of special circuits. With slight circuit modifications,

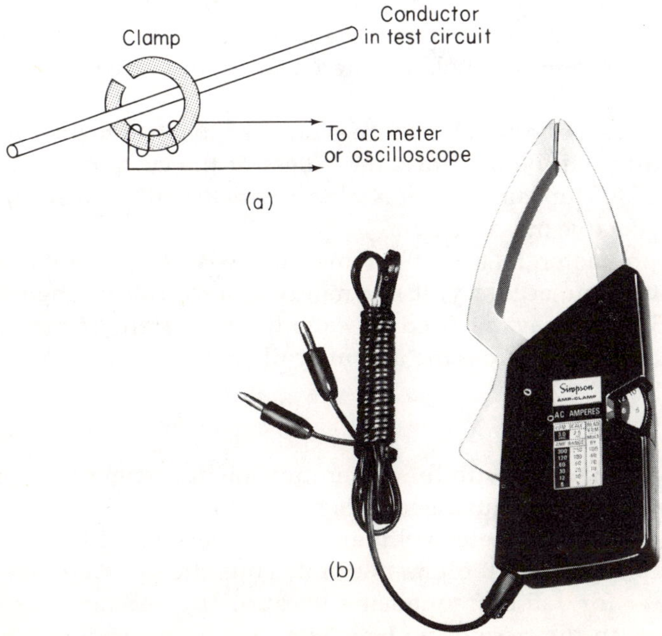

Figure 1–13 Clip-on Probe (Courtesy Simpson Electric Company)

it can be used in an ammeter, a voltmeter, a wattmeter, a power factor meter, a frequency meter, or a reactive power meter. When used for voltage and current, this meter can be calibrated on dc and will directly equate dc and ac (effective) measurements.

When used to measure current, the coils are in series as shown in Figure 1–14. All three coils are in series with the test circuit; the current being measured furnishes the energy for all three coils.

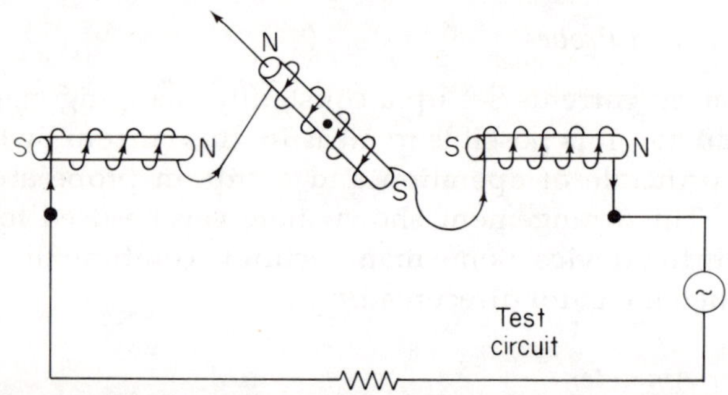

Figure 1–14 Dynamometer Ammeter

The direction of current is indicated for one alternation of the ac input. On the next alternation, the direction of current and magnetic poles will reverse. The pointer gives the same reading on both alternations, and meter damping holds it there as the sine wave amplitude changes.

When used as a dc ammeter, there is no need to observe polarities. The same reading will be obtained regardless of the current direction through the coils. The same meter scale provides a direct reading of either dc or ac effective measurement.

The dynamometer is useful for measuring small currents; and because of its extreme accuracy, it is a handy tool for calibrating other instruments. Shunts may be used to extend the current range. The shunt is normally placed across the moving coil only.

Dynamometer Voltmeter

Placing a resistor in series with the meter movement changes the dynamometer to a voltmeter as illustrated in Figure 1–15.

At best the dynamometer voltmeter is not very sensitive; sensitivity ranges up to about 30 ohms per volt. Thus the dynamometer is virtually useless for radio frequencies because the reactance of the coils increases with the frequency. However, it is so accurate at power frequencies that it may be used as a secondary standard.

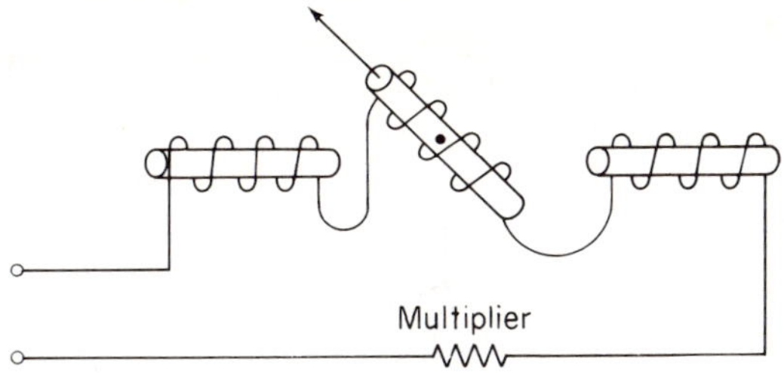

Figure 1–15 Dynamometer Voltmeter

The chief disadvantage of the dynamometer for either ammeter or voltmeter is its loading effect on the circuit being tested. In either case, energy from the circuit must be sufficient to activate all three coils.

ELECTRONIC MULTIMETERS

The electronic multimeter serves the same function as the electric multimeter, but it is engineered for the more precise measurements of the laboratory environment. It uses the same movement (D'Arsonval or taut band) found in its electric counterpart, but it has several improvements. It is more sensitive, and its high input impedance greatly decreases the loading effect on the test circuit.

Figure 1–16 is a simplified circuit for a transistorized electronic multimeter. The combination selector and range switch (SW) selects the portion of the voltage divider that will be used by the input. On voltage functions, high resistance values are used. This resistance is in parallel with the section of the test circuit being measured. The high resistance allows a very small current even with large voltage drops, and it minimizes loading of the test circuit.

On current functions, very small values of resistance are used. This resistance is in series with the test circuit and uses a minimum of energy even when measuring large values of current.

The meter movement, with a zero adjustment, is located between the emitters of two balanced amplifiers. Input through the field-effect transistor (FET) unbalances the amplifiers and causes meter deflection. The FET input provides for high input impedance on all ranges. Also, the FET is sensitive to extremely small inputs and greatly increases the sensitivity for small values.

In practice, the input circuit to the FET is much more complicated. One reason for this is the problem of drift when measuring dc. Any change in power supply voltage tends to cause an erroneous reading.

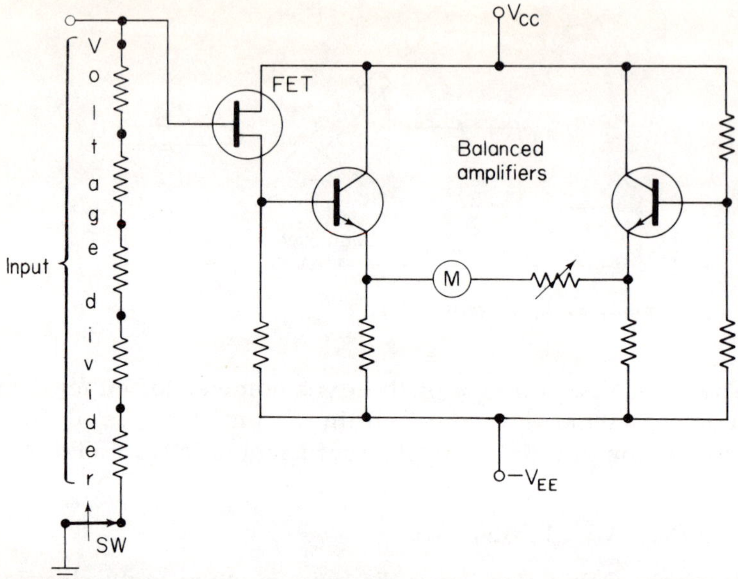

Figure 1–16 Electronic Multimeter Circuit

Another item for consideration is the fact that inputs can be either ac or dc. Figure 1–17 is a block diagram of some functional components that commonly appear between the input jacks and the FET.

When measuring dc, the voltage from the voltage divider is fed to a chopper, which changes it to a pulsating dc. The ac is not subject to drift, so it bypasses the chopper. In either case, the voltage is rectified and filtered before application to the FET.

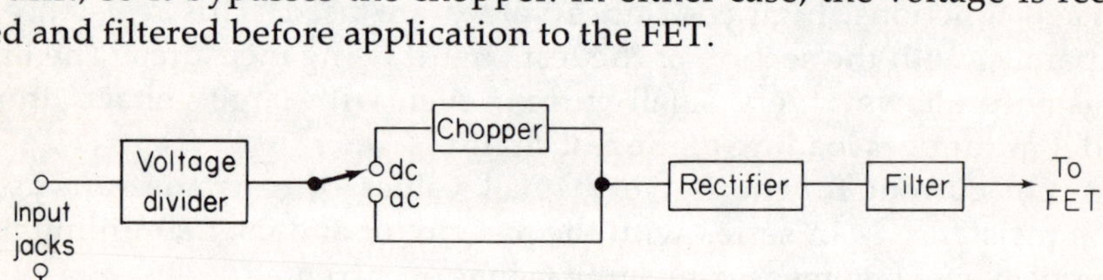

Figure 1–17 Input to FET

Most solid-state electronic meters use batteries as a power supply. This eliminates the dc drift problem and enables circuit simplification. Figure 1–18 is a view of such a meter. This meter is used to measure both alternating and direct current from 1 μA to 3 A, both alternating and direct voltage from 1 mV to 1 kV, dB relative to 1 mW, temperature,

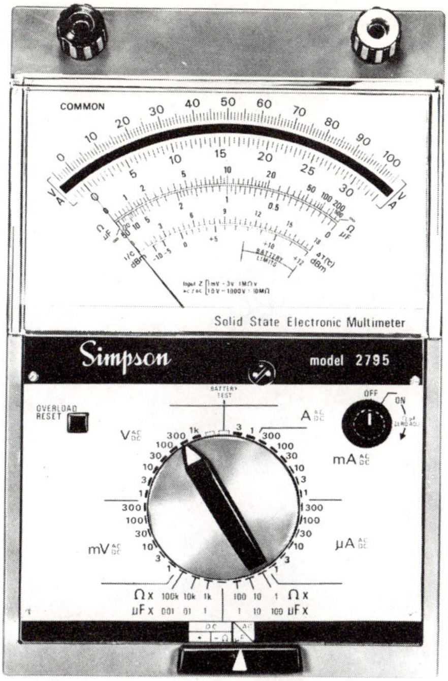

Figure 1–18 Modern Solid-State Electronic Multimeter (Courtesy Simpson Electric Company)

and resistance and capacitance. The voltmeter sensitivity is from 1 MΩ/V to 10 MΩ/V. There are six ranges for resistance and six for capacitance measurements.

Notice that the ohms scale is direct—zero on the left. This indicates that the ohmmeter is of the shunt type. The zero-adjust is set for full-scale deflection with the external leads open.

The μF scale is inverse—zero on the right. Again there is no need to short the leads together, and the zero-adjust is set for full-scale as before.

The dark line between the V and A scale is an antiparallax mirror. This particular meter does not have a scale for peak-to-peak readings, but the peak-to-peak scale is available on many models of electronic voltmeters and multimeters.

chapter two

INSTRUMENTS FOR REACTIVE CIRCUITS AND POWER

This chapter is concerned with instruments that are designed to measure capacitance, inductance, reactance, impedance, and power. Certain basic checks of capacitors can be made with an ohmmeter or voltmeter. Basic circuit power can be estimated using the dB scale on a multimeter or measured on a simple wattmeter. However, actual analysis of reactive circuits and measurement of high-frequency power require special equipment.

BRIDGE INSTRUMENTS

Bridges are among the most accurate of measuring devices, and they are used to measure a variety of quantities, including resistance, capacitance, inductance, impedance, and power.

dc Bridges

There are many types and variations of *bridge* circuits. This discussion includes several of those most frequently encountered. Figure 2–1 is a sample bridge circuit that illustrates the general principles of bridge

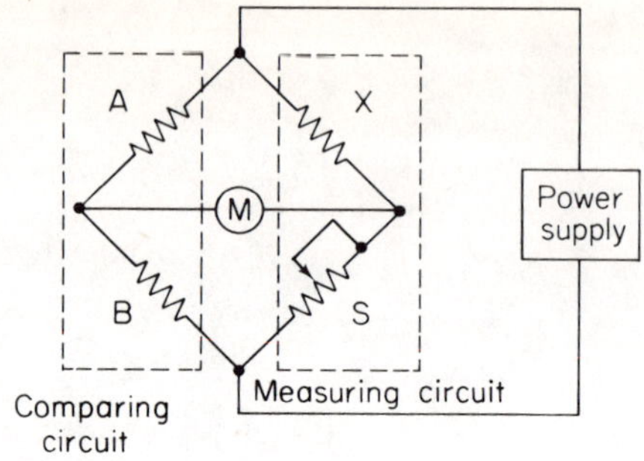

Figure 2–1 Sample Bridge Circuit

measurements. Notice that branches *A* and *B* are comparing circuits, while branches *X* and *S* are measuring circuits. The ratios of resistance in either circuit, in respect to the other, can be changed. Branch *X* contains the component to be measured, and branch *S* contains the variable control for balancing the bridge.

The meter is usually a zero-center galvanometer, but other forms of detection are sometimes used. Electron-ray tubes and headsets are types of less sensitive detectors. When the bridge is balanced, there is no current through the meter. The null (no-current) meter condition indicates the following:

$$I_A = I_B$$

$$I_X = I_S$$

$$\frac{R_A}{R_B} = \frac{R_X}{R_S}$$

$$R_X = \frac{R_A R_S}{R_B}$$

Figure 2–2 illustrates the four basic bridges most frequently encountered.

ac Bridges

Most basic bridge circuits can be adapted to ac measurements. Complex impedances are used, and two balance conditions must be satisfied.

One very versatile ac bridge is actually a flexible combination of several bridges. It is commonly identified as the universal bridge. The typical universal bridge will contain an audio driving source, a dc sup-

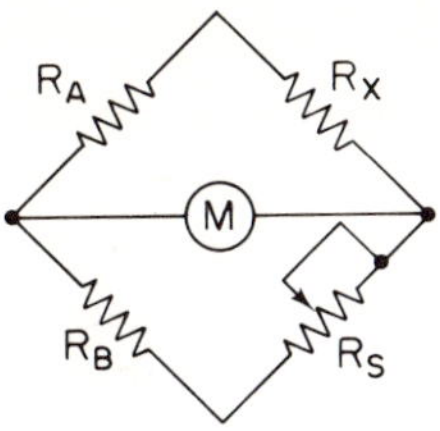

Wheatstone bridge for resistance

$$R_X = \frac{R_A R_S}{R_B}$$

(a)

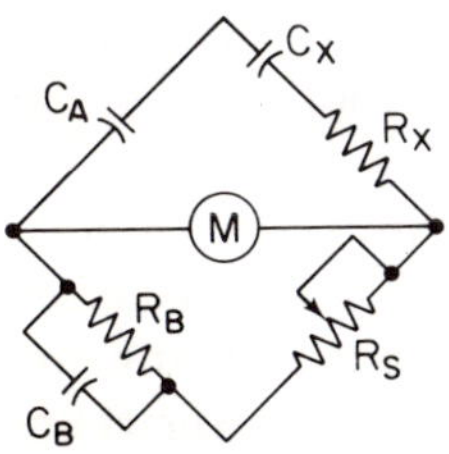

Schering bridge for capacitance

$$R_X = \frac{C_B R_S}{C_A} \qquad (R_X = \text{Resistive component of } C_X)$$

$$C_X = \frac{R_B C_A}{R_S}$$

(b)

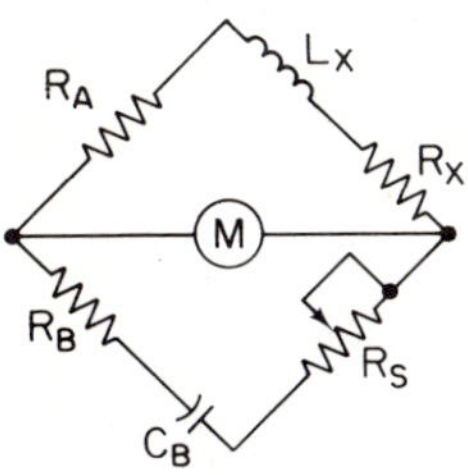

Hay bridge for inductance

$$L_X = \frac{R_A R_S C_B}{1+(\omega R_B C_B)^2} \qquad (R_X = \text{Resistive component of } L_X)$$

$$R_X = \frac{R_A R_B R_S (\omega C_B)^2}{1+(\omega R_B C_B)^2} \qquad (\omega = \text{Angular velocity, } 2\pi f)$$

(c)

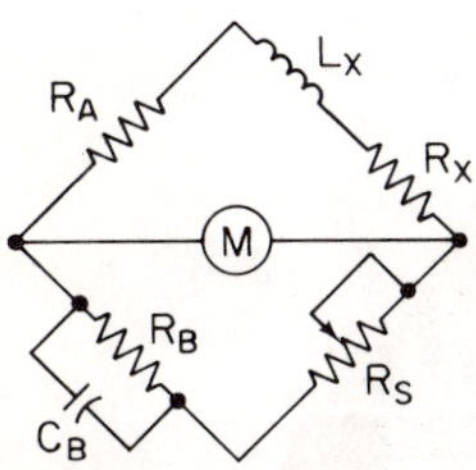

Maxwell bridge for inductance

$$L_X = R_A R_S C_B$$

$$R_X = \frac{R_A R_S}{R_B}$$

(d)

Figure 2–2 Common dc Bridge Circuits

ply, a null meter circuit, and the four arms of a bridge. Components in the arms can be switched to form the bridge needed for the problem at hand. The audio driving source is generally adjustable from about 50 Hz to 10 kHz. The universal bridge can be used to measure resistance, capacitance, inductance, impedance, conductance, quality (*Q*), and dissipation factor (*D*).

Automatic Bridges

A technician with a few highly stable precision components can construct any of the basic bridges and can use them to make measurements. However, such a self-help project is not necessary. Many instrument companies offer a variety of bridge instruments. Some of these instruments employ a universal bridge that can make various types of measurements. Components switch in and out to form the bridge circuit needed at any time. Others use various combinations of the dc bridges or other bridge arrangements. The *Q* meter, the RX meter, the admittance meter, the LC meter, and the vector impedance meter are a few applications of automatic bridges.

Figure 2–3 is a photograph and a simplified circuit of a *Q* meter. The simplified circuit illustrates the principles of operation in a series arrangement. The meter consists of an RF oscillator and a measuring circuit. The *Q* meter measures reactive electronic parts by resonating them with a capacitor of known characteristics.

The meter shown in Figure 2–3 is composed of a stable, continu-

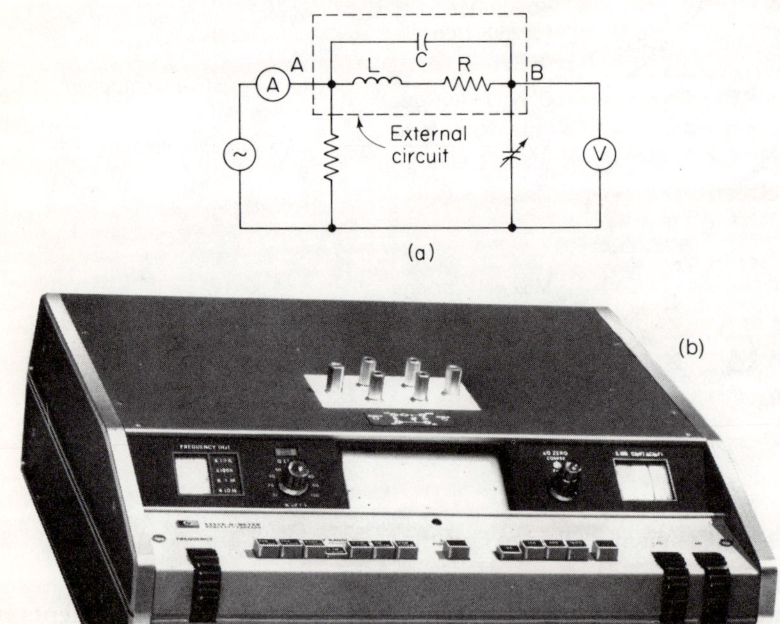

The Hewlett-Packard Model 4342A Q-Meter covers the Q range from 5 to 1000, and simplifies measurements with pushbutton controls.

Figure 2–3 *Q* Meter (Courtesy Hewlett Packard)

ously variable oscillator with automatic level control. The output is applied in series with an external unknown and an internal variable capacitor. This is a solid-state meter capable of measuring Q from 5 to 1000, and it can detect very small changes in Q. The expanded scale and push button operation make for easy, rapid measurements.

LOW-FREQUENCY POWER MEASURING INSTRUMENTS

Measuring power in most *electric* circuits is a simple operation accomplished with relatively simple instruments. In electronic circuits the problem grows more complex because of the wide range of circuits requiring power measurements. It is necessary to measure power in all types of circuits from dc power supplies to microwave equipment. This includes transmission lines, antennas, waveguides, amplifiers, oscillators, etc.

Dynamometer

As previously discussed, the dynamometer movement can be used for many types of measurement. It can be used for power in either ac or dc circuits, and it is not restricted to a purely sinusoidal waveshape. Figure 2–4 illustrates coil connections for a wattmeter. The fixed current coils are in series with the line current (total current), and the movable voltage coil is in parallel with the load. With coils in this arrangement, the deflection is proportional to the average power, which is

$$EI \cos \theta$$

There is a slight loss because a small amount of energy is consumed in

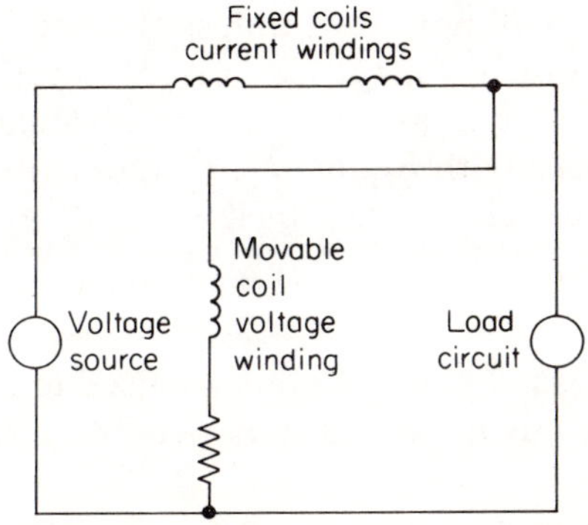

Figure 2–4 Dynamometer Wattmeter

maintaining the magnetic fields. However, the loss is so small that it can be safely ignored in most cases.

Notice that with the load circuit of Figure 2–4 disconnected, the meter movement still provides a complete path for current. This will cause a simple meter such as this to indicate some power even with the load removed. This fallacy can be overcome by using a meter with compensated windings as illustrated in Figure 2–5. The current windings are composed of heavy wire, and they carry the total current. The compensating windings are made of light wire and carry only the voltage coil current. The compensating coils are wound inside the current coils. The field resulting from the current in the compensating coils opposes the field of the current coils. This small cancellation effect causes this meter to indicate zero when the load is disconnected.

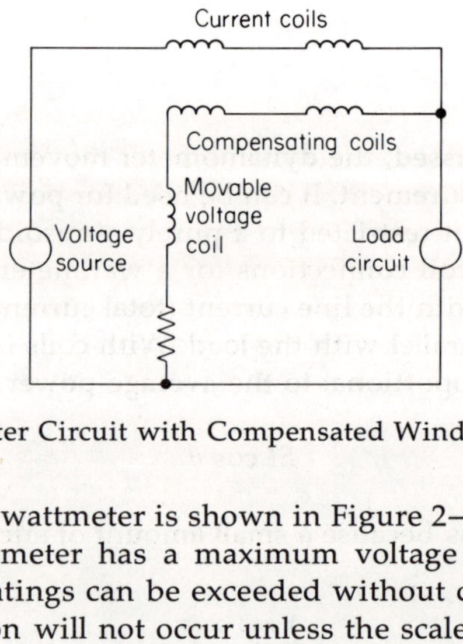

Figure 2–5 Wattmeter Circuit with Compensated Windings

The face of a wattmeter is shown in Figure 2–6. Notice in the picture that the wattmeter has a maximum voltage and current rating. Neither of these ratings can be exceeded without danger to the meter. Excessive deflection will not occur unless the scale value of wattage is exceeded, but too much voltage or current could damage the coils.

Dynamometer meters are used for frequencies up to about 2500 Hz, but error increases with frequency. Meters with rated accuracy may be obtained for frequencies up to 800 Hz.

Reactive Power Meter

The previously discussed power meters are used to measure true power: $EI \cos \theta$. But in certain instances it is necessary to measure reactive power.

The dynamometer movement is easily adapted to a reactive power meter. A large inductance in series with the movable voltage coil re-

Figure 2–6 Wattmeter (Courtesy Simpson Electric Company)

places the resistor and changes the meter to read reactive power. This is illustrated in Figure 2–7. The inductor in series with the voltage coil causes a phase difference between the current in the current coils and the current in the voltage coils. Compensating networks are used to cause the voltage coil current to lag exactly 90°. The meter reading is proportional to the reactive power of the load circuit.

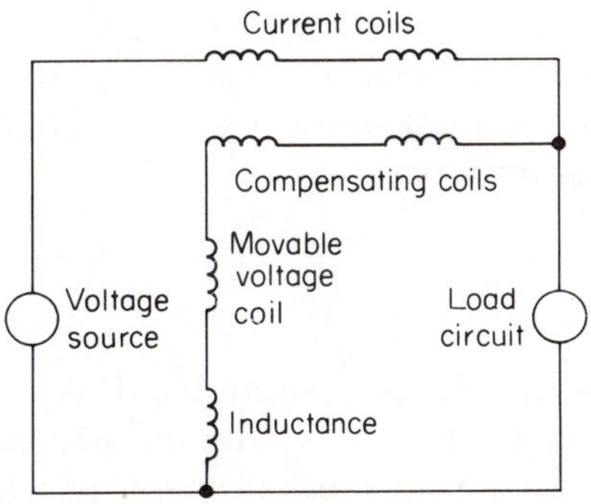

Figure 2–7 Reactive Power Meter

Thermocouple Power Meter

At audio and low-radio frequencies, power meters often use thermocouples. Line current is used to heat the thermoelements. This heat is converted to EMF by the thermocouple, which causes current through the meter movement. The movement is generally a D'Arsonval or taut band. Figure 2–8 illustrates such a meter.

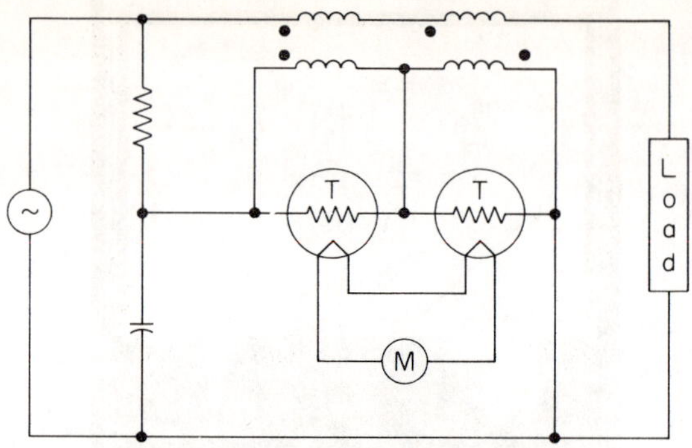

Figure 2–8 Thermocouple Power Meter

The meter circuit uses two thermoelements that are heated individually by current from transformers. The transformers are in series with source and load, and they are oppositely polarized. A meter of this type may be calibrated with dc power. The indication will remain independent of frequency until stray capacitances become a problem. A given meter may have a precision of 0.1% and cover frequencies from 500 Hz to 2000 kHz.

The thermocouples produce a voltage that is proportional to the sum and the difference of the heater currents. Since matched thermocouples are used and connected in opposition, the resulting voltage is proportional to the combined energy consumed in the load and primary windings of the transformers.

Output Power Meters

The output power wattmeter is designed to directly measure the output power in an arbitrary load. The instrument provides a set of resistive loads to be selected for power measurement. In addition to power, the output meter can be used to measure impedance and frequency response characteristics. A sample circuit is illustrated in Figure 2–9.

The input impedance network consists of two tapped resistors and a coil. The input impedance can be varied in steps from 2.5 Ω to 40 kΩ. At low impedance values the coil shunts a portion of R_1; an increase in resistance results in fewer turns on the coil. This arrangement keeps the meter reading proportional to the energy dissipated by the resistor. At high impedance values, the coil is replaced by another tapped resistor, R_2.

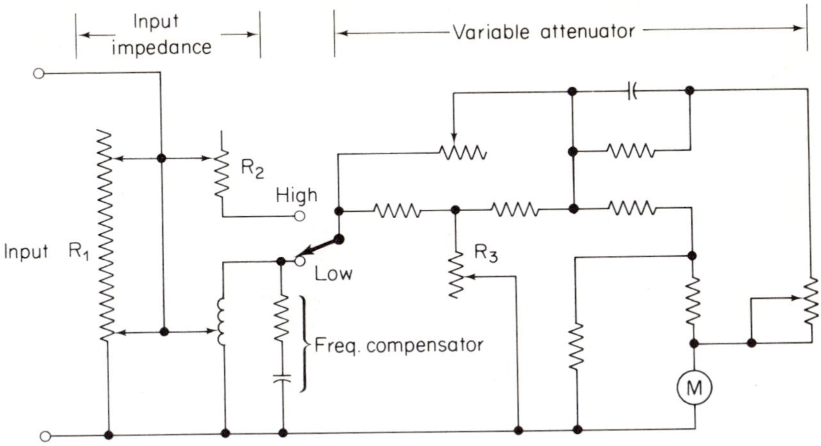

Figure 2–9 Output Power Meter Circuit

The RC frequency compensator is parallel to the coil. At low frequencies X_C is high, and X_C decreases as frequency increases. This compensates for losses of the coil at low frequencies. The frequency range is generally 20 Hz to 20 kHz.

R_3 is the control in a variable T-network. This amounts to a variable meter shunt, which is used to extend the range of the meter. The lowest meter range is normally 5 mW, but it can be extended in decade steps: 50 mW, 500 mW, etc. The remainder of the circuit is a combination calibration and frequency compensation network. A meter of this type may have a midscale accuracy of ±2% at 1000 Hz. Over a frequency range of 20 Hz to 15 kHz the accuracy may be within ±2% of the 1000 Hz value.

The input meter is subject to waveform error when the input is other than sinusoidal. This is not a grave error in this band of frequencies—audio to low frequency radio. These are communication frequencies, and the waveform will be sinusoidal with perhaps some harmonic distortion. The meter can indicate true rms with up to 20% second and third harmonics present in the input signal. A sample output power meter is shown in Figure 2–10. This General Radio instrument can be used to measure audio power (20 Hz to 20 kHz). Its input impedence is a tapped transformer with 48 impedance settings. This is a practical instrument for measuring power output of oscillators, amplifiers, transformers, transducers, and low-frequency lines. It can be used to measure output impedance by adjusting for maximum power. It can also be used to check frequency response characteristics of audio-frequency devices.

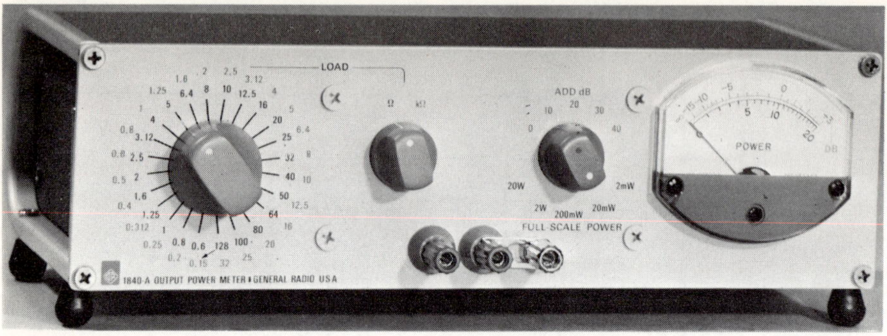

Figure 2–10 Output Power Meter

Photometric Power Meter

Photometric power meters are used to directly measure RF power from the medium range up to approximately 30 watts. These meters use a barrier layer *photocell* in series with a resistor. The photocell causes current through the resistor proportional to the intensity of the illumination of the photosensitive area of the cell. The illumination is provided by a load lamp in series with the RF power source. A suitable meter, calibrated in watts, completes the circuit as illustrated in Figure 2–11.

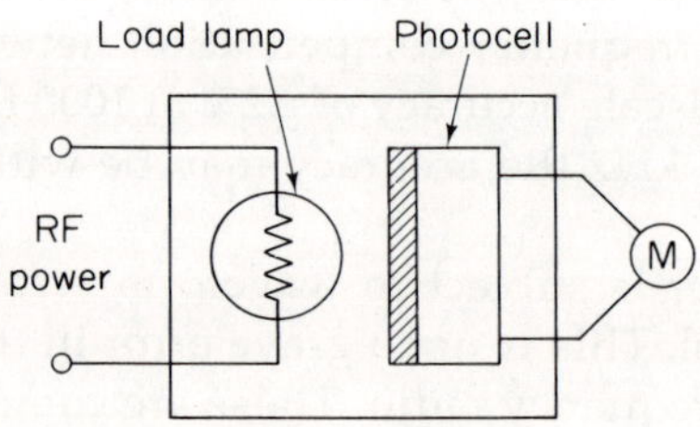

Figure 2–11 Photometric Power Meter

The load lamp emits color or light intensity from a hot wire element that serves as an RF load. The element is enclosed in a glass envelope that is filled with hydrogen gas. Very sensitive meters can be made by using very little gas; higher wattage capabilities can be obtained by increasing the gas pressure.

Since these meters are calibrated with dc power, the inherent error from use of RF increases with frequencies. Other factors that contribute to error are RF energy losses in the glass seals and reactance of the wire element.

Neon Lamp Power Meter

Neon lamp power meters are used to indicate relative power from pulse-modulated RF transmitting equipment. The general appearance of such an instrument is shown in Figure 2–12. This is not a meter as such, but it is a power indicating instrument. When the feed horn is held a specified distance in front of a radar antenna, RF power, picked up by the horn, lights a neon lamp. The attenuator knob is then adjusted until the lamp is just barely extinguished. The attenuator dial is calibrated in relative power. Maximum transfer of power is an important factor, and this can be obtained by using the proper sand load to terminate the waveguide.

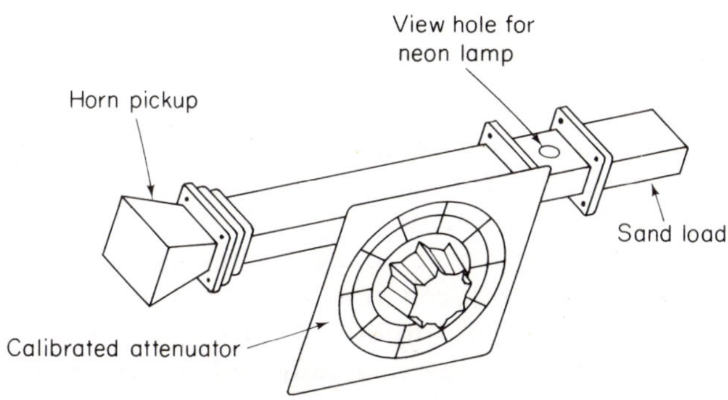

Figure 2–12 Neon Lamp Power Meter

This type of instrument can be constructed with excellent sensitivity and used with CW transmitters. If the lamp is removed and the sand load replaced by a matched crystal detector, the instrument becomes adaptable to AM signals. Of course, this latter variation is not exactly a neon lamp power meter, but it may be used in conjunction with other instruments to obtain power indications.

HIGH-FREQUENCY POWER MEASURING INSTRUMENTS

At high radio and microwave frequencies, power instruments become sophisticated and as a rule more sensitive. Some high-frequency devices also radiate energy at relatively high-power levels. Among the most common devices are calorimeters and bolometers.

Calorimeters

Calorimeters are popular devices in high-frequency and/or high-power instruments. They are rated among the most accurate of all high-power measuring instruments. Power measurement is based solely on temperature, mass, and time. The type of waveshape and frequency of the signal are minor considerations, except for observation of frequency range applications.

The calorimetric principle calls for complete conversion of the input electromagnetic energy into heat. A number of methods are employed to accomplish this conversion. Direct heating requires measurement of the heating effect on the load that terminates the line. Indirect heating requires measurement of the heating effect on a medium other than the original energy absorbing load. The substitution method uses a known low-frequency power to produce the same heating effect as that of an unknown RF power that is being measured.

Static Calorimeters

A static calorimeter uses a thermally shielded body to minimize the loss of heat to the surrounding medium. The RF power causes an increase in temperature of the isolated body. The product of the rate of temperature rise in the calorimetric body, and its heat capacity equals the applied power. One type of static calorimeter is illustrated in Figure 2–13. This model is a generalized version of an adiabatic calorimeter using low-frequency power substitution. In adiabatic-type calorimeters, power is applied directly to the thermally isolated body. The rate of temperature rise is determined by the change in temperature per unit time.

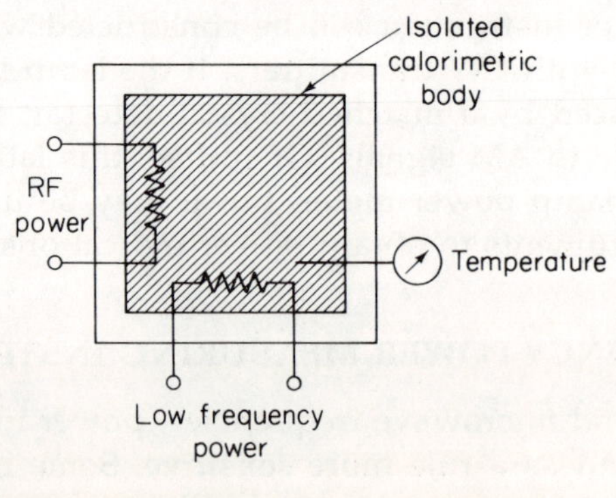

Figure 2–13 Static Calorimeter

Figure 2–14 shows additional details of the calorimetric body. The calorimetric body in this case is water. The open-ended waveguide contains a sealed, inclined glass partition to create a wedge-shaped water load with a low-reflection coefficient. The stirrer agitates the water and assures that the entire body has a uniform temperature (adiabatic means even heat). A sensitive thermometer measures the rise in temperature. The heating coil is wound around the waveguide inside the calorimetric body. Low-frequency power can be applied to the heater coil for calibration. A meter of this type can withstand peak power up to 50 kW, and can be used to measure average power of several hundred watts.

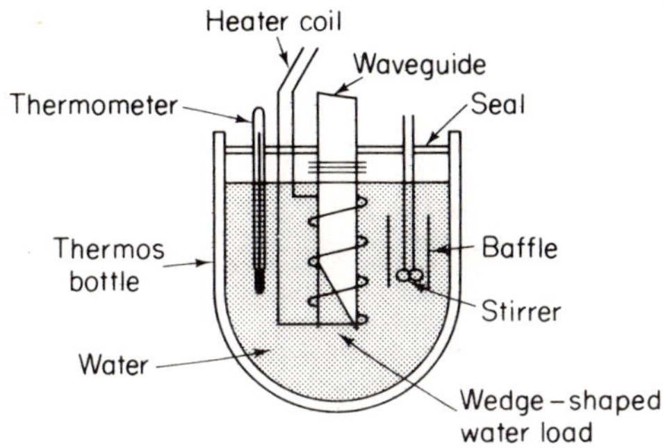

Figure 2–14 Dewar Flask Calorimetric Body

The nonadiabatic calorimeters function on the principle of a heat differential. They have RF lines terminated in loads formed from lossy dielectric materials or resistive film strips. Temperature indications may be obtained by any of a number of heat indicators; some examples are thermometers, thermopiles, thermocouples, thermisters, and bimetallic strips. These calorimeters are calibrated against a known low-frequency power or a power standard.

A coaxial calorimeter of the nonadiabatic type may be used for measuring power at frequencies from 0 to 1200 MHz. It needs to have good sensitivity and a short 50 Ω resistive film on a center conductor. The outer conductor is thin-walled and tapered. The RF termination is electrically connected to a mounting plate by a short section of silvered lava coaxial line. The high thermal resistance provides thermal isolation between the RF termination and the plate. The steady-state temperature rise of the outer casing of the load with respect to the mounting plate is generally detected by a differential resistance thermometer in a Wheatstone bridge.

Another type of static calorimeter is called the twin calorimeter. It is generally used with microwave frequencies and provides a high sensitivity for measurements in the milliwatt range. This method uses two calorimetric bodies thermally shielded against ambient temperature variations. The power to be measured is applied to one calorimetric body, while the other calorimetric body is used for temperature reference. The steady-state temperature difference between the two bodies is used as a measure of RF power. Figure 2–15 illustrates this device. This method of power measurement is extremely accurate.

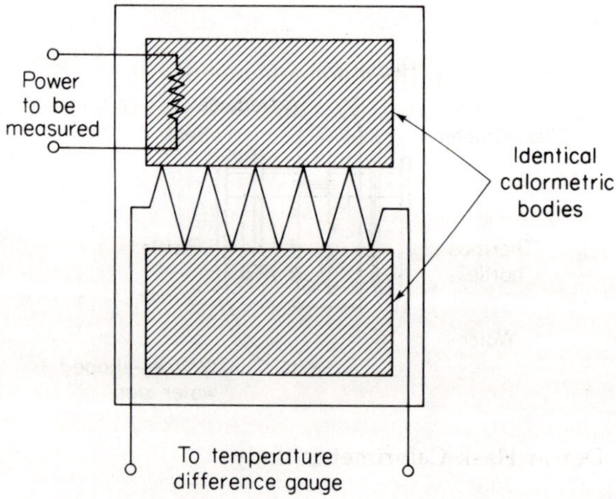

Figure 2–15 Twin Calorimeter

One variation of the twin calorimeter is known as the differential air thermometer. It is illustrated in Figure 2–16. This instrument is used to measure microwave power in the 0 to 100 mW range. It consists of two similar glass cells connected by a capillary tube. Each glass cell has a carbon strip calorimetric body surrounded by air. A liquid pellet in the capillary tube separates the air mass from the two cells. A heat differential between the two bodies results in a difference in air pressure between the two cells. This difference in air pressure is registered by a displacement of the liquid pellet. The pellet is viewed through an aperture in the waveguide wall. A highly sensitive instrument may be fitted with a microscope for more precise viewing. Power measurement is accomplished by using dc power to balance the RF power and center the pellet. The RF power is then replaced with a known dc power to maintain the balance. This procedure permits an accuracy of ±2% at 10 mW.

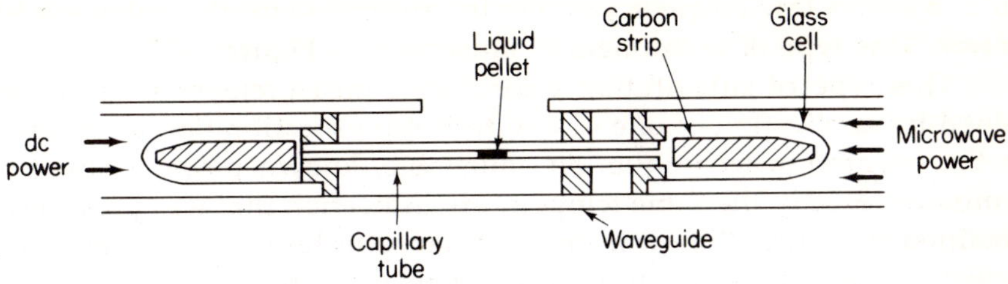

Figure 2–16 Differential Air Thermometer

Flow Calorimeters

Flow calorimeters provide the primary standards for the measurement of high-power levels. They are also used in conjunction with calibrated directional couplers, attenuators, power dividers, and other similar devices to standardize medium- and low-power instruments.

There are several classifications of flow calorimeters: (1) type of circulation (open or closed), (2) type of heating (direct or indirect), and (3) type of measurement performed (true calorimetric or substitution).

In an open system the calorimetric fluid (water or oil) is used only once. An overflow system is used to maintain a constant rate of flow.

The closed system recirculates the fluid by means of a pump. A cooling system restores the fluid to the ambient temperature before it is returned to the calorimeter. This self-contained method enables construction of portable, as well as practical, instruments.

In the indirect heating method, the circulating fluid is used solely as a coolant. In direct heating, the fluid is both a coolant and a dissipative medium. In both heating methods, water is generally used at frequencies of 1000 MHz or higher because it has excellent thermal properties and a high dielectric loss. Below 1000 MHz, the dielectric losses of water are insufficient and oil is generally used. The indirect heating method offers a wider frequency and power range than the direct heating method, and it can be readily used in substitution-type measurements.

In true calorimetric measurements several imperfections have a bearing on accuracy. Nonuniformity of flow rate, air bubbles, inaccuracies of flow rate measurement, and variations in temperature rise—all are factors that may contribute to appreciable errors in power measurements. These errors are minimized by devices such as bubble traps, flow regulators, and thermal insulation.

In the substitution method of measurement, the accuracy depends on constant parameters and the exactness of the substitution power determination. Instead of measuring direct heat dissipation of a moving fluid, a *known* low-frequency power is substituted for the unknown RF power. This type of instrument is illustrated in Figure 2–17.

This type of substitution calorimetric measurement may be obtained by application of one of two techniques: calibration or balance. In the calibration technique, an adjustable, known power is used to reproduce exactly the same temperature indication that was previously obtained from the RF power. In the balance technique, low-frequency power is used to establish a steady-state temperature *rise*. When RF power is applied, low-frequency power is reduced to maintain the same temperature indication. The power removed is then equal to the RF power.

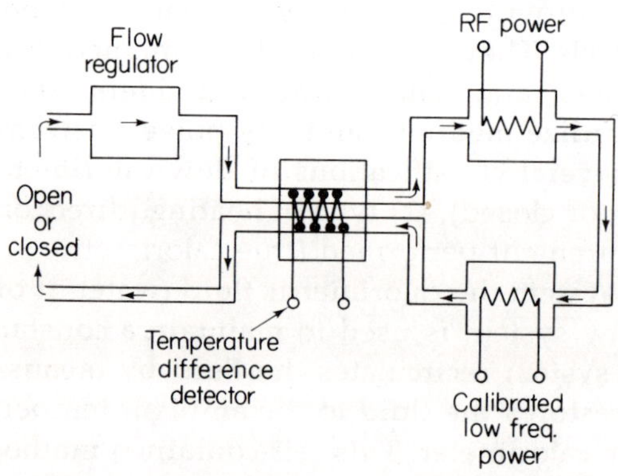

Figure 2–17 Flow Calorimeter with Substitution Measurement

A variety of instruments make use of the calorimetric principle. Figure 2–18 presents a practical example of one of these instruments.

This Hewlett Packard meter features ease of operation, rapid response, accuracy, and a wide range in both frequency and power. It employs a self-balancing bridge and a high-efficiency heat transfer system. The full-scale response time is five seconds or less, which enables the meter to follow small power changes. It covers a frequency range from 0 to 12.4 GHz and measures power up to a 10 W average. The accuracy is within ±5% of full-scale, and there are seven meter ranges. The calorimetric body is oil which is passed through a parallel-flow heat exchanger.

Bolometers

A *bolometer* is a temperature-sensitive resistive element. In the form of a thin wire, the active material has a positive temperature coefficient and

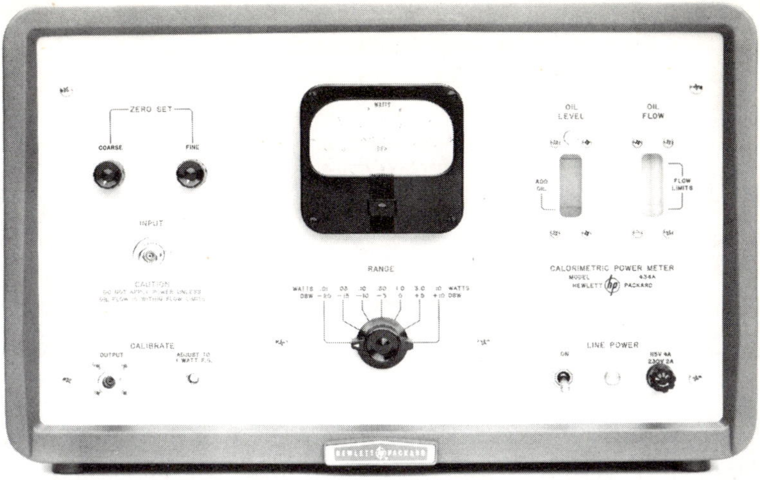

Figure 2–18 A Practical Calorimetric Power Meter (Courtesy Hewlett Packard)

is called a *barretter*. In the form of a semiconductor bead, it has a negative temperature coefficient and is known as a *thermister*. Below 10 mW, power is generally measured with bolometers in conjunction with a balanced bridge. The barretter is rather delicate, both electrically and physically; therefore, the thermister is the more popular type.

When RF power is applied to a bolometer, energy is absorbed, and the element is heated. This causes a change in the electrical resistance. The bolometer is generally used as one leg in a bridge circuit so that small resistance changes can be detected. Power measurement is accomplished by substituting a known low-frequency power to produce an equivalent heating effect. One method uses a bridge which is initially balanced by low-frequency bias power. When RF is applied, the bias power is gradually removed until the bridge is again balanced. The RF power is then equivalent to the bias power removed. Another method uses a bridge balanced as before; but after application of RF, the bridge is not rebalanced. The unbalanced indicator reading is converted directly into power. A basic bolometer bridge is illustrated in Figure 2–19.

Bias low-frequency power, is controlled by the variable resistor, *R*. The RF power is applied across the bolometer, *B*. This is the balanced bridge application. *R* is calibrated to indicate power removed in order to maintain a balanced condition.

A bolometer must be small for high sensitivity; it must be equally responsive to both low-frequency and RF power; and it must be matched to the RF power input line. The cross-sectional dimension of a bolom-

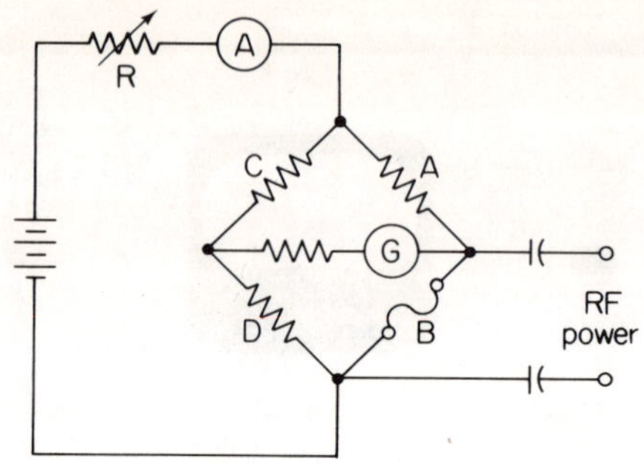

Figure 2–19 Bolometer Bridge

eter is approximately equal to the depth of RF current penetration at the highest operating frequency. An air-mounted bolometer is more than a hundred times as sensitive as the average static calorimeter. Additional sensitivity is obtained by mounting the bolometer in an evacuated envelope to eliminate convective heat loss.

Self-Balancing Bridges

Bolometer power instruments frequently employ a bridge circuit which automatically rebalances itself when RF power is applied. Such a bridge is illustrated in Figure 2–20.

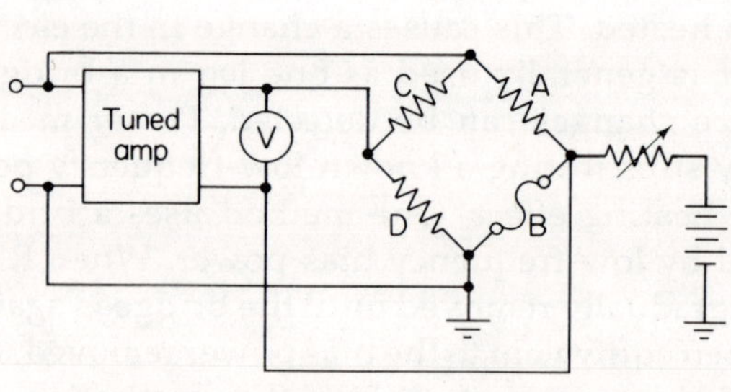

Figure 2–20 Self-Balancing Bridge

The tuned amplifier uses feedback to maintain proper amplitude and phase to keep the bridge in a balanced condition. Self-balancing bridges are available with both barretters and thermisters. They are also designed to cover a wide range of both power and frequency.

Figure 2–21 shows two self-balancing power meters which use temperature compensated bolometers. These two instruments serve

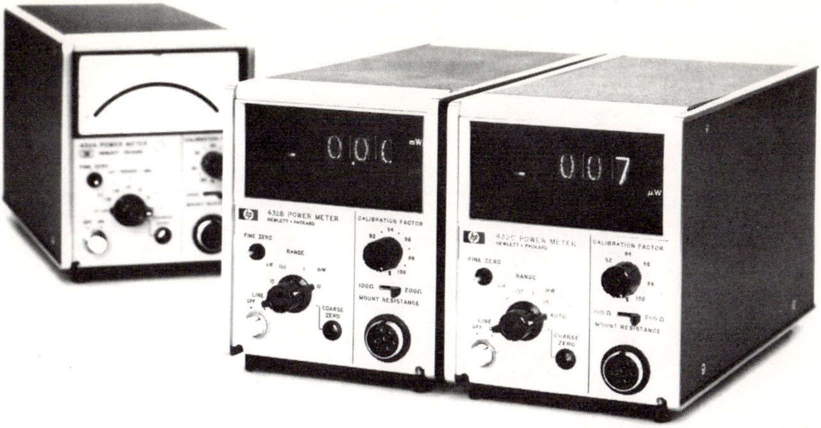

MICROWAVE POWER METERS GET DIGITAL READOUT AND BCD OUTPUT, OPTIONAL AUTORANGING AND PROGRAMMABILITY.

Figure 2–21 Self-Balancing Bolometer Bridges (Courtesy Hewlett Packard)

the same purpose; the main difference is the type of readout. One has an analog scale and the other a digital indicator. Both may be fitted with a variety of bolometer mounts to cover a frequency range from 10 MHz to 40 GHz. The analog meter has seven power ranges with full-scale values of 10, 30, 100, and 300 μW, and 1, 3, and 10 mW. The digital meter has four ranges with maximum readings 10 μW, 100 μW, 1 mW, and 10 mW.

chapter three

FREQUENCY MEASURING INSTRUMENTS

It is of prime importance that frequency measuring instruments be accurately calibrated against a reliable frequency standard. The absolute standard of frequency is the mean sidereal year as recorded for the U. S. National Bureau of Standards.

SOURCES OF STANDARD DATA

The U. S. National Bureau of Standards broadcasts time and frequency standards from WWV at Fort Collins, Colorado, and from WWVH at Maui, Hawaii. These stations broadcast continuously except for short periods of scheduled silence. Radio broadcasts from WWV are on frequencies 2.5, 5, 10, 20, and 25 MHz. WWVH transmits on frequencies of 2.5, 5, 10, and 20 MHz.

Data Available from Broadcasts

WWV and WWVH each broadcast time signals, radio propagation notices, and standards. Standards are broadcast against which audio frequencies, radio frequencies, time intervals, and musical pitch may be calibrated. A further breakdown of this information is shown in Figure 3–1.

The propagation forecasts are transmitted in International Morse Code and consist of a letter followed by a number. The letters are *N* for normal, *U* for unsettled, and *W* for disturbed. The numbers are 1–9. The significance of letters and numbers is further illustrated in Figure 3–2.

Transmission	WWV	WWVH
RF signal	2.5, 5, 10, 15, 20 and 25 MHz	2.5, 5, 10, and 15 MHz
Frequency stability	1 part in 10^9	5 parts in 10^9
Frequency deviation	2 parts in 10^{10} per day	4 parts in 10^{10} per day
Time signal	1000 Hz	1200 Hz
Duration	5 mS	6 mS
Accuracy	1 part in $10^8 \pm 1\ \mu S$	
Audio tones accuracy	440 and 600 Hz	
	1 part in 10^9	
Propergation forecast	19.5 and 49.5 min past the hr.	9.4 and 39.4 min past the hr.

Figure 3–1 Frequency Standards and Time Transmission

Letter	Digit	Condition
W	1	Impossible
	2	Very poor
	3	Poor
	4	Fair to poor
U	5	Fair
N	6	Fair to good
	7	Good
	8	Very good
	9	Excellent

Figure 3–2 Propagation Forecast

Two other National Bureau of Standards (NBS) stations are WWVL on a frequency of 20 kHz, and WWVB on a frequency of 60 kHz. All of these NBS stations, including the two mentioned previously, broadcast a great deal of information not covered here. Complete details may be obtained from the Superintendent of Documents, Government Printing Office, Washington, D.C. 20402. This office publishes a pamphlet on NBS Frequency and Time Broadcasts.

FREQUENCY STANDARDS

Frequency standards are classified in two categories: primary and secondary. A primary standard requires no other reference for calibration. Secondary standards must be calibrated during manufacturing and at periodic intervals during use. There are four types of frequency standards in common use:

1. Atomic hydrogen maser.
2. Cesium atomic beam-controlled oscillator.
3. Rubidium gas cell-controlled oscillator.
4. Quartz crystal oscillator.

Of these four, the first two are primary standards and the latter two are secondary standards. Since time and frequency are interrelated, a standard for one is a standard for both.

Atomic Hydrogen Maser

The atomic hydrogen maser produces an RF signal at the transition frequency of hydrogen atoms. This is a highly stable frequency of 1.42 GHz which should remain constant indefinitely. This maser is generally used as a master oscillator which controls a slave oscillator at a much lower frequency. The slave oscillator generally has a frequency of either 1 MHz or 5 MHz. The output of the slave oscillator is rigidly controlled and serves as a primary frequency standard.

Cesium Atomic Beam-Controlled Oscillator

Cesium beam standards are in use in environments where the requirement is a highly accurate primary standard. The NBS frequency standard is a cesium atomic beam-controlled oscillator. The oscillator consists of an atomic resonance device which produces an unvarying frequency of 9.192 GHz. This is the transition frequency of cesium atoms.

This primary standard may be obtained in portable equipment with the same levels of accuracy and long-term stability as large-scale laboratory models. Figure 3–3 is one version of a cesium beam frequency standard.

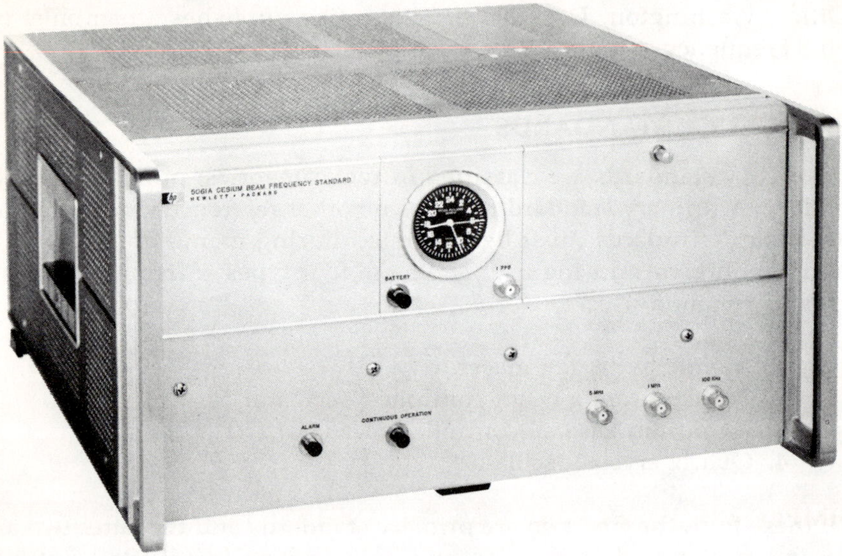

Figure 3–3 Cesium Beam Frequency Standard (Courtesy Hewlett Packard)

Rubidium Gas Cell-Controlled Oscillator

This oscillator is a secondary standard, but it features a high degree of both short-term and long-term frequency stability. It is generally portable equipment designed to be compact and light in weight.

The atomic resonant element controls the frequency of a quartz oscillator through a frequency lock loop. The frequency of the gas cell is dependent on the gas mixture and the gas pressure in the cell. It must be calibrated and is subject to a small degree of drift. However, the drift is approximately 100 times less than that of the best quartz oscillator. The controlled output frequency is that of rubidium atom transition, 6.834 GHz. Figure 3–4 is a photograph of such a standard.

Quartz Crystal Oscillator

A quartz crystal oscillator is used in nearly all frequency control applications. In addition to being an integral part of the atomic standards, it is used extensively as an independent frequency source where the applications are less demanding. Quartz oscillators may be obtained in portable, lightweight, rugged, and relatively inexpensive equipment.

Figure 3–4 Rubidium Frequency Standard (Courtesy Hewlett Packard)

An inherent characteristic of crystal oscillators is that their resonant frequency changes with time. After a few days aging, the frequency drift is almost constant. This drift can accumulate into a serious error, but with periodic calibration and frequency checks, these oscillators can be maintained as reliable frequency standards.

SIMPLE FREQUENCY MEASURING INSTRUMENTS

Vibrating-Reed Meters

Power supply frequencies can be measured with very simple devices. One of the simplest frequency meters is the vibrating-reed type. It is composed of an electromagnet and an array of thin metal strips. Each strip is fastened at one end and is free to vibrate. The vibration of these reeds is used to detect frequencies over a narrow range. Figure 3–5 illustrates the arrangement of reeds and magnet. The alternating current is passed through the windings of the electromagnet. The polarity of the resulting magnetic field reverses each time the current changes

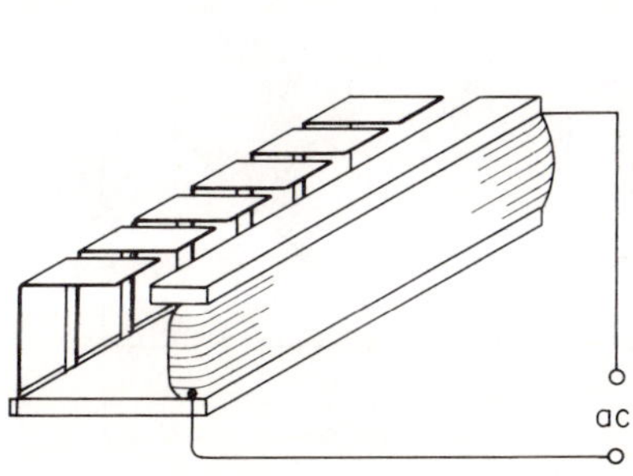

Figure 3–5 Vibrating-Reed Frequency Meter

direction. An extension of the magnet core forms a base plate for both the magnet and the reeds. The cyclic reversal of the magnetic polarity transmits a vibratory rate to the metal reeds.

The reeds are firmly attached to the base plate at one end, and the other end is free to move. The free end of the reed has a metal weight, and the weight on each reed is different. Even though all reeds are the same length, each vibrates at its own rate because of the weight. All reeds vibrate, but the greatest amplitude of vibration is exhibited by the reed which has a natural vibration period nearest to the energizing frequency.

When two adjacent reeds vibrate at the same amplitudes, the energizing frequency is midway between the two vibration frequencies. Frequencies can be estimated to the nearest half cycle by the relative amplitudes in the two affected reeds.

Crossed-Coil Meters

The crossed-coil frequency meter is a variation of an iron vane meter movement. It is designed to measure the frequency of a single-phase power source. The frequency range is rather narrow, usually about 5 Hz to either side of a designated power supply frequency. An accuracy within ±0.3% may be expected. Figure 3–6 illustrates this meter movement. Notice that the current path for winding *A* is resistive, and the path for winding *B* is inductive. Current through the resistive path energizes coil *A* on one half cycle. During the next half cycle, coil *B* is energized by current through the inductive path. Frequency has no effect on the resistive path but is directly proportional to the reactance of the inductive path.

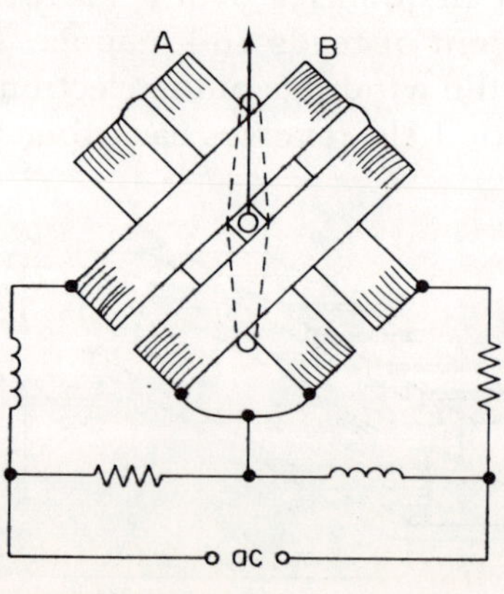

Figure 3–6 Crossed-Coil Frequency Meter

The indicator is secured to an iron vane which reacts to the ratio of the controlling and deflecting forces. The meter is calibrated in terms of frequency. When this meter is not in use, a caging device is used to keep the vane in a neutral position. Otherwise, it is floating free and the indicator can point to any spot on the scale.

Tuning Forks as Timing Instruments

A simple tuning fork, illustrated in Figure 3–7 is a useful timing instrument. The principal use of this tuning fork is checking the speed of rotation of motors and generators. The fork is set in motion by striking it against the hand, and it vibrates at a predetermined frequency. A slotted plate is attached to the end of each tine. As the tines vibrate, the slot opens and closes. The target area on the motor or generator is viewed through these slots. Viewing is possible only during discrete intervals. The target consists of black and white segments painted on the outer circumference of the flywheel. No special lighting is required, and speed of rotation can be checked in a matter of seconds.

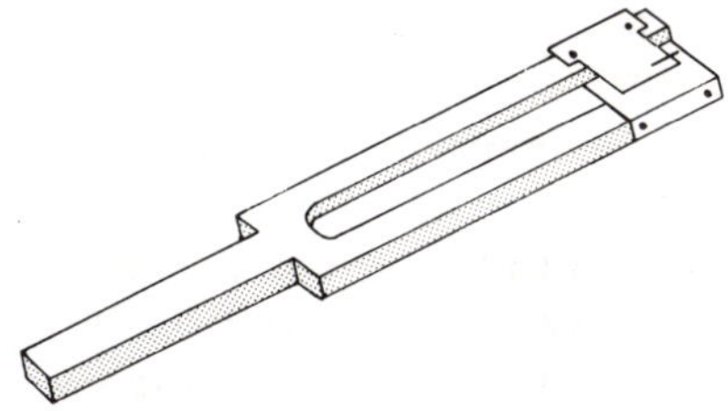

Figure 3–7 Tuning Fork

WAVEMETERS

There are two basic types of wavemeters: reaction and absorption. Both measure frequency, and both are passive in that they absorb energy from the circuit under test. The reaction type is preferred in low-power circuits because it uses less energy. In high-power circuits the absorption type is preferred because of greater accuracy. General principles of both types are developed, then some sample meters are discussed. When discussing the actual meters, the function will be emphasized instead of the basic classification.

Wavemeters are not noted for a high degree of accuracy. They tend to detune self-excited oscillator circuits to which they are coupled. However, they are extremely valuable instruments for preliminary frequency adjustments and general experimental work. No better tool exists for checking neutralization of an amplifier, detecting parasitic oscillations, checking transmitters for correct fundamental frequency, and for measuring radiated field strength in relative terms.

Reaction Wavemeter

The functional area of a reaction wavemeter is a tank circuit which can be tuned to resonate with the test frequency. This is illustrated in Figure 3–8.

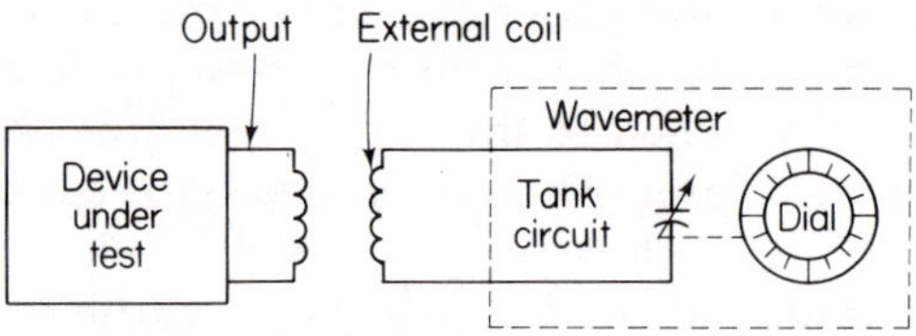

Figure 3–8 Reaction Wavemeter Circuit

The external coil is a plug-in unit, and the frequency range can be changed by using different coils. The dial varies the tunable capacitor until a condition of resonance is reached. The indicator is generally a headset or an ammeter. The condition of resonance is signaled by either a maximum or a null indication, depending on how the indicator is connected in the circuit. When the condition of resonance is obtained, the tank-circuit frequency matches the frequency of the device under test. The dial setting may then be converted to a frequency by reference to a frequency calibration chart.

The coupling between the wavemeter and the device under test should be loose and is gradually reduced as resonance is approached. The sharpest indication of resonance is obtained when the signal is barely strong enough for use.

Absorption Wavemeter

In actual construction there is little difference between the reaction wavemeter and the absorption wavemeter. This is readily apparent when Figures 3–8 and 3–9 are compared.

The lamp indicator in the absorption wavemeter circuit is bypassed by a large fixed capacitor. At resonance, voltage will be dropped across this capacitor to light the lamp, but this has negligible effect on the resonant frequency of the tank.

Coupling should be just close enough to produce a faint glow of the indicator when resonance is reached. Dial reading is converted to

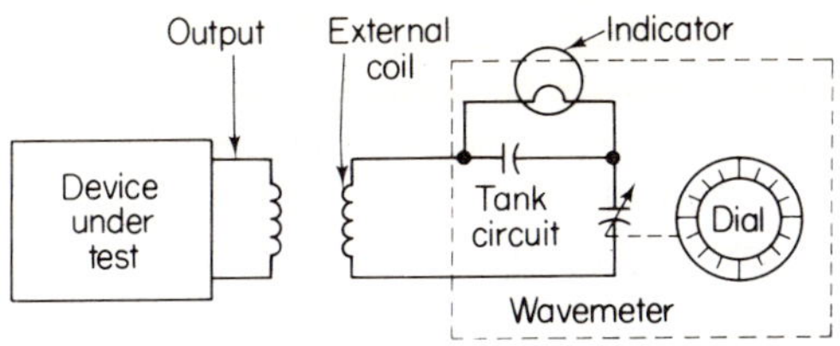

Figure 3–9 Absorption Wavemeter Circuit

frequency by reference to a chart. A good absorption wavemeter will provide accuracy of 0.25 to 2%. Since vibration, temperature, or moisture causes a change in capacitance and inductance, calibration should be checked frequently.

Resonant Cavity Wavemeters

At frequencies above 300 MHz it is necessary to use either a resonant cavity or a resonant coaxial line if wavemeters are to be employed. Figure 3–10 is a mechanical drawing of a cut-away resonant cavity wavemeter.

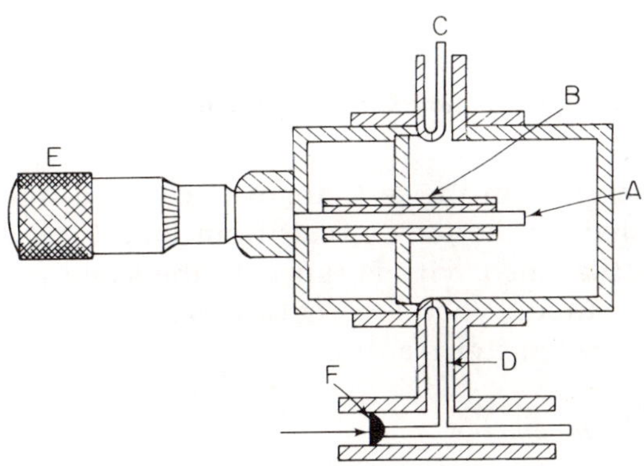

Figure 3–10 Resonant Cavity Wavemeter

The fundamental frequency of a resonant cavity is the resonant frequency with the longest wavelength. The resonant frequency of a cavity resonator can be changed by altering the mechanical dimensions. Large changes are possible with sliding members inside the cavity. Oscillations may be coupled in to, and out of, the resonant cavity by means of coupling loops.

The coaxial line in this resonator is a quarter-wave length with one end open and the other end shorted. Point *A* is the open end; a shorting bar at point *B* shorts the line.

The quarter-wave section can be lengthened or shortened by turning the micrometer thimble, *E*. The unknown signal may be coupled into either *C* or *D*. Inputs at *C* make use of the transmission method of measurement. Inputs at *D* use the absorption method. *F* is a diode rectifier to rectify the meter current. The equivalent circuit of this resonator is shown in Figure 3–11.

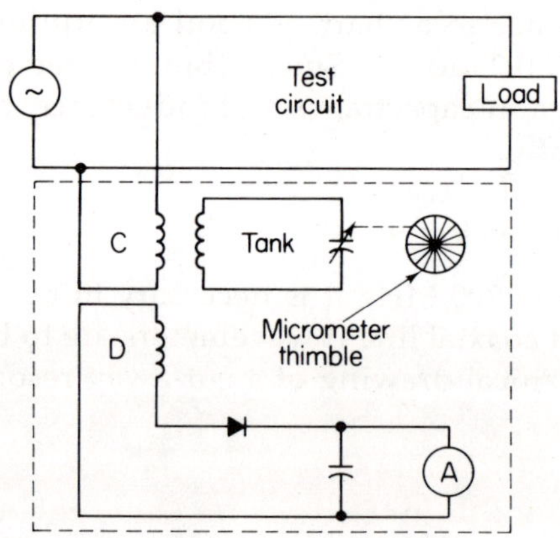

Figure 3–11 Circuit of Resonant Cavity Wavemeter

The RF energy from the test circuit is coupled into the resonant cavity, represented by the tank circuit from loop *C*. A dip in the ammeter indicates that the cavity is tuned to the same frequency as the signal being measured. When loop *D* carries signal, resonance is indicated by a dip in the meter reading.

Resonant Coaxial Wavemeter

Resonant coaxial wavemeters are similar in construction and function to the resonant cavity wavemeters.

Lecher Lines

Lecher lines use the absorption principle and provide a simple and direct means of measuring frequencies in the UHF band. They are resonant devices using direct measurement of the length of a standing-wave pattern on a short section of transmission line.

Lecher lines are generally composed of copper tubing and separated by a distance of 1 to 1.5 in. An accuracy as high as 0.1% is possible.

Open-End Lecher Lines

The function of an open-end Lecher line is illustrated in Figure 3–12. Ideally the ammeter is a thermocouple type, and its contacts slide along the line. Either minimum or maximum current may be used for checkpoints. The distance between checkpoints (two minimum or two maximum) is a half wavelength. This distance can be read from a calibrated scale. Frequency in megahertz is obtained by dividing 150 by the distance of a half wavelength in meters.

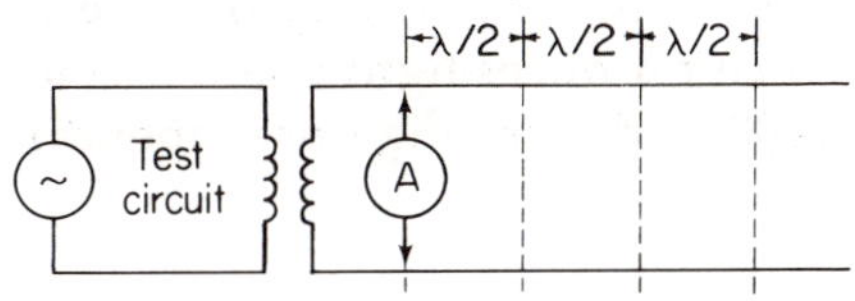

Figure 3–12 Open-End Lecher Line

Shorted-End Lecher Lines

The function of a shorted-end Lecher line is illustrated in Figure 3–13. This line has a shorting bar which slides along the line making contact with both conductors. The line terminates at the position of the shorting bar. The bar will reach points of resonance at half-wave distances. These resonant points are indicated by a dip in the meter. Frequency is determined as explained for open-end Lecher lines.

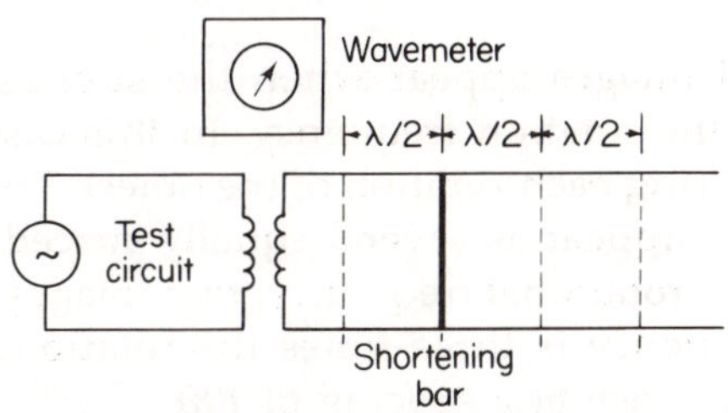

Figure 3–13 Shorted-End Lecher Line

Stroboscopes

The stroboscopic principle uses a high-intensity light, which flashes at precise intervals. This light may be directed upon a rotating or vibrating object. The stroboscopic effect is apparent when rotational or vibratory speed is in proper ratio to the frequency of the light flashes.

Most stroboscopes are composed of an oscillator, a reed, and a flasher as illustrated in Figure 3–14. The oscillator provides trigger pulses to the flasher mechanism to control the flashing rate. The oscillator is generally an externally synchronized multivibrator. The vibrating reed serves as a reference for accurately calibrating the stroboscope. The reed is driven from the ac lines and vibrates 7200 times each minute. This steady rate is used to standardize the calibration scale over a narrow range. The flasher produces the illumination for the measurements. The tube is fired by a capacitor discharge, which is, in turn, controlled by trigger pulses from the oscillator. The tube is filled with a suitable inert gas which produces light when it is ionized. The tube life ranges from 200 to 1000 hr depending upon operating conditions.

Figure 3–14 Block Diagram of Stroboscope

When frequency of movement exactly matches the stroboscope frequency, the moving object is viewed clearly only once during each revolution. This causes the moving object to appear as a single stationary image. A stationary image also appears when the speed of rotation is some exact multiple of the stroboscope frequency. The highest scale reading that produces the single, still image is the fundamental frequency.

Multiple still images appear when the stroboscope frequency is some multiple of the rotation frequency. In this case the light flashes more than once during each rotation of the object. The radial line on the end of a shaft may appear as several equally spaced lines. If lamp frequency is twice the rotational frequency, two images are produced 180° apart. If lamp frequency is three times the rotational frequency, three images will appear, each at a spacing of 120°.

Moving images are obtained when the light frequency and the rotational frequency are not synchronized. When the image appears to rotate in an opposite direction to that of actual rotation, the rotation frequency is less than the flasher frequency. When it appears to rotate in the same direction as actual rotation, the rotation frequency is higher than the flasher frequency.

A stroboscope may be used to check motor or generator speeds ranging from 60 to 1,000,000 revolutions per minute. The stroboscope is highly versatile, uses no power from the circuit being measured, and when carefully calibrated, accuracy can be as close as 0.1%. General appearance is illustrated in Figure 3–15.

This scope from General Radio uses the line frequency for calibration. The flash-lamp and reflector assembly rotate 360° for maximum flexibility. The case may be mounted on a tripod. The flash rate is 110 to 150,000 flashes per minute, enabling measuring speeds up to 1,000,000 revolutions per minute. Light output varies with flash rate from 3 μs to 0.5 μs.

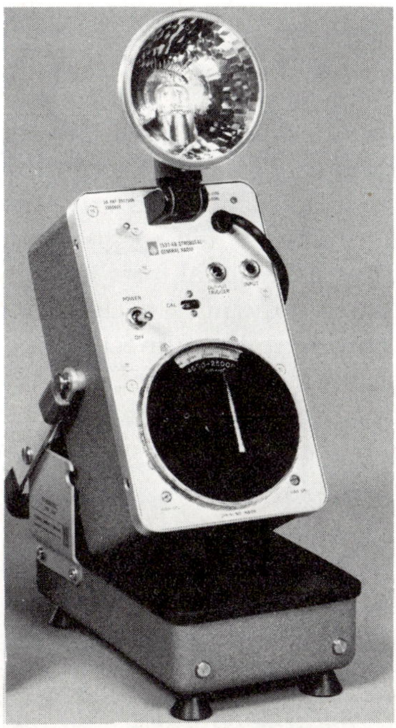

Figure 3–15 Portable Stroboscope (Courtesy General Radio)

Frequency Counters

A frequency counter is a digital frequency meter. It is a high-speed electronic counter with an accurate, crystal-controlled time base. It automatically counts the number of events in a precise time interval and displays these in the form of a digital readout. The counter is independent of input waveform and is easy to operate. It generally consists of a time-base, signal input, gating, display, and reset sections. The arrangement is illustrated in Figure 3–16.

The signal input section is primarily a Schmitt trigger. It produces standard-shaped pulses at the same frequency as the input signal. The

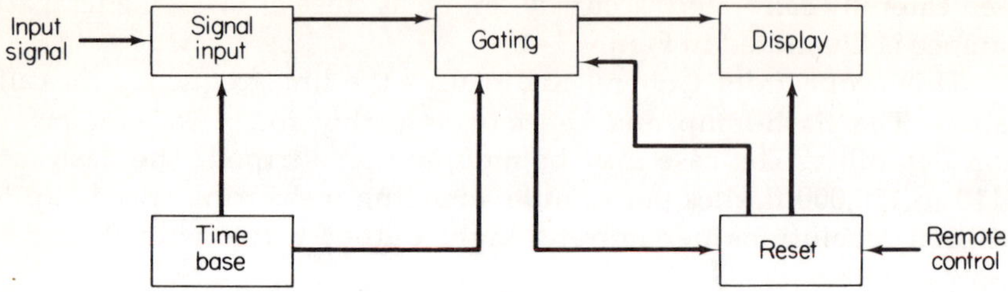

Figure 3–16 Functional Diagram of Frequency Counter

time-base section consists of a crystal-controlled oscillator and frequency divider multivibrators. A 100 kHz crystal oscillator produces time bases of 10, 1, and 0.1 Hz. The outputs are gate pulses which gate input signals to the display section and drive the reset section.

A new cycle is initiated when a signal from the time-base section occurs at the same time as a signal from the reset section. Signals pass through the gating section during a cycle. The display section is composed of decimal counting units connected in cascade. These counting units total and display the Schmitt trigger pulses which occur during each counting period. The display time is variable, and all counters are reset to zero at the end of each display period. One model of a frequency counter is illustrated in Figure 3–17.

Figure 3–17 Frequency Counter (Courtesy Hewlett Packard)

This digital readout counter features automatic measurement from 0.125 Hz to 20 MHz, using any gate time up to 4 seconds. It also contains a multiplier which converts from hertz to revolutions per minute when desired.

Grid Dip Meters

The *grid dip* meter is not a precision instrument, but it is a
strument because it is small, portable, easy to operate, and r
accurate. It is composed of an RF oscillator, a null indicator, ar
priate coupling circuits. One example of circuit design is illus
Figure 3–18.

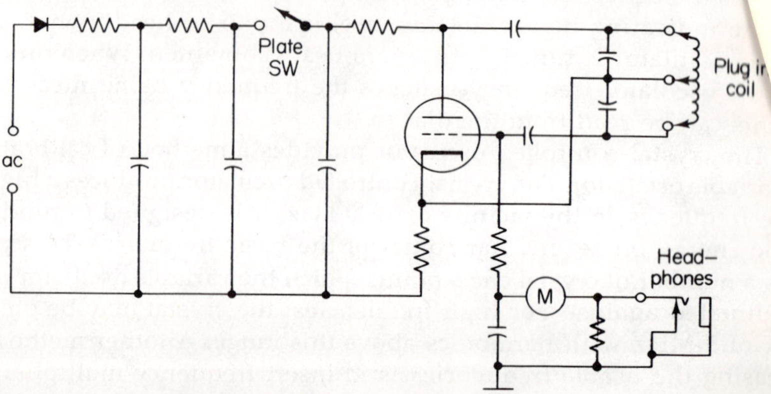

Figure 3–18 Grid Dip Meter

The grid dip meter functions in either an active or passive fashion depending on the type of source of the signal being measured. Input signals are loosely coupled to the tuned circuit of the oscillator. When the oscillator is tuned to the frequency of the test circuit, energy is absorbed from the grid circuit. This causes a null (dip) in the meter indication. Minimum coupling provides the most accurate measurements.

When more precise frequency measurements are required, the grid dip meter may still be employed. In this case it should be heterodyned with a more accurate frequency standard. It may also be used as a passive wavemeter by removing the plate voltage from the oscillator. As a passive device, the grid dip meter indicates resonance by a maximum deflection instead of a null.

Heterodyne Frequency Meters

A frequency meter of the heterodyne type can be extremely accurate. The degree of accuracy is largely dependent on the sensitivity of the indicating device. This meter is used to measure the frequency of RF signals by matching the test signal (or some harmonic thereof) with a

y generated signal. Basically a heterodyne frequency meter con- of a variable oscillator, a mixer, a detector, an audio amplifier, and ndicator. A *second* oscillator is included for calibration purposes. e stages are functionally arranged as illustrated in Figure 3–19.

The RF signal to be measured is generally coupled from a meter nna, through the input jack to the mixer. It mixes with the signal n the variable oscillator to produce an audio signal. The detector cesses the difference frequency and passes it on to the audio ampli- r. The indicating device, in this case, is a pair of headphones. The riable oscillator is tuned until a zero beat is obtained. When this oc- urs, the oscillator frequency matches the frequency being measured, nd this can be read from the dial.

The crystal-controlled oscillator provides a method of calibrating e variable oscillator. The crystal-controlled oscillator produces a highly able frequency in the vicinity of 1000 kHz. It is designed to produce sable energy on several harmonics of the basic frequency. This pro- ides a number of crystal checkpoints which the variable oscillator may be calibrated against. For high frequencies, the crystal may be on the order of 5 MHz with harmonics above this range. Another method of increasing the usable frequencies is to insert frequency multipliers at the output of the crystal-controlled oscillator. A sturdy high-grade instrument will hold calibration for a long period of time.

The headphone is the least sensitive type of indicator. When the zero beat is properly attained, the error may be off from 30 to 100 Hz. Other types of indicators are electronic voltmeters, AF voltmeters, neon lamps, electron-ray tubes, and cathode-ray tubes. Accuracy when using headphones ranges from 0.02 to 0.04%. The oscilloscope indicator enables accuracy up to 0.001%.

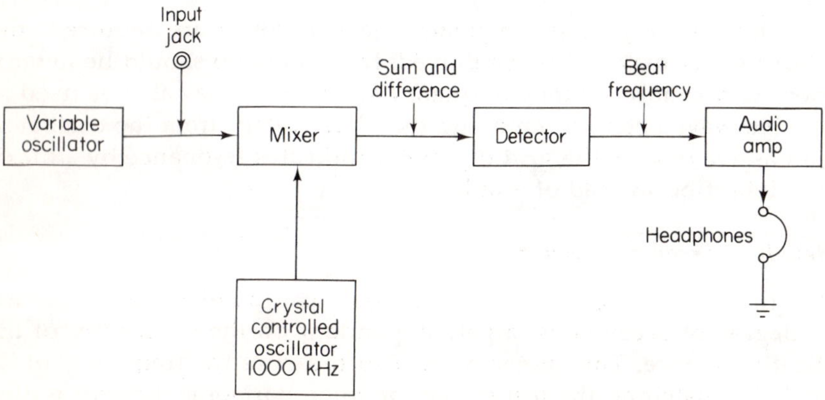

Figure 3–19 Heterodyne Frequency Meter

In the ultrahigh-frequency range, the principles are the same, but the components of the meter appear quite different. Resonant cavities become frequency-determining devices with micrometer tuners.

With some frequency meters, the calibration book is an important part of the frequency meter. The book is intended to match a particular meter; book and meter carry the same serial number. The book contains a listing of dial settings and corresponding frequencies. It also contains precise operating instructions and crystal checkpoints.

chapter four

WAVEFORM MEASURING INSTRUMENTS

A waveform is a graphic representation of a varying signal plotted against time. An unknown waveform may be plotted by using a system of coordinates with a linear relationship between time and amplitude. Many instances in electronic work require careful analysis of waveforms.

A cathrode-ray tube provides the basic component in the structure of all waveform measuring instruments. The cathode-ray instrument common to most electronic shops is the *oscilloscope*. The oscilloscope is the most versatile of all waveform measuring instruments. There is an oscilloscope in a size, shape, and price range to fit nearly all applications.

Other less common waveform measuring instruments which use the cathode-ray tube are: waveform monitors, vectorscopes, characteristic curve tracers, scan converters, graphic computer terminals, and spectrum analyzers.

This chapter develops the operational principles of waveform measuring instruments, with particular emphasis on the oscilloscope.

CATHODE-RAY TUBE

The cathode-ray tube (CRT) has rightly earned its place as the *picture tube* of the electronics industry. It is used for computer displays, television picture tubes, radar scopes, and for observation of electronic waveshapes. There are two standard types of CRT: electromagnetic and electrostatic.

The type of CRT is determined by its deflection system. The electromagnetic uses electromagnets. Here, deflection is accomplished by varying the field strength of deflection coils. This type is seldom, if ever, used in test equipment.

The electrostatic CRT accomplishes deflection by varying the potential on deflection plates. This creates electrostatic fields proportional to the amplitude of the voltages used. This is the CRT used in electronic instruments. Since the electrostatic CRT is encountered continuously in the field of electronics, a review of its structure and function is in order.

Electron Gun

Each CRT contains one or more *electron guns.* A gun consists of a heater, cathode, grid, and two anodes. The gun generates, directs, and focuses a sharp beam of electrons onto a phosphor (fluorescent) coated screen. The relationship of these parts is illustrated in Figure 4–1.

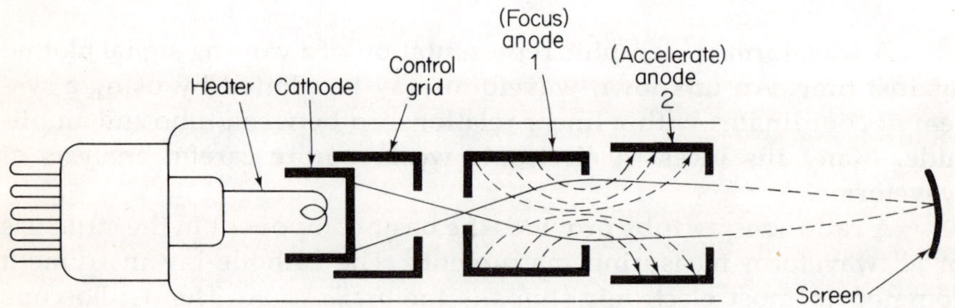

Figure 4–1 Electron Gun

Thermionic emission drives electrons from the end of the cathode. A portion of these electrons is directed through a small opening in the grid which surrounds the cathode. Varying the potential on the grid determines the quantity of electrons allowed to pass.

The first anode (anode 1) is a hollow cylinder with a high negative potential. The potential repels the electrons in all directions, squeezing them into a tight beam. Turning the focus control varies this potential and determines the focal point of the beam. A focal point produces a sharp bright spot exactly where the beam touches the screen.

The second anode is the most positive part of the electron gun; it is about 300 V positive. In respect to anode 1, it is about 1000 V positive. This difference in potential drives the electrons toward the screen with a high acceleration. The relative potential between the first and second anode is variable and is controlled by the focus control.

All CRTs have one electron gun, and some CRTs have several. A single-beam oscilloscope uses a single electron gun; a dual-beam CRT has two electron guns, and some picture tubes use five electron guns.

Envelope and Screen

The CRT employs an evacuated glass envelope, and the viewing screen is created by a phosphor coating on the inside front of the tube. The electrons strike the screen with considerable force and cause the material to fluoresce. The emitted light produces a bright spot which is visible from the outside front of the screen. The envelope is illustrated in Figure 4–2.

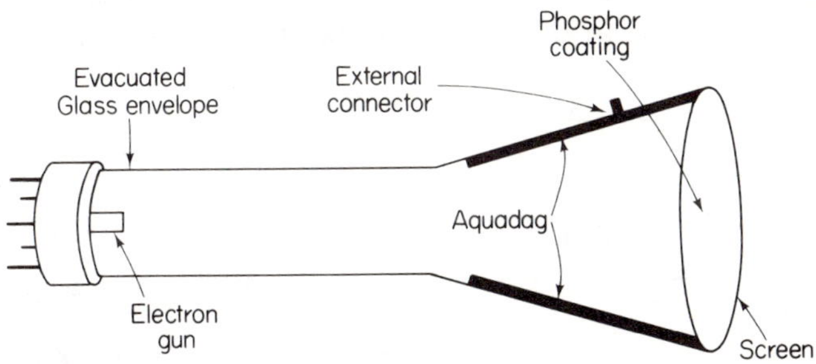

Figure 4–2 CRT Envelope

As the electrons bounce from the screen, they are collected by the aquadag coating and returned to the power supply. The aquadag is a metal coating which covers all the inside of the forward portion of the CRT. A very high positive is connected to the aquadag external connector. This is a dangerously high potential, and when the cover is removed, the connector can easily be contacted.

Electrostatic Deflection System

The deflection system consists of two sets of plates for each electron gun, as illustrated in Figure 4–3. The electron beam is the stylus, and the deflection plates move it about on the screen to draw an oscillograph of the applied signals. The vertical plates move the beam up and down, while the horizontal plates move it right and left. The two sets of plates are physically separated far enough to prevent interaction of

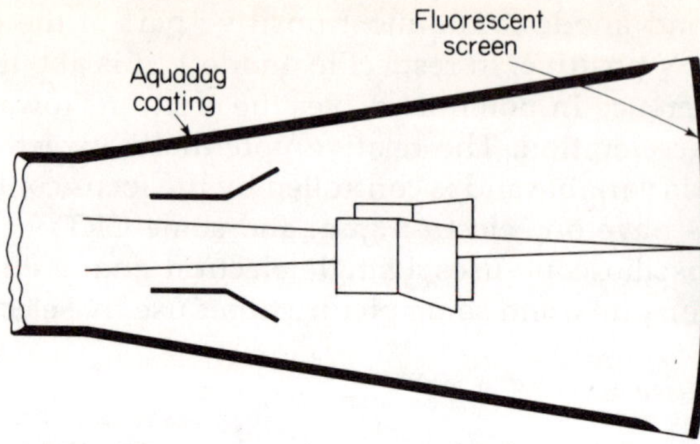

Figure 4–3 Electrostatic Deflection Plates

the fields. This enables vertical and horizontal movements to be independent of one another.

Signals, or potentials, are applied to opposite plates of a pair in push-pull fashion as illustrated in Figure 4–4. The dc potential on the plates is balanced to place the dot exactly in the center of the screen.

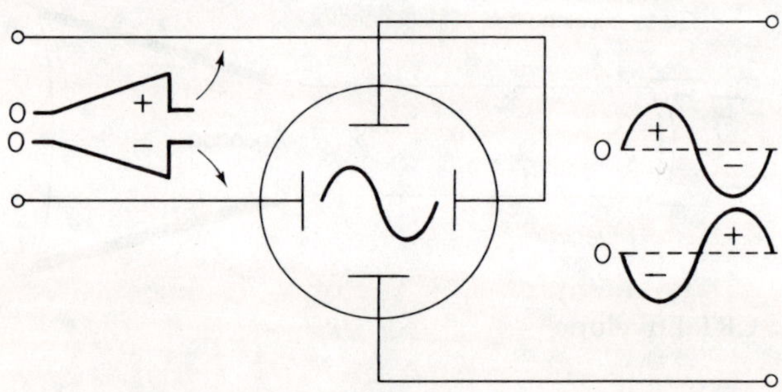

Figure 4–4 Push-Pull Deflection

An ac signal of any shape on the horizontal plates will produce a horizontal trace. The length of the trace will be proportional to the signal amplitude. The same is true of an ac signal on the vertical plates, except that this moves the dot vertically and causes a vertical trace. The position of the dot at any instant is a resultant of the potentials on the two sets of plates at that instant.

BASIC OSCILLOSCOPE CIRCUITS

Each type of oscilloscope has certain special circuits and features, but all contain the basic circuits covered here. These circuits are covered from an operational viewpoint.

Functional Analysis

Figure 4–5 is a simplified functional diagram of a basic oscilloscope. Both power supplies are operated from the ac line voltage. The low-voltage power supply furnishes dc potentials for most of the circuits, including centering voltages for the deflection plates. The high-voltage power supply provides the high potentials required by the electron gun and aquadag coating.

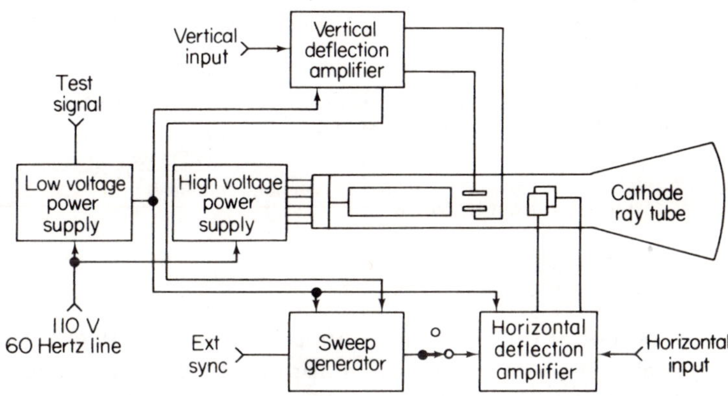

Figure 4–5 Basic Oscilloscope

The horizontal channel, sometimes called the *X*-axis, consists chiefly of a sweep generator and horizontal deflection amplifiers. The sweep generator provides a linear sawtooth signal as a sweep voltage. The sawtooth is amplified and placed on the horizontal plates in push-pull. This is generally the signal used for horizontal deflection. It provides the trace, and its duration is the time base. When desired, the sweep generator can be switched out of the circuit. In this case, an external signal is used to replace the sawtooth wave.

The vertical channel (*Y*-axis) is mainly composed of the vertical deflection amplifiers. The external signal, which is to be displayed, is applied through the vertical input jack. After appropriate amplification, the external signal is placed on the vertical deflection *plates* as push-pull signals.

CRT Potentials

Figure 4–6 illustrates the relative dc potentials as applied to the CRT. This drawing shows a low-voltage power supply which applies a fixed,

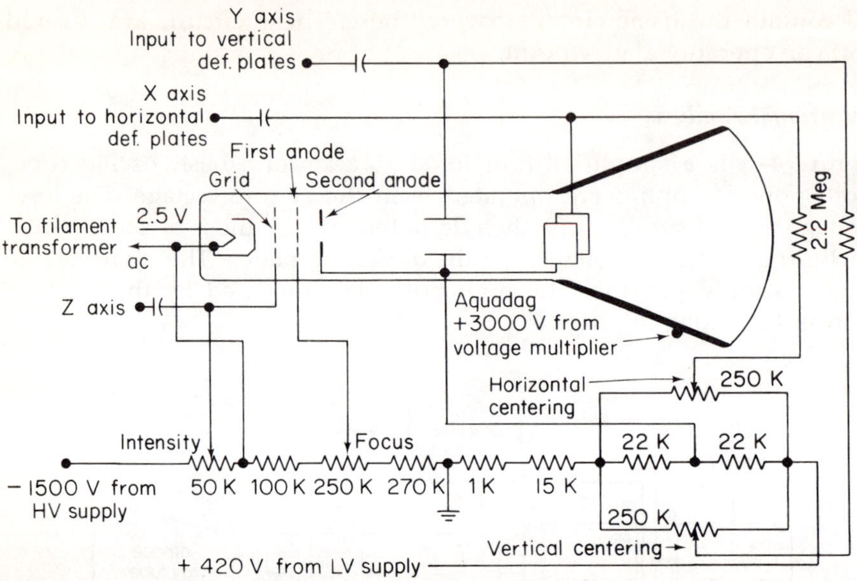

Figure 4–6 CRT Potentials

positive dc potential on the second anode, bottom vertical plate, and left horizontal plate. A variable dc is applied to the top vertical plate and right horizontal plate. The respective centering controls vary the deflection voltages for positioning of the picture on the screen.

The −1500 V (high-voltage) source furnishes the high potentials for cathode, control grid, and first anode. The cathode has a fixed potential of slightly more than −1400 V. The intensity control places a variable potential on the grid, which is slightly more negative than the cathode. The intensity control exercises complete control of the electron stream from shut-off to a very intense beam.

The focus control varies the first anode potential to either side of a −750 V center point. This control varies the focal point of the electron beam, and the more negative the setting, the shorter the focal point.

Not all scopes use such high voltage on the aquadag coating. In fact many of them use about 300 V from the low-voltage power supply. Regardless of the source or value of this potential, it is high enough to be dangerous, and its location makes it easy to touch if proper care is not exercised.

External inputs may be applied to the *X*-, *Y*-, and *Z*-axes. The accompanying diagram is not intended to show the actual coupling of these inputs. For instance, the *X* and *Y* inputs each appear to be applied to only one deflection plate when they are actually coupled to both plates. The Z-axis provides a means of applying intensity modulation to the electron beam. The effect of intensity modulation is illustrated in Figure 4–7.

Inputs to the Z-axis provide frequency or time-scale markers. In the examples shown, each negative alternation of the input cuts off the electron beam and leaves a blank spot on the waveform being observed. The same thing can be accomplished by applying a series of negative trigger pulses. This is essentially the same type of blanking action that occurs during flyback time. A series of positive pulses to the Z-axis results in a bright spot on the presentation for each pulse of the input. This idea is utilized for range markers.

Figure 4–7 Intensity Modulated Display

TYPES OF OSCILLOSCOPES

There is a wide range of oscilloscopes available today. They range from simple, general-purpose types to complex, highly sophisticated laboratory models. Some of the simpler models are scarcely more complicated than the basic oscilloscope described in the preceding paragraphs, whereas a top of the line laboratory model will have multiple electron guns, multiple presentations, digital readout, memory circuits, pushbutton controls, and many other special features.

The price of course is proportional to the degree of sophistication. The intended use of the oscilloscope should dictate the model selected. Many very excellent oscilloscopes are available for $200 or less. These models are great for general-purpose shop work. In the laboratory class, however, it is not unusual to find price tags in excess of $3000 with plug-in options to raise the price well above this figure.

There is no sharp dividing line between general-purpose and laboratory oscilloscopes. Except for a discussion of this type, there would be no need for the two categories.

DISTINGUISHING FEATURES

All oscilloscopes have more features in common than they have differences. This section will treat features in a general fashion and point out the differences which normally differentiate general-purpose from laboratory oscilloscopes. All modern oscilloscopes, except the very basic types, have many advanced features. Design emphasis on one or more features frequently makes the difference between general-purpose and laboratory instruments.

Screen Coating

Various phosphor chemicals are used for forming the screen on the inside of the CRT; each type having its specific advantages for the application intended. All types have the property of fluorescence; they emit light when bombarded by electrons.

The intensity of the spot on a screen is determined by the number of electrons striking that spot per unit time and the speed of collision. The amount of light per unit area which the phosphor is capable of emitting, of course, is a limiting factor. The intensity control varies the number of electrons and controls the intensity up to the saturation point of the phosphor.

The *afterglow* is the light that remains after bombardment ceases. The *persistency* of a screen is the time required for the afterglow to diminish by a specified amount. The general classifications are short, medium, and long persistency.

The persistence of vision is a crucial factor with short-persistence phosphor. The human eye sees a continuous pattern when it is retraced at least 16 times each second. In some cases, it is essential to view waveforms at frequencies below 16 Hz, but the combination of screen persistency and eye persistency gives the illusion of a continuous picture.

At medium- and high-phosphor persistency, the human eye is no longer a factor. The afterglow remains long enough to allow solid pictures at slower rates. However, the human eye is not always the viewer. In many applications, a camera observes the screen, and personnel observe the photographs. Cameras are often employed in laboratory environments, and some oscilloscopes are designed primarily for making photographic records. Such oscilloscopes generally have a white or bluish white phosphor with a short to very short persistency.

Screens designed primarily for direct viewing are usually coated with a green phosphor which has a medium to high persistence. This type of screen can be expected in all general-purpose oscilloscopes as well as in the majority of laboratory scopes. In laboratory models, the purchaser is usually offered a choice of phosphors. In permanent installations, the position and lighting of the screen should be taken into consideration when this choice is made.

Vertical Channel

The vertical amplifiers determine the sensitivity, frequency response, and to a large extent, the cost of an oscilloscope. The deflection system is relatively insensitive; it may require several hundred volts on the deflection plates to produce full-scale deflection. Except for very strong signals, a certain amount of amplification is essential in the vertical channel. Provisions are made for applying strong signals directly to the deflection plates, and attenuator probes are frequently used to limit the amplitude of extremely strong signals.

It is a well-known fact that for a given stage of amplification, sensitivity and bandwidth are diametrically opposed. Further, an amplifier faithfully reproduces a signal only when the signal meets strict limits of both frequency and amplitude. Wide frequency response is attained at the expense of lowering the signal gain. Less gain in a given stage means that more stages are necessary in order to produce the required amplification. This leads to some highly complex amplifier arrangements with price tags to match.

At higher frequencies, additional problems are encountered. Input impedance and gain control may have adverse effects. A variable resistor may cause different effects on the same frequency as the resistance is changed over a wide range. This leads to distortion of both phase and frequency. Laboratory oscilloscopes generally employ a step-type switch as a frequency compensated attenuator which precedes the first amplifier and the gain control. This switch reduces the effect of overloading as well as the phase and frequency distortion. The same switch often provides direct coupling to the first amplifier to enable observation and measurement of outputs from dc devices.

General-purpose oscilloscopes can be obtained for either medium or high frequency. Medium-frequency scopes have a satisfactory response from zero up to about 5 MHz. A 0-to-10 MHz response may be expected in a high-frequency model. A 3 MHz bandwith with a deflection sensitivity of 100 mV/cm is satisfactory for most general-purpose applications. Laboratory oscilloscopes generally provide a choice of gains, sensitivities, and bandwidths.

Horizontal Channel

Most of the discussion on vertical amplifiers also applies to horizontal amplifiers. One exception is the fact that attenuators rarely appear in the horizontal channel. There are two reasons for this: (1) the horizontal channel is less sensitive because the plates are closer to the screen, and (2) most procedures use the internal sweep generator. The amplifiers provide gain for either the internal sweep voltage or for an external signal coupled to the horizontal channel. The horizontal gain control works through the amplifiers to provide the control on the width of the presentation.

The sawtooth sweep generator is a switching device controlling the charge and discharge of a capacitor. A small portion of the charge waveform provides the ramp voltage which rises reasonably linearly with time. The discharge of the capacitor is the flyback of the sweep. A sweep voltage and flyback blanking are illustrated in Figure 4–8. The *blanking* waveform is applied to the grid of the CRT. The negative alternation cuts off the electron stream and blanks out the flyback portion of the sweep.

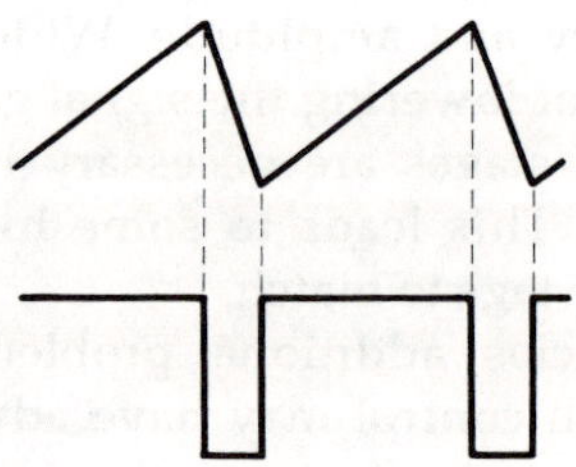

Figure 4–8 Sweep and Flyback Blanking

Most general-purpose oscilloscopes have a frequency range from about 10 Hz to perhaps 500 kHz. TV service requires a sweep frequency of about 1 MHz. A course sweep range control steps the sweep frequency, and a fine control varies the frequency within the steps. The general-purpose oscilloscope has about six calibrated steps within a range of 1μs to 100 ms/cm. The laboratory oscilloscope may have more than 20 calibrated steps within a range of 1μs to 5 s (or more)/cm.

The sweep must be synchronized with the signal under investigation in order to obtain a stationary pattern. There are three common sources of synchronization, and a selector control is used to select the type of sync desired. These are: internal, external, and line. The internal position selects a portion of the test signal from the vertical amplifier and uses it to trigger the sweep generator. The external position is used when a convenient external signal is triggering the signal being observed. This uses the same signal to trigger the sweep generator, and it conveniently synchronizes the sweep with the test signal. Line synchronization uses the 60 Hz time signal to trigger the sweep generator. This is convenient in cases where the waveform under observation is also triggered by a signal at the line frequency. The blanking waveform previously mentioned is generated in the sweep generator circuits, so there is no problem in synchronizing the blanking signal with the sweep signal.

It is true that a stationary display can be attained without synchronization by varying the frequency of a free-run sweep generator. Each time the sweep frequency matches the test signal frequency, one stationary cycle of test signal is visible. Each time the sweep frequency matches a harmonic of the signal frequency, a stationary pattern is visible. In the latter case, there is one complete cycle of the signal for each harmonic; two complete cycles of the signal for the second harmonic, three for the third, etc.

Some general-purpose oscilloscopes may be encountered which do not have the driven sweep synchronization capabilities. When purchasing an oscilloscope for any purpose, the extra cost for a driven sweep over a free-run sweep is a good investment. Over the expected useful period of the instrument, the synchronized sweep will save many hours of frustration.

Nonlinear Time Bases

Oscilloscopes are not limited to linear, straight-line time bases. Laboratory oscilloscopes intended for specialized engineering and research applications are sometimes equipped with circular- , spiral- , or radial-type sweeps. One reason for this is a longer time base on a given CRT screen. Another reason is elimination of flyback time which can cause loss of a portion of the test signal. These oscilloscopes have special circuits for producing such sweeps.

The engineer or technician can have some of the advantages of circular and elliptical sweeps even with ordinary general-purpose oscilloscopes. A circle can be obtained by applying sine waves to the deflection plates. Two sine waves of the same frequency must be applied in a manner that causes 90° phase difference between adjacent plates and 180° phase difference between opposite plates. This is illustrated in Figure 4–9.

With the waveshapes shown, the dot appears at the bottom center of the screen at each time t_1 and rotates counterclockwise. For a perfect circle, the two signals must be of equal amplitudes as well as having the phase relations just described. Unequal amplitudes will produce an elliptical sweep.

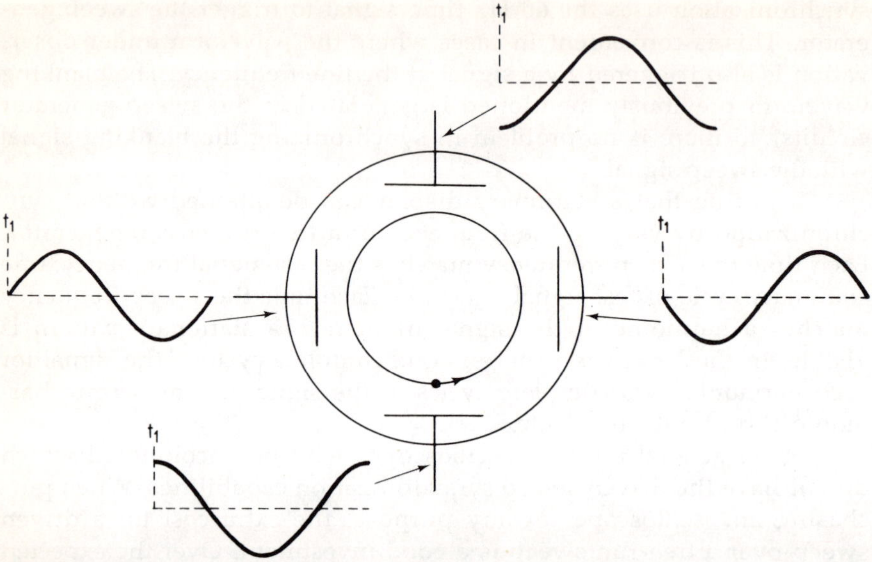

Figure 4–9 Production of a Circular Sweep

Signals fed into the vertical channel will now be displayed on the circle. The test signal amplitude modulates the sine wave on the vertical deflection plates to produce this display. Figure 4–10 illustrates the results of a square-wave input. The same synchronization problem exists as with any other sweep. For a stationary display, the signal being observed must be some exact harmonic of the sweep frequency.

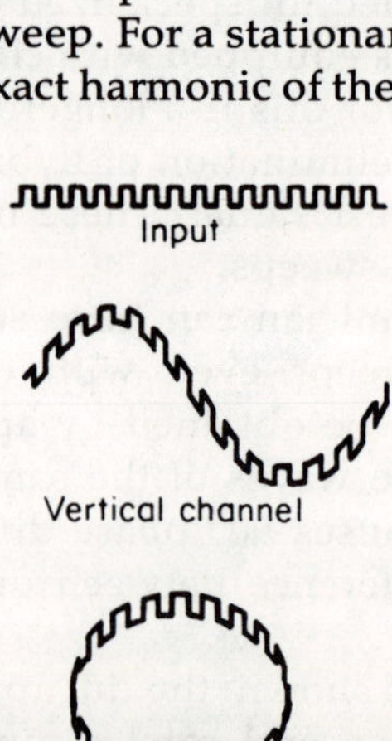

Figure 4–10 Display on Circular Sweep

Sweep Calibration

In ordinary wave observation, the length of the sweep waveform is not vitally important. In radar maintenance and other pulse applications, pulse duration, pulse recurrence time, pulse rise time, and pulse positioning are primary considerations. These applications require calibration of the time base.

Laboratory and some general-purpose oscilloscopes provide methods of calibrating the sweep. An internal oscillator of a known, stable frequency furnishes an output which can be injected into the vertical channel. The sweep can then be adjusted until the reference signal on the screen covers a specified distance. The sweep velocity can be expressed in distance per unit time. For instance; when a 1-MHz calibration signal covers a horizontal distance of 10 cm, the sweep is calibrated to 1 μs/10 cm or 100 ns/cm.

When the oscilloscope does not contain calibration for the sweep, calibration is still possible by inserting a signal from a standard-frequency source. It should be kept in mind that the full sweep includes flyback time. This can cause a minor error which is usually safely ignored.

Marker Generators

Marker generators are another way to calibrate the sweep on an oscilloscope. These markers are standard on laboratory oscilloscopes and are found on many general-purpose oscilloscopes. The markers are produced by an accurately adjusted tuned circuit. They are spaced at definite time intervals, and the time is selected by a switch. Marker intervals are 0.2, 1, 10, 100, and 500 μs, as a typical example.

One type of marker generator is a tuned tank which is triggered by the start of the sweep. The circuit then oscillates at its own frequency, which is determined by the marker interval selected. The first markers from this circuit are strong, and they gradually grow weaker as the oscillations decay. A sample output of such a circuit is illustrated in Figure 4–11.

These markers can be applied directly to the CRT cathode, or they can be inverted and placed on the grid. Either way provides intensity

Figure 4–11 Markers from a Tuned Tank

modulation of the electron beam, and each marker places a bright spot on the screen presentation.

Voltage Calibration

Most oscilloscopes have an internal voltage calibrator circuit to enable calibration of the vertical deflection. This is vitally important if signal amplitudes are to be measured.

One popular type of calibrator converts the 60-Hz signal from the CRT filament circuit into square waves. These square waves are changed to zero reference and are fixed at a specified peak-to-peak amplitude. This signal is injected into the vertical channel, and the gain control is set for the desired deflection. If three divisions are selected, each division represents one volt at the input. On this basis, exact amplitude of input signals can be *measured.*

Another type of calibrator has a variable control. This enables measuring a signal amplitude by comparing it with the calibrator signal of known amplitude. The amplitude can be read directly from the calibration control when the two signals are the same amplitude. If different amplitudes are used, ratio calculations are required.

When there is no provision for amplitude calibration, the same thing can be accomplished by using an external source for obtaining a signal of known amplitude. A voltage calibrator may be purchased as a separate piece of test equipment, but any stable signal of known amplitude may be used.

Expanded Sweep

Laboratory oscilloscopes and some general-purpose oscilloscopes have circuits which permit expansion of selected segments of the sweep. This enables close-up inspection of signals for greater detail. A potentiometer control enables selection of the desired portion of the sweep. Extra amplification is provided for the sawtooth voltage, beginning at the selected point. The expansion factor may be some fixed amount, or it may be under a multiplier control which enables the operator to select the desired expansion. This is sometimes referred to as *delayed sweep* because the sweep starts at a later time.

Polarity Reversal

Generally a positive potential in the vertical channel causes a deflection upward in respect to the horizontal axis. Some oscilloscopes have a display polarity reversal switch in the vertical amplifier section. This switch enables the operator to select either positive or negative deflection when either polarity signal is applied.

SAMPLE GENERAL-PURPOSE OSCILLOSCOPE

There are many oscilloscopes that are similar in nature, and yet each model has so many individual characteristics that it is impossible to feature a typical oscilloscope. The RCA WO-535A* dual-mode oscilloscope shown in Figure 4–12 has been selected as a sample general-purpose oscilloscope.

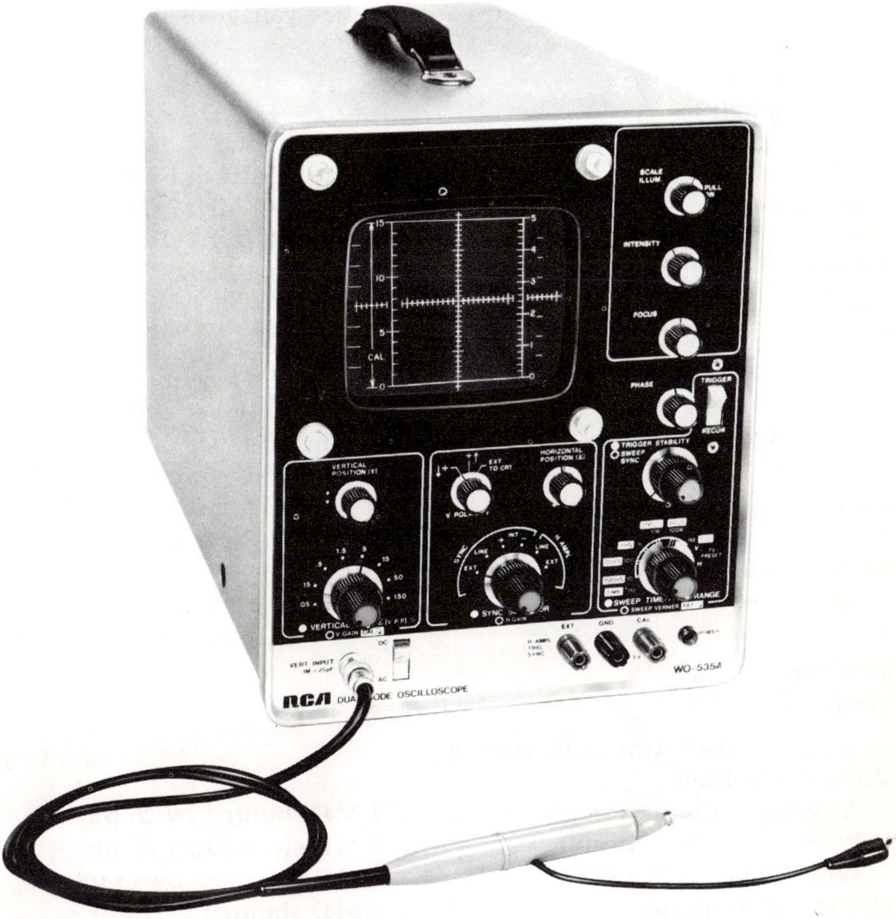

Figure 4–12 General-Purpose Oscilloscope (Courtesy RCA)

*Pictures and explanations concerning this instrument reprinted by special permission of RCA.

Description

The WO-535A is a solid-state, high-performance oscilloscope designed for general-purpose electronic servicing and industrial applications. It is a lightweight portable instrument with a wide frequency response and high gain. In addition to the recurrent sweep mode, it also has a trigger sweep mode with expansion capabilities.

Specifications

NOTE: Performance figures are for a line voltage of 120 V, 60 Hz.

FREQUENCY RESPONSE	
Vertical Amplifier	
dc Function	dc to 10 MHz (dc to 5 MHz, ±1 dB)
ac Funtion	5 Hz to 10 MHz (Hz to MHz, ±1 dB)
Horizontal Amplifier	3.5 Hz to 2 MHz, ±3 dB
DEFLECTION SENSITIVITY	
Vertical Amplifier	
Direct Probe	15 mV P-P per in. (5.9 mV P-P per cm)
Lo-Cap Probe	150 mV P-P per in. (59 mV P-P per cm)
Horizontal Amplifier	0.4 V P-P per in. (0.16 V P-P per cm)
Vertical Deflection Plate Terminals	40 V P-P per in. (15.7 V P-P per cm)
RISE TIME	
Vertical Amplifier	35 ns
INPUT RESISTANCE AND CAPACITANCE	
Vertical Amplifier	
At V Input Connector	1 MΩ shunted by 25 pF
With Probe on "Direct"	1 MΩ shunted by 75 pF
With Probe on "Lo-Cap"	10 MΩ shunted by 13 pF
Horizontal Amplifier	1 MΩ shunted by 20 pF
Sync Input Terminal	1 MΩ shunted by 20 pF
SYNC	Internal, positive or negative External 60 Hz Line

SWEEP OSCILLATOR	
Recur Mode	
Frequency Range	10 Hz to 1 MHz (1 μs to 100 ms per sweep interval) continuously adjustable
Preset Positions	30 Hz (TV V) 7875 Hz (TV H)
Trigger Mode	
Frequency Range	5 Hz to 1 MHz, in six continuously adjustable, noncalibrated ranges
Triggering Sensitivity	5 mV
MAXIMUM AC INPUT VOLTAGE	600 V P-P (in presence of 400 V DC)
Phase Control Range	0 to 160°
Z-Axis Input	
Minimum Input Voltage for Blanking	12 V rms
Frequency Response	3 Hz to 500 kHz
Power Requirements	
Voltage	105–130 V
Frequency	50–60 Hz
Average Power Consumption	30 W
MECHANICAL	
Height	11⅜ in.
Width	9 in.
Length	16½ in.
Weight (Net)	25 lb

Functions of Controls and Terminals

The following figures show and describe the functions of each control and terminal. These controls are logically grouped and are relatively simple to operate.

Operation

which mode to use—trigger or recurrent Either recurrent or trigger sweep can be used to lock in most waveforms, and the mode used is simply a matter of individual preference. For most general-purpose applications, many will prefer to use the recurrent sweep mode because most patterns can be locked in a little faster and easier using recurrent sweep.

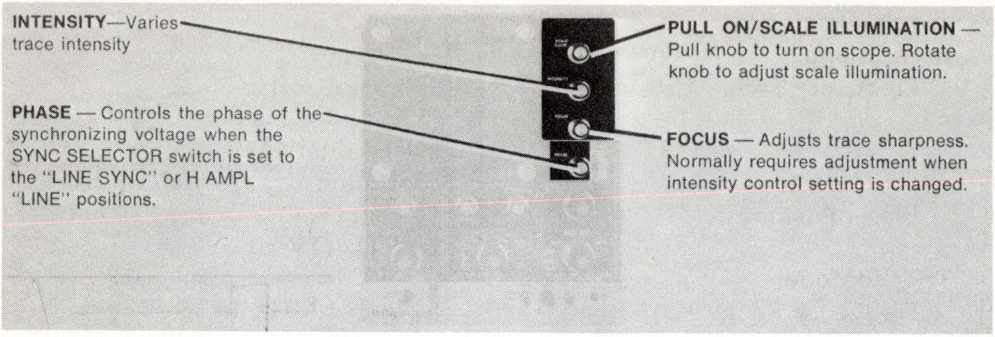

Figure 4–13 Set-Up Controls

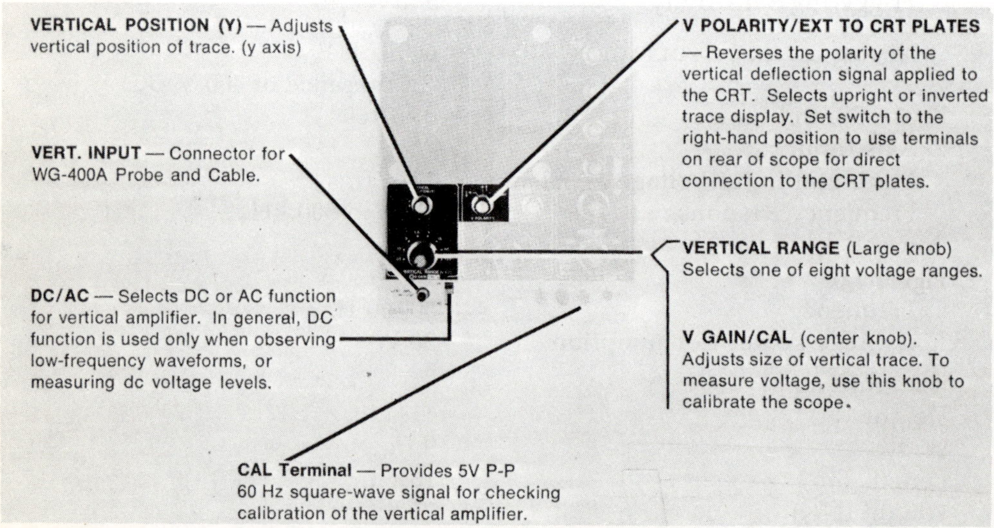

Figure 4–14 Vertical Controls

The basic advantage of triggered sweep is that it can lock in waveforms and patterns that cannot easily be viewed using recurrent sweep. For example, the triggered sweep is used to view a short-time event (narrow pulse, etc.) that occurs in a long-time period, to view a random or nonrecurrent pulse or marker that does not occur at each waveform interval, or to expand a small segment of a complex waveform.

A good example of a pattern that can be viewed only by using the trigger sweep mode is the television VITS (vertical-interval test signal). This signal occurs only during the vertical interval of the televised video waveform. Recurrent sweep will easily lock in the total composite signal, but cannot lock in on the VITS. Using the trigger

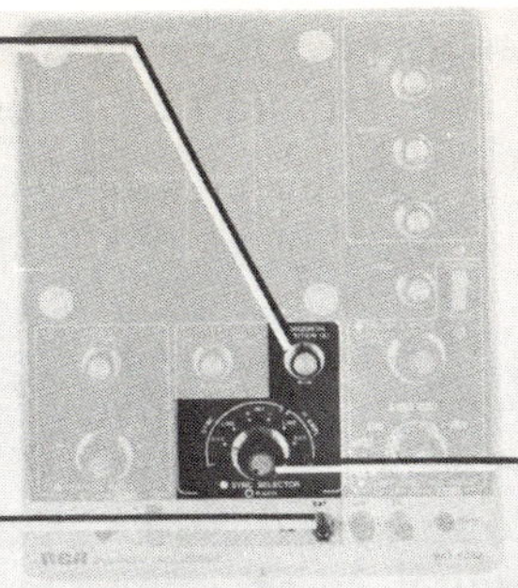

Figure 4–15 Horizontal Controls

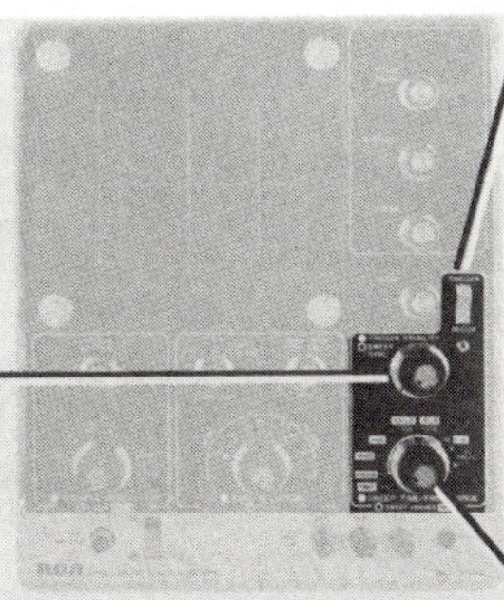

Figure 4–16 Sweep Controls

mode, the oscilloscope can be set so that the sweep oscillator will trigger on the VITS, and display and expand that brief event in the total composite signal.

Initial Procedure

1. Connect the power cord at the rear of the instrument to an ac outlet supplying 105–130 V at 50–60 Hz.

2. To turn on the scope, pull the SCALE ILLUMINATION/PULL-ON knob. Set the ac/dc switch to ac, unless monitoring the dc voltage level of the signal is desirable.

3. Adjust the INTENSITY control until either a spot or a horizontal line appears on the screen. It may be necessary to adjust the vertical or horizontal position control to locate the trace. The spot or line should increase in brilliance as the intensity control is turned clockwise.

> *NOTE*: *Do not* allow a small spot of high brilliance to remain stationary on the screen for an appreciable length of time because discoloration or burning of the screen may result.

4. Adjust the FOCUS control for an image of maximum sharpness.

5. Set the VERTICAL POLARITY knob to " ↑+"; this will provide an upright trace. To invert the trace, set to "+ ↓ ".

Waveform Lock-in Using Recurrent Mode

1. Connect the probe to the test signal. Set the following controls as indicated:

MODE	RECUR
SWEEP SYNC	FULLY COUNTERCLOCKWISE
V. GAIN	FULLY CLOCKWISE
H. GAIN	MIDRANGE

2. Set the VERTICAL RANGE switch to a position which provides suitable pattern height—about one inch. Unless voltage measurements are to be made, adjust the V. GAIN control to obtain any pattern height.

3. Turn the SYNC SELECTOR control to the "+ INT" or "− INT" position to obtain sync on the positive or negative pulse of the input signal. If in doubt, use the position that provides the best pattern stability.

4. Set the SWEEP TIME/FREQ. RANGE switch and the SWEEP

VERNIER control to the position that provides the desired sweep rate or pattern. Turn the SWEEP SYNC control clockwise just far enough to provide solid lock-in.

Example of Recurrent Sweep Operation Using the Cal Signal

Connect the WG-400A probe to the CAL terminal on the panel. Set the oscilloscope controls as described above, with the SWEEP TIME/FREQ. RANGE switch in the 10–100 Hz range. Turn the SWEEP VERNIER fully clockwise, then counterclockwise just to the point where one complete cycle is locked in. Turn the SWEEP VERNIER slowly counterclockwise, and note that the pattern will go out of sync. Continue turning the control counterclockwise and note that the pattern will lock in with two cycles, then with three cycles. In this example, the frequency of the input signal is 60 Hz. Therefore, when two cycles are locked in, the sweep rate is 60/2, or 30 Hz. When three cycles are locked in, the sweep rate is 60/3, or 20 Hz, etc.

Waveform Lock-in Using Trigger Mode

1. Connect the probes to the test signal. Set the following controls as indicated:

MODE	TRIGGER
TRIGGER STABILITY	MIDRANGE
SWEEP SYNC	FULLY COUNTERCLOCKWISE
SYNC SELECTOR	+ INT (Use –INT to lock in on a negative pulse)
H. GAIN	¼-TURN CLOCKWISE
SWEEP TIME	FULLY
FREQ. RANGE	COUNTERCLOCKWISE

NOTE: To become familiar with the operation of the trigger mode, use the 5 V P-P square-wave calibrating signal by connecting the probe to the CAL terminal.

2. Adjust the VERTICAL RANGE switch, VERTICAL GAIN control, and VERTICAL and HORIZONTAL POSITION controls to obtain a vertical line trace about one inch high. Use the HORIZONTAL POSITION control to locate the trace toward the right side of the screen.

3. Adjust the SWEEP SYNC knob slowly clockwise just to the point where the horizontal trace appears. Adjust HORIZONTAL GAIN for convenient width. Readjust HORIZONTAL POSITION to center the trace.

4. Set the SWEEP TIME/FREQ. RANGE switch and adjust the SWEEP VERNIER control as required to obtain the desired pattern. Alternately adjust SWEEP SYNC and TRIGGER STABILITY to obtain a stable trace. If a stable pattern cannot be obtained, reset the SYNC SELECTOR switch to "—INT".

> *NOTE*: If the TRIGGER STABILITY or SWEEP SYNC control is turned clockwise, the sweep circuit may eventually revert to the recurrent mode, and the horizontal pattern will be maintained even if there is no signal input.

Trace Expansion Using Trigger Mode

If you wish to observe a small segment of an overall waveform, first adjust the SWEEP TIME/FREQ. RANGE switch to a higher frequency (faster speed) position; then adjust the SWEEP VERNIER control and the SWEEP SYNC controls to obtain solid lock-in on a partial waveform that contains the area to be expanded. Slight readjustment of TRIGGER STABILITY may be required to stabilize the pattern. Expand the trace by turning the HORIZONTAL GAIN control clockwise.

Use of WG-400A Direct/Low Capacitance Probe and Cable

The WG-400A direct/low capacitance probe and cable are designed especially for use with the WO-535A oscilloscope. This single unit probe is equipped with a sliding switch in the probe housing which permits setting the probe for direct measurements or for automatically connecting a built-in high impedance network in series with the test point and the probe cable. When the switch is set to the "LOW CAP X10" position, the input capacitance of the cable and scope is reduced to 10 MΩ. These high impedance characteristics permit use of the WO-535A in high impedance circuits, such as those found in TV sync-separator and video amplifier stages. These circuits would not operate properly if loaded down by a conventional scope probe and cable.

Whenever the probe is used in its low capacitance position, however, the signal is attenuated by a factor of 10. Therefore, when voltage measurements are made with the WO-535A, the indicated voltage should be multiplied by 10.

Voltage Measurement

The vertical amplifier of the WO-535A can easily be calibrated so that peak-to-peak waveform amplitude can be measured.

The graph screen has two scales for measuring voltage. The scale used depends on the setting of the VERTICAL RANGE switch. For reference purposes, centimeter marks are located along the outer edges of the graph screen, and scales calibrated in inches are located along the horizontal and vertical center lines of the screen. See Figure 4–17.

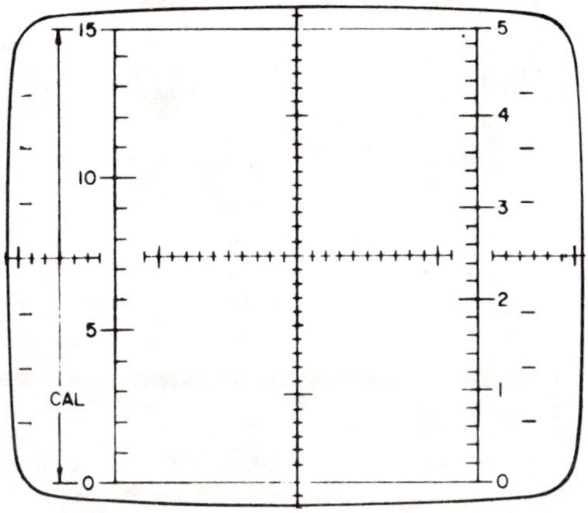

Figure 4–17 Graph Screen with Voltage Scales

Calibration

Calibrating the VO-535A is a simple matter of turning the vertical gain control fully clockwise. As long as the control is in this position, the instrument is calibrated in all ranges. The calibration may be checked by connecting the probe to the cal terminal and using the 5 V P-P square-wave signal as a reference voltage. If the calibration is not correct, it can be adjusted internally.

Back Plate Terminals

Four additional terminals are located on the back of the oscilloscope case: one is ground; one is Z-axis input; and the remaining two provide direct inputs to the vertical deflection plates.

SAMPLE LABORATORY OSCILLOSCOPE

A laboratory oscilloscope may be so versatile as to be useful in all manner of oscilloscope applications, or it may be so highly specialized that it is useful for only one purpose. The sample oscilloscope featured here is in the highly versatile category. This is the Tektronix 7704* shown in Figure 4–18.

*By permission of Tektronix, Inc.

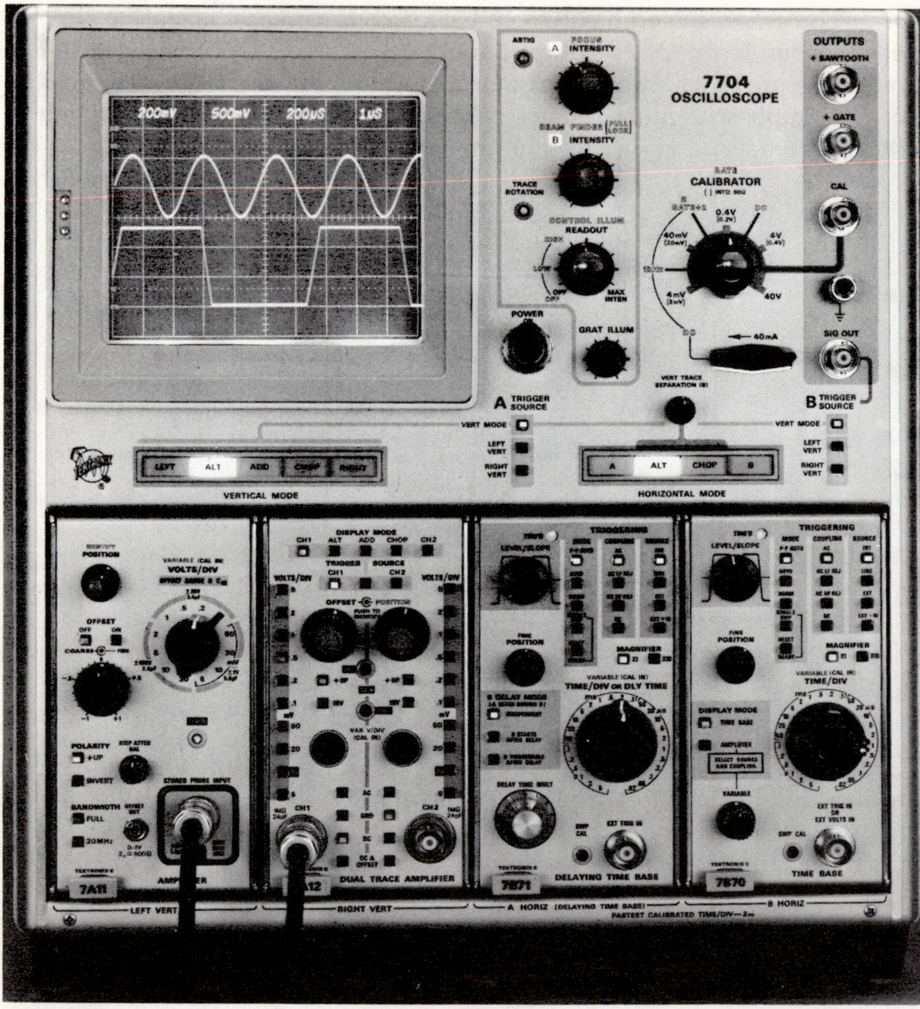

Figure 4–18 Sample Laboratory Oscilloscope (Courtesy Tektronix, Inc.)

Description

The Tektronix 7704 oscilloscope is a solid-state, high-performance instrument designed for general-purpose applications. It is also a satisfactory laboratory measuring instrument. The variety of plug-in units enables the 7704 to be a complete, highly flexible measurement system.

The 7704 has four plug-in compartments. The left pair of plug-ins is connected to the vertical deflection system; the right pair is connected to the horizontal deflection system. Electronic switching between the plug-ins of each system allows a dual-trace vertical display and a dual-sweep horizontal display.

The CRT features a small spot high resolution with writing speeds up to 7000 centimeters per microsecond. The graticule area is 8 × 10 centimeters marked into divisions of one-centimeter squares. The graticules have internal illumination with variable edge lighting. The instrument includes a readout system providing CRT display of alphanumeric information from the plug-ins. Readout includes deflection factor, sweep rate, and encoded parameters.

Specifications

The choice of plug-in units produces a great variety of specifications. The following list represents a small percentage of the total number.

Characteristic	Performance Requirement
Vertical Deflection	
Deflection accuracy	Less than 1% difference between vertical compartments.
Low frequency linearity	No more than 0.1 compression or expansion of a center screen two-division signal when positioned anywhere within the graticule area.
Bandwidth	dc to 190 MHz.
Isolation between compartments	At least 100:1 from dc to 150 MHz.
Vertical display modes	LEFT: Left vertical unit only. RIGHT: Right vertical unit only. ALT: Dual trace, alternate between vertical units. ADD: Added algebraically CHOP: Dual trace, chopped between vertical units.
Horizontal Deflection	
Deflection accuracy	Less than 1% difference between compartments.
Fastest calibrated sweep rate	2 ns/division.
Horizontal display modes	A: A horizontal unit only. B: B horizontal unit only. ALT: Dual sweep, alternate between horizontal units. CHOP: Dual sweep, chopped between horizontal units.

Calibrator	
Waveshape	Positive square wave with baseline at OV.
Output Voltage	
Open	4 mV, 40 mV, 0.4 V, 4 V, 40 V.
Into 50 ohms	2 mV, 20 mV, 0.2 V, 0.4 V.
Amplitude accuracy	0° C to 50° C within 2%. 15° C to 35° C within 1%.
Frequency (1 kHz) accuracy	0° C to 50° C within 0.25%.
	CRT
Phosphor	Choice available
Resolution	At least 12 lines/division
Geometry	0.1 division or less bowing or tilt of displayed line
Photographic writing speed	Up to 7000 cm/μs, depending on type of phosphor and camera used.
Power Supply	
Line Voltage	
115 V	90 V to 136 V
230 V	180 V to 272 V
Line frequency	48 to 440 Hz

VECTORSCOPE

The vectorscope is a specialized CRT instrument for checking color response in a color television receiver. The color bar generator produces ten color signals with spacing of 30° between signals. These signals are displayed on the receiver as a rainbow or ten vertical stripes. The color reproduction in the receiver depends on the phase relation between the color signal and the color reference burst. Both of these signals have a frequency of 3.58 MHz.

The vectorscope can be connected directly to the color grids of the picture tube for monitoring the color signals. The phase relation and amplitude of the color signals are then displayed in a single pattern. A vectorscope is composed of a conventional CRT with special deflection circuits. Some oscilloscopes have built-in vectorscope features. A combination of control settings and connections to the deflection channels enables these oscilloscopes to be used as vectorscopes.

The vector pattern is formed by applying the color signals from the picture tube of the receiver to the deflection system of the CRT in the vectorscope. The red luminance is connected to the vertical deflection system, and the blue luminance is connected to the horizontal deflection system. The envelope of the video signals forms a sine wave pattern. If these were really sine waves, a circle would form on the screen. Since they are pulses of varying amplitudes, each pulse forms a petal. This petal is the resultant of the horizontal and vertical deflection. The phase relationship of the signals in the two channels determines the size and shape of the petals. This principle is closely related to the production of Lissajous patterns as described in Chapter 11 and illustrated in Figure 4–19. The vector pattern is used to adjust both frequency and phase. The proper method is to compare the patterns to pictures of ideal patterns. The receiver is then adjusted to obtain the closest resemblance to the ideal.

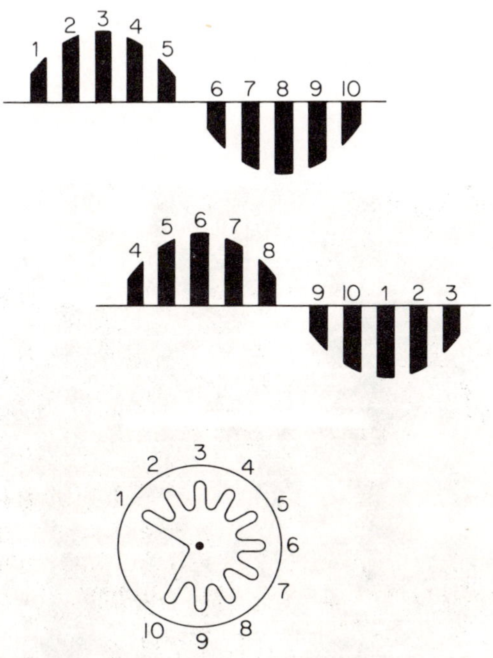

Figure 4–19 Principle of Vector Deflection

Figure 4–20 is a photograph of a sample vectorscope. This is an advanced vectorscope with many special features. It is designed to measure luminance amplitude and chrominance phase and amplitude of the composite color signal.

Figure 4–21 is a vector display with full color bars using a test signal generator as a signal source. This vectorscope features a selection of two graticules: one for vector amplitude and phase; the other for differential gain and phase.

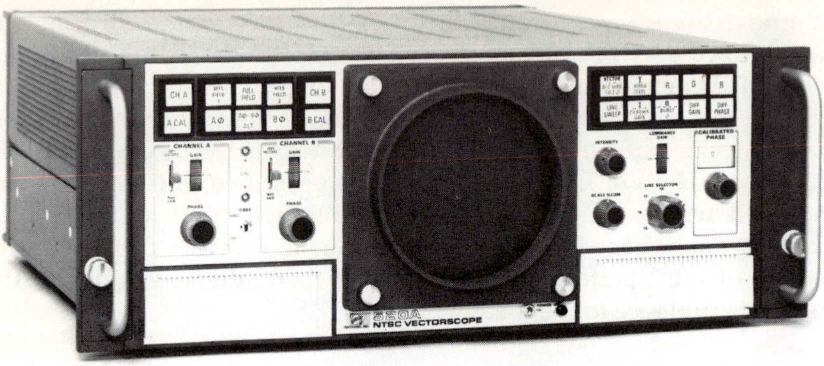

Figure 4–20 Vectorscope (Courtesy Tektronix, Inc.)

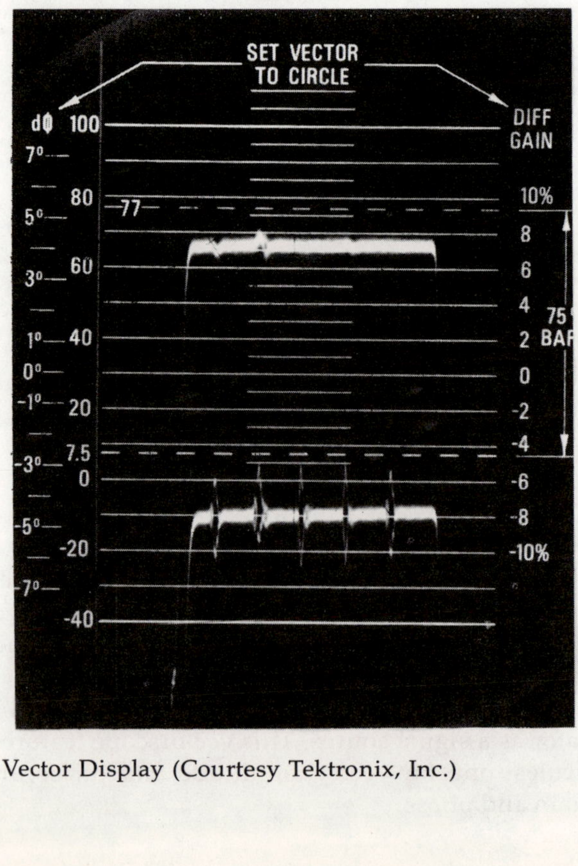

Figure 4–21 Vector Display (Courtesy Tektronix, Inc.)

The vector display (Figure 4–21) has points which correspond to the proper phase and amplitude of the primary and complementary colors.

CHARACTERISTIC CURVE TRACERS

The characteristic curve tracer is a specialized device used primarily by circuit designers. One version of this device is used for electron tubes, and another is for use with solid-state devices. The following description applies to curve tracers for solid-state devices.

The solid-state characteristic curve tracer will display a selected characteristic curve or a family of curves for the solid-state device that is plugged into the recepticle. Figure 4–22 shows a sample curve tracer with automatic readout features.

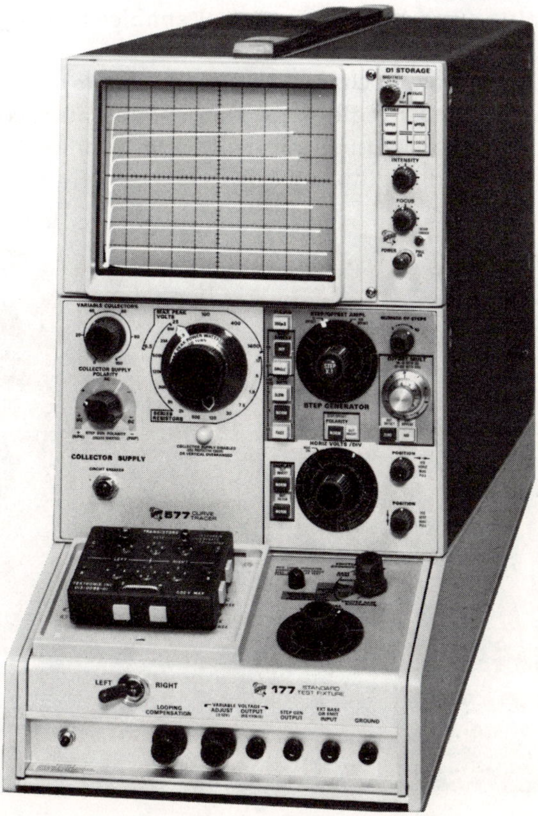

Figure 4–22 Semiconductor Curve Tracer (Courtesy Tektronix, Inc.)

Figure 4–23 illustrates a CRT display of the family of curves from a field-effect transistor. The readout gives the following information:

Vertical current/division	2 mA
Horizontal voltage/division	2 V
Voltage/step	1 V
Beta or transconductance/division	2 m

The curve tracer displays dynamic characteristic curves, and two sets of curves may be superimposed. This enables direct comparison of the characteristics of a standard transistor and of a transistor under test.

The test transistor may be inserted into either a common base or a common-emitter test circuit. The designer can then view the curves and determine the behavior of this transistor in circumstances similar to its anticipated use. A selector control enables testing of either PNP or NPN transistors.

In addition to tracing characteristic curves, the curve tracer enables many specialized measurements of diodes, selinium-controlled rectifiers, power transistors, tunnel diodes, Zener diodes, field-effect transistors, and integrated circuits.

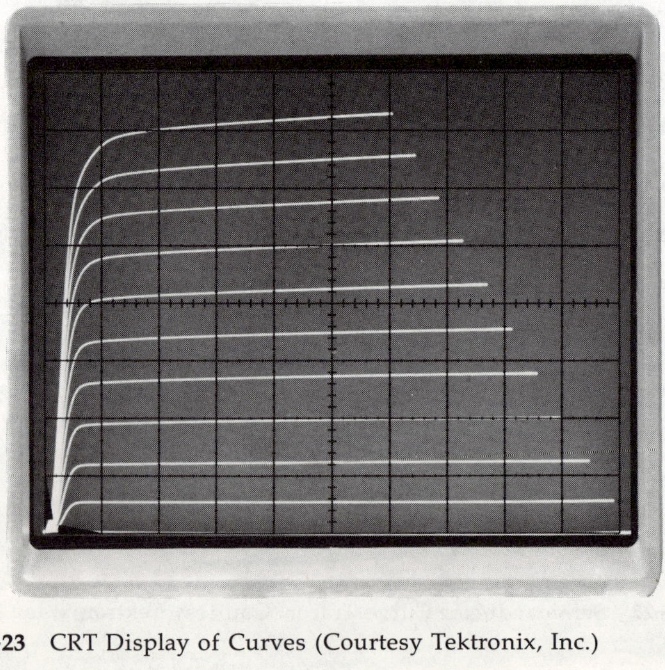

Figure 4–23 CRT Display of Curves (Courtesy Tektronix, Inc.)

SPECTRUM ANALYZERS

A spectrum analyzer is a swept receiver that provides a CRT display of amplitude versus frequency. It displays the Fourier components of a given waveform. It is used to measure frequency response of mixers, doublers, and other frequency conversion components. It can be used to measure signal purity and the bandwidth of a given signal.

The spectrum analyzer is particularly useful in measuring pulse-modulated signals of short duration such as those encountered in radar. These analyzers are available in various sizes and shapes, from small portable units to fixed automatic monitoring installations. A sample spectrum analyzer is shown in Figure 4–24.

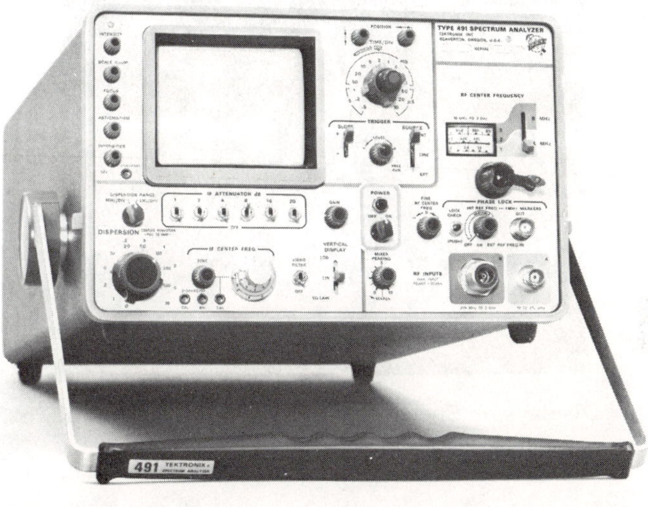

Figure 4–24 Spectrum Analyzer (Courtesy Tektronix, Inc.)

This portable spectrum analyzer has a center frequency range of 1.5 GHz to 40 GHz. It selects a portion of the electronic frequency spectrum for display. Any signal that occurs within the selected portion of the spectrum is plotted as a series of pips on the CRT. Calibrated dispersion permits detailed study of the signal, with frequency difference read directly from the CRT. Figure 4–25 contains photographs of spectrum analyzer displays.

Figure 4–25(a) is a spectrum of a 90 ns pulse with a PRT of 1 μs. The frequency span is from 0 to 20 MHz. Figure 4–25(b) is a pulse of RF at 200 MHz. Horizontal deflection is set to 5 ms/cm.

(a)

(b)

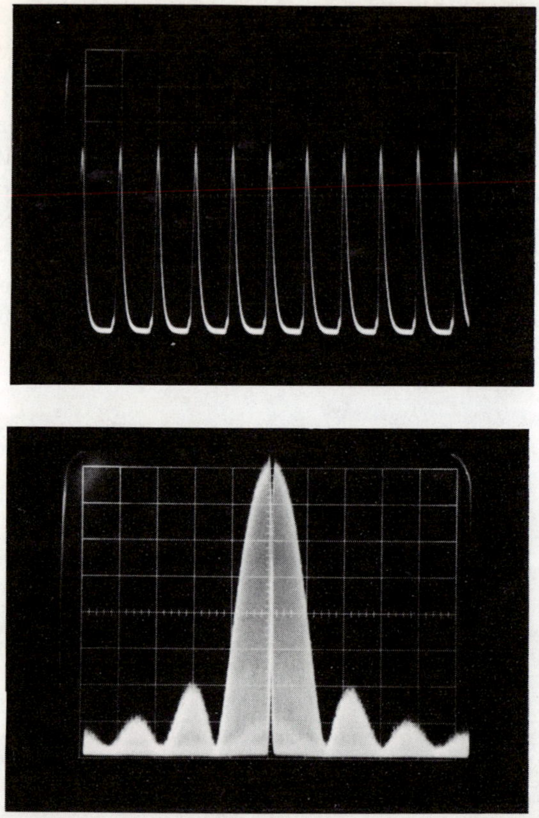

Figure 4–25 Displays on Spectrum Analyzer (Courtesy Tektronix, Inc.)

chapter five

SIGNAL GENERATING INSTRUMENTS

A signal must be present before it can be measured or observed on a CRT. This may be too obvious to require comment, but it is still an indisputable fact. Most test circuits are either incomplete or a part of inoperative equipment. In most cases then, the signal must be artificially produced and injected into the circuit. This fact establishes the signal generator as an indispensable item of electronic test equipment. Although signal generating devices are not measuring instruments, they must be used in many cases in order to have a signal to measure.

SIGNAL GENERATORS

Uses

A signal generator is a radio transmitter constructed to generate a test signal at a desired frequency. It is not a frequency meter, and it cannot be relied upon as a frequency standard. Signal generators are useful in aligning tuned circuits, signal tracing, measuring sensitivity, measuring field intensity, and aligning color television receivers.

Classification

Signal generators are classified as electromechanical or electronic. In the electromechanical category there are two types of generators: the transmitter distributor and the noise generator. The *transmitter distributor* generates and transmits an electric code for checking teletypewriters. The *noise generator* produces a wide variety of random frequencies by producing a high voltage spark between two conductors.

The electronic category contains frequency generators of four types: audio frequency, radio frequency, noise frequency, and video frequency.

Audio-frequency generators produce signals with a frequency range from 20 Hz to 20 kHz. Most audio signal generators provide a choice of output waveshapes between a sine wave and a square wave.

Radio-frequency generators can be obtained to cover selected portions of the radio-frequency spectrum from 10 kHz to 10 GHz. This spectrum is too broad for complete coverage by a single instrument, but most shops have no need for complete coverage. Most RF signal generators provide a separate audio-frequency signal at 400 and/or 1000 Hz. The RF generators provide signals at a choice between a fundamental radio frequency and a fundamental with modulation. When modulation is selected, there is a choice of amplitude modulation, frequency modulation, or pulse modulation.

Noise-frequency generators provide a wide range of test signals at frequencies from 20 Hz to something beyond 40 GHz.

Video-frequency generators are most useful in testing television circuits. The signals are roughly a square wave composed of a variety of both low and high frequency harmonics.

TRANSMITTER DISTRIBUTOR

The transmitter distributor is an electromechanical signal generator used with teletype mechanisms. It is used to determine the operating margins of a start-stop teletypewriter, and also to check the selection mechanism efficiency.

Distributor

The *distributor* is a disk fitted with four concentric rings. The two inner rings are solid contacts, and the two outer rings are divided by insulator segments. Contact brushes ride on the rings and provide contact between pairs of rings. Starting with the outer ring, the pairs are: 1 and 3, and 2 and 4. This arrangement is illustrated in Figure 5–1.

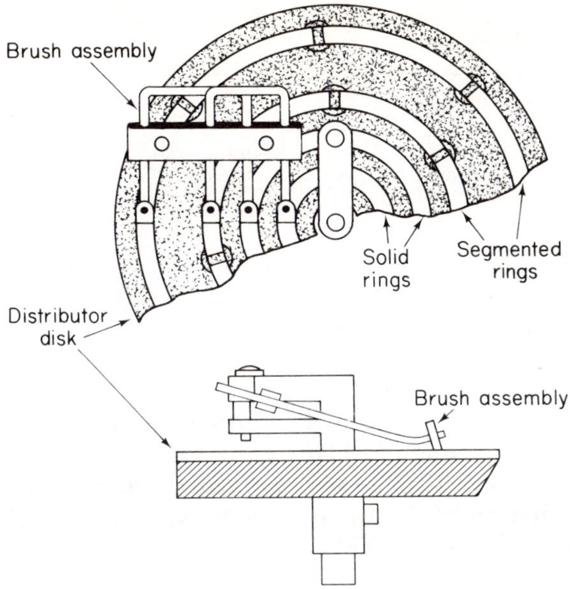

Figure 5–1 Distributor Disk

The main shaft of the mechanism rotates and draws the brushes along the rings. The rings are electrically connected through the brushes and character selection switch to the signal line. This arrangement transmits signals at the proper speed and sequence to the equipment being tested.

The segmented rings correspond to the impulses of a teletypewriter code. This disk has a start segment for the start pulse, five code segments for intelligence pulses, and a stop segment for the stop pulse. One complete revolution is required to transmit the code combination for each character or function.

Transmitter

The *transmitter* is also a code disk assembly consisting of five sets of code disks, levers, and contacts. This is illustrated in Figure 5–2. The five contact levers are mounted on a common shaft. The code disk associated with each lever has a cammed surface which pivots the contact lever. The contact tongue is moved by the lever between two sets of contacts. The high portion of the cam pushes the arm up and makes contact between the tongue and the upper contact. The low portion of the cam allows the spring to pull the arm down and make contact between the

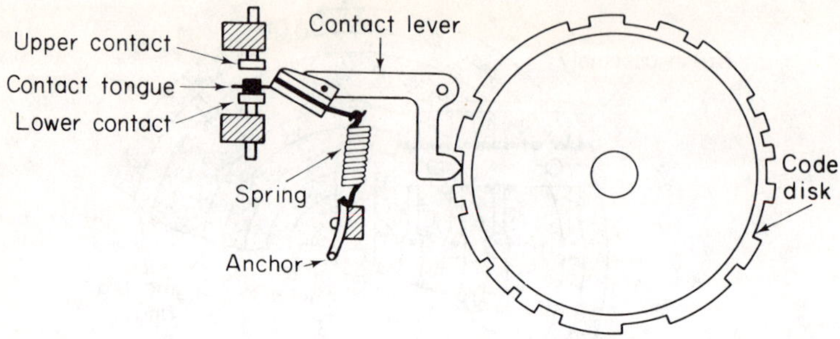

Figure 5–2 Transmitter Assembly

tongue and the lower contact. The spring completes the electrical connection between the tongue and the terminal box.

The five disks are bolted together and rotate with a fixed relationship to the rotation of the distributor brush assembly. One complete revolution of the disk assembly transmits a short test message which contains all alphanumeric characters on the teletypewriter. The message is usually, *"The quick brown fox jumped over a lazy dog's back 1 2 3 4 5 6 7 8 9 0 testing."*

AUDIO-FREQUENCY SIGNAL GENERATORS

Circuit testing and troubleshooting in many types of audio equipment require a source of controlled audio frequencies. Voice-frequency apparatus must have audio signals with little or no harmonic content, and for flexibility the frequency must be adjustable. An audio signal generator with these capabilities is illustrated in Figure 5–3.

Sample Audio-Frequency Signal Generator

Figure 5–3 is the RCA WA-44C sine/square-wave signal generator. Frequencies are controlled by a vernier dial and a range selector switch. The generator can be set to any frequency from 20 Hz to 200 kHz. Frequency stability is within ±2%. Dial calibration accuracy is within ±5% over the entire audio-frequency band, 20 Hz to 20 kHz.

The sine wave is developed by a bridged-tee oscillator. The square wave is produced by shaping and limiting the sine wave. Signal amplitude variation is no more than ±1.5 dB, and total harmonic distortion is limited to 0.25%. Operation is simple and straightforward. Figure 5–4 illustrates the function of each operating control.

Figure 5–3 Audio-Frequency Signal Generator (Courtesy RCA)

Oscillator Circuits

The sine wave portion of most AF signal generators consists of an oscillator circuit. The oscillator is generally one of three types: bridge, phase-shift, and bridged-tee. However, there are other types of oscillators, including the tuning-fork and beat-frequency oscillators.

The most common oscillator found in AF signal generators is the Wien bridge oscillator. Its frequency can be adjusted over a relatively wide range by varying the capacitance in the frequency-determining bridge network.

The ladder-type phase-shift oscillator is more likely to be used in a fixed frequency oscillator of 400, 600, or 1000 Hz. However, the output frequency can be made variable over a wide range by simultaneously varying either all three capacitors or all three resistors.

The tuning-fork oscillator functions on the magnetostrictive principle. When used in AF oscillators, it has a bimetallic temperature-compensated tuning fork. Feedback to one coil drives the fork, and pickup on another coil feeds the amplifier element. Tuning is accomplished by varying the inductances.

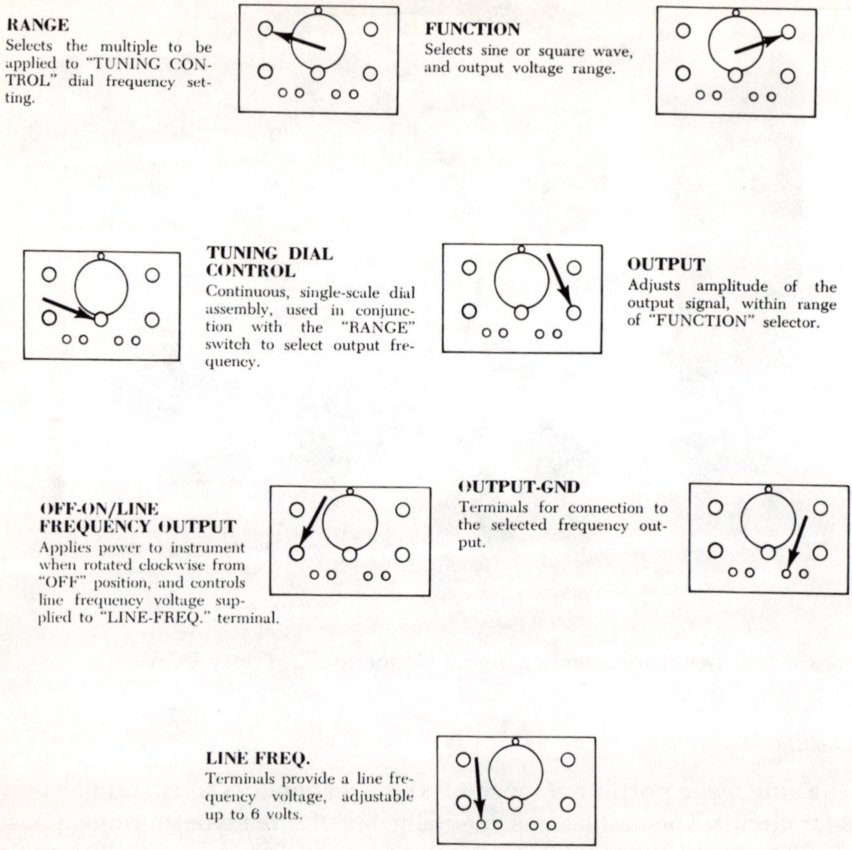

Figure 5–4 WA-44C Controls and Terminals (Courtesy RCA)

The beat-frequency, or heterodyne, oscillator is very common in AF oscillators. It consists of a fixed-frequency oscillator, a variable-frequency oscillator, and a mixer with a filter. The difference frequency can be adjusted over the entire audio-frequency range.

The square-wave portion of an AF generator may be either sine wave amplifier-limiters or free-run multivibrators. Either type produces output square-wave signals at the selected frequency.

Square waves can be produced from sine waves by limiting both positive and negative excursions of the signal. This is easily accomplished by using the sine wave to overdrive an amplifier. The only requirements are an amplifier operating near the center portion of its linear characteristic and a strong sine wave input. The positive excursion drives the amplifier into saturation and limits the positive peak.

The negative excursion drives the amplifier below cutoff and limits the negative peak. The result is a square wave at the sine wave frequency.

There are several types of multivibrators. The Eccles-Jordan multivibrator is a free-running multivibrator that can be either free run or synchronized by an input signal. It is a stable oscillator and can be made to produce rectangular outputs of almost any shape. Obtaining a symmetrical square wave from this type of oscillator is no problem, and it is frequently used in audio signal generators for this purpose. The frequency of a free-running multivibrator can be changed by varying the time constants in the resistor-capacitor coupling circuits. The synchronized multivibrator follows the frequency of the input trigger, and it is only necessary to change the time constants for wide bands of frequencies.

Buffer Amplifiers

Frequency stability is a very important feature in a signal generator. This is accomplished by circuit isolation through buffer amplifiers. The buffer amplifiers present a constant load to the oscillator, isolate frequency generator circuits from the other circuits, and produce added amplification of the output signal.

Attenuators

Attenuator networks provide an adjustment of output level. They can be resistance-capacitance networks ranging from a simple variable control to ladder attenuators. The more complex attenuator is usually accompanied by a metering circuit to indicate decibels of output level into a specified load.

RADIO-FREQUENCY SIGNAL GENERATORS

Radio-frequency signal generators provide practical outputs ranging from 10 kHz to 10 GHz. A particular instrument is designed to cover a specified segment of this broad radio-frequency spectrum. The RF signal generator is particularly useful in aligning radio and television receivers.

Sample Radio-Frequency Signal Generator

A sample of an RF signal generator is shown in Figure 5–5. This RCA WR-50B RF signal generator is a general-purpose instrument designed primarily for aligning and servicing radio and television receivers. Its crystal-controlled oscillator produces tunable RF outputs from 85 kHz to

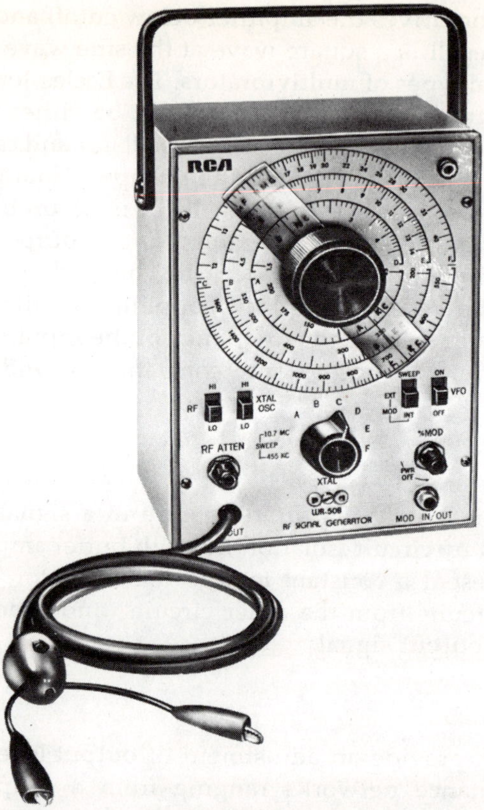

Figure 5–5 Radio-Frequency Signal Generator (Courtesy RCA)

40 MHz. It also provides a 400 Hz audio-frequency signal and sweep outputs of 455 kHz and 10.7 MHz, and it has provisions for using an external crystal with the oscillator for either calibration or frequency control.

The variable-frequency oscillator produces the total range of frequencies from 85 kHz to 40 MHz in six ranges. Frequencies above 40 MHz are available by using second or third harmonics of the fundamental of the high frequencies.

The RF output on all ranges is a minimum of 0.05 V effective. The dial calibration accuracy is within ±2%. The internal modulation frequency is 400 Hz and provides adjustable modulation up to 30%.

This portable instrument is lightweight and features simplified operating controls. Figure 5–6 explains the function of each control.

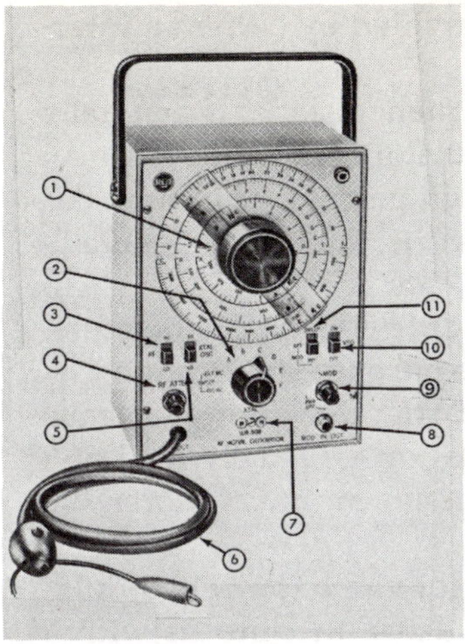

① TUNING DIAL — Tunes through all six ranges (A through F) of the variable rf oscillator.

② RANGE SWITCH — Selects 455 kHz or 10.7 MHz sweep, and VFO ranges A through F.

③ RF HI/LO — Attenuates rf (VFO, crystal, and sweep) output when set to LO position.

④ RF ATTEN — Provides fine rf attenuation adjustment.

⑤ XTAL OSC HI/LO — Provides additional attenuation of crystal output.

⑥ RF OUT — RF output cable.

⑦ XTAL — Crystal socket. Accepts crystals with HC-6U type base. Inserting crystal activates crystal oscillator.

⑧ MOD IN/OUT — External modulating signal can be applied through this jack when MOD switch is set to "EXT".
Provides 400 Hz output when MOD switch is set to "INT".

⑨ PWR OFF/% MOD — Applies power when turned clockwise from "PWR OFF" position. Varies modulation level. Varies level of 400 Hz output.

⑩ VFO ON/OFF — Turns variable frequency oscillator on or off.

⑪ SWEEP/INT MOD/EXT MOD — Three position switch.
SWEEP — Provides sweep output with retrace blanking when range switch is set to one of the sweep positions.
EXT MOD — Permits external modulation of VFO or crystal oscillator. Removes blanking when used with sweep function.
INT MOD — Modulates VFO or crystal oscillator with 400 Hz signal.

Figure 5–6 WR-50B Controls and Terminals (Courtesy RCA)

Oscillator Circuits

Crystal-controlled oscillators are favorites in RF oscillators. The special cut crystals serve as electrical equivalents of tuned circuits. The piezo-electric effect of the crystal results in physical vibration of the crystal, and this generates an alternating voltage at the fundamental frequency of the crystal. The crystal-controlled oscillator is very stable and produces an output relatively free from harmonics.

Some other oscillators that may be used, especially for the lower end of the RF spectrum, are Hartley oscillators and Colpitt oscillators. These are less stable and are more subject to harmonics that the crystal controlled oscillator. However, they are reasonably reliable and produce usable sine waves at radio frequencies.

In the very high to ultrahigh radio-frequency band, special types of oscillators are used. At these frequencies, skin effect and transit time are serious problems. Also stray coupling between circuits becomes more pronounced. The oscillators in signal generators at these frequencies are probably resonant cavity oscillators with light house tubes as amplifiers.

Modulator Circuits

The modulation circuits are used to change either the frequency or the amplitude of the output RF signal. An audio- or video-frequency signal is imposed upon the RF signal and causes it to vary at the audio (or video) rate. The modulating signal may be provided by either an internal or an external audio-frequency oscillator.

In amplitude modulation, the audio-frequency signal is generally a sine wave. Direct modulation of the RF oscillator is likely to cause an undesirable frequency shift, especially if a high percentage of modulation is used. For this reason, the modulation percentage is generally kept to a value of about 30%.

There are various types of frequency-modulated signal generators, but only the sweep generators have general-purpose applications. A signal generator provides control over the RF carrier deviation by means of a sweep width control. The rate of modulation is usually held at a fixed frequency. The signal causing the frequency modulation may be either a sine wave or a sawtooth.

Some signal generators have provisions for video (pulse) modulation. These instruments produce pulse-modulated RF outputs with a wide range of pulse repetition frequencies and pulse widths. This is accomplished by feeding the RF carrier and the modulating pulse into a mixer stage. The positive excursion of the pulse allows the mixer to conduct and pass on a burst of the RF carrier. The negative excursion of the pulse drives the mixer to cutoff and allows no output.

Output Circuits

The output circuits of RF signal generators are composed of a calibrated attenuator and usually an output-level metering circuit. The meter permits precise control of the output level of the oscillator. When the attenuator is adjusted to unity, the meter provides a direct indication of the output in microvolts. At other settings, the decimal value of the indication is multiplied by the attenuator setting. This calculation produces the output value in microvolts.

NOISE-FREQUENCY SIGNAL GENERATORS

A *noise generator* is a test instrument that provides a random (white) noise signal suitable for a number of tests and measurements. Conventional noise generators cover frequencies from about 20 Hz to 40 GHz and beyond.

Applications of Noise Generators

Noise is a common form of a disturbing and interfering signal that limits the threshold of detectability. The noise generator provides controlled noise signals for testing receivers, public address systems, communications equipment, and detection circuits for susceptibility to interference. The noise generator is a useful training device where operators must learn to communicate through noise interference. The wide frequency range of noise generator outputs is particularly useful in calibrating audio equipment. A burst of noise can be used as a reference signal for determining frequency response, signal-to-noise ratio, harmonic distortion, and flutter noise.

Sample Noise Generator

A sample noise generator is pictured in Figure 5–7. This is the Hewlett Packard 3722A noise generator. It uses digital techniques to synthesize binary and Gaussian noise patterns of known content and duration. These patterns are repeated over and over without interruption.

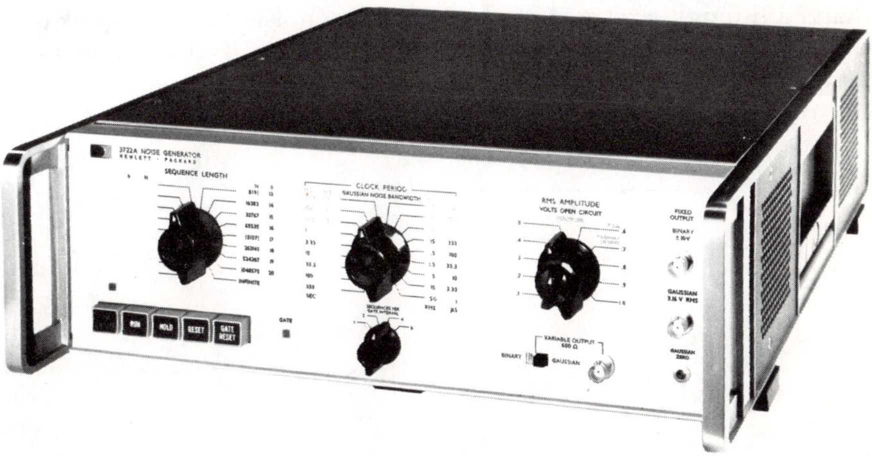

Figure 5–7 Noise Generator (Courtesy Hewlett Packard)

The 3722A has a binary waveform generator—a shift register, which operates under the control of either a feedback mechanism or a random-noise source. The pseudo-random mode selects the feedback control; the random mode selects the random-noise source control. The binary output has a shaped spectrum, and the Gaussian output spectrum is almost rectangular.

The bandwidth is controllable, and the output power remains constant regardless of bandwidth. This feature is needed in very narrow bandwidths. The frequency of the first null in the binary spectrum is selectable from 0.0003 Hz to 1 MHz. The bandwidth of the Gaussian noise is selectable from 0.00015 Hz to 50 kHz at the −3 dB point. A precision amplitude control provides a variable output from 0.1 V to 3.16 V Gaussian and to ±10 V binary. (Voltage in effective values.)

Types of Noise Generators

The simplest noise generator is electromechanical, but there is a wide variety of types to choose from. Some are solid-state, some electron-tube, and some are constructed from sections of waveguides. A noise generator may be a very simple device, or it may be a rather complex instrument. Some are designed for special purposes and others for general applications in tuning and troubleshooting.

Electromechanical Noise Generator

When a section of coaxial line is charged up from a dc source and discharged through a relay contact, it generates noise. This is a spark discharge, and an electrical spark contains harmonics of all frequencies. A sample of such an arrangement is illustrated in Figure 5–8.

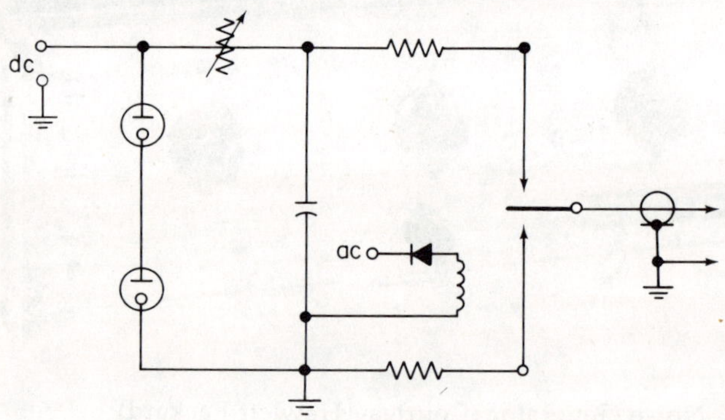

Figure 5–8 Electromechanical Noise Generator

Each spark discharge constitutes an impulse of random noise. Such a generator would have very limited applications because of a lack of control over the amplitude and frequencies of the generated signals.

Solid-State Noise Generator

The solid-state noise generator has the advantages of being lightweight, small in size, and with low power requirements. The circuits of such a random-noise generator may be as simple as the illustration in Figure 5–9. This circuit uses a Zener diode as the noise source and a single transistor as an amplifier. The Zener diode produces the noise signal current, and at the same time stabilizes the collector operating point for the transistor. The output signals across the transistor are amplified random-noise signals. The Zener diode is a satisfactory noise source in the frequency range from a few MHz to several hundred MHz.

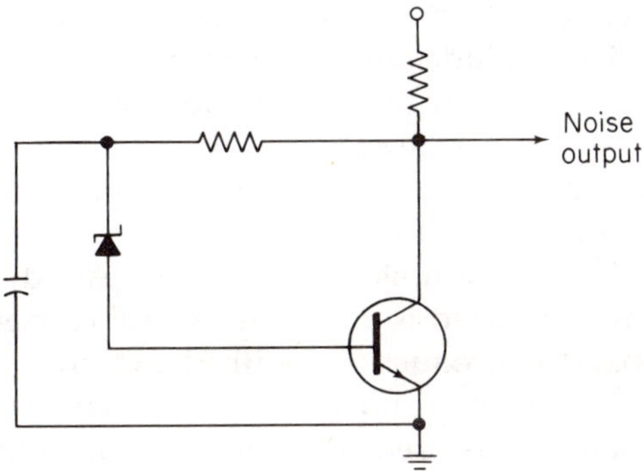

Figure 5–9 Solid-State Noise Generator

Special Applications of Random Noise Generators

A random-noise signal is a signal whose amplitude and frequency are completely unpredictable at any given point in time. The completely random noise is known as *white noise*.

At high frequencies, a type of noise is encountered which has an increase in power level that is directly proportional to an increase in frequency. This type of noise signal is known as *blue noise*.

At low frequencies, there is a type of noise signal which has power levels inversely proportional to frequency changes. This type of noise signal is known as *pink noise*. Pink filters may be installed on the output of a random-noise generator to convert the white noise to pink noise output. Pink noise has the advantage of a constant energy factor per octave.

Random noise is a close simulation of the natural interference encountered. For this reason the output of random-noise generators is an important signal for masking, interference tests, audiometric tests, and determination of critical bandwidths.

The study of how sounds affect the listener is called *psychoacoustics*. Noise control must often be applied for the protection of personnel who work in noise-polluted areas. Over a long period of exposure to noise levels in excess of 85 dB above 0.0002 dynes/cm^2, brain and eardrum damage can occur. Even below the levels of actual damage, environmental noises cause annoyance, fatigue, nervousness, and decreased production. Since few work areas are free of noise, random-noise signals are used to blank out the audibility of sounds. This is noise masking, and it is a very important procedure in psychoacoustics.

Electroacoustics refers to the sound levels of devices used to convert sound energy to electrical energy, or electrical energy to sound energy. These devices include speakers and microphones. The random-noise signals from noise generators enable a wide variety of electroacoustical tests. To name a few, there are reverberation tests, panel transmission measurements, measurement of space irregularities, and measurement of steady-state transmission.

When high-frequency noise signals are desired, the output of a noise generator may be used to modulate an RF carrier. This can generate a 10 MHz band of frequencies with the carrier as the center frequency. A series of carriers and modulators may be used to obtain bandwidths as broad as desired. For some applications it is advantageous to use suppressed carriers and/or balanced-type modulators.

TELEVISION SIGNAL GENERATORS

Television signal generators are a special category of instruments. They must simulate the signals transmitted by the television stations. This includes color, black and white, audio, video, and test patterns. Some signal generators incorporate all these capabilities, while others provide only portions of them. These signals are applied to television receivers and provide a means of checking the response of the receivers.

COLOR TRANSMISSION

A television camera feeds three basic colors into a signal matrix. The matrix separates the signal into brightness and color. The brightness portion is commonly referred to as the *luminance* signal, or simply as the *Y* signal. The luminance signal contains all the details of the picture. The color portion is the *chrominance* signal, and it contains the hue and saturation of the picture.

Any color can be reproduced with a proper mixture of red, green, and blue. The camera has a separate image pickup tube for each of the basic colors. It divides the scene into basic colors while recording the relative intensity of the details in the scene.

After separation of the luminance and chrominance signals,

the chrominance signal is impressed on a 3.58 MHz subcarrier. The chrominance signal causes a phase shift of the subcarrier in accordance with the color information.

The luminance signal is varying in amplitude in accordance with the shades of brightness and darkness which compose the picture details.

After amplification, these signals are again mixed and transmitted as a composite signal. A functional diagram of a transmitter is illustrated in Figure 5–10.

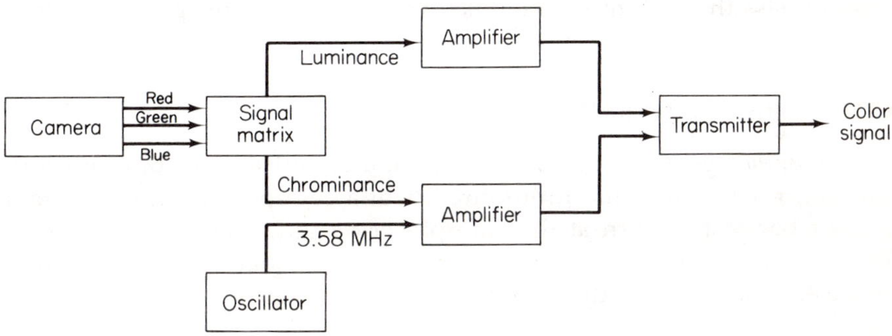

Figure 5–10 Color Transmitter

TELEVISION RECEPTION

On the receiving end, black and white receivers are concerned only with the luminance signal. The signal composition is such that luminance and chrominance can be separated without interference. The color subcarrier was selected to be 3.579545 MHz (3.58) above the video carrier for this purpose.

The color receiver has a luminance channel and a chrominance channel. The composite video signal is detected, amplified, and processed through the luminance channel. The audio is separated, detected, amplified, and applied to the speakers. The luminance signal is applied to the CRT cathodes.

The composite color signal is amplified, demodulated through the chrominance channel, and applied to the color grids of the picture tube. Synchronization between channels is accomplished so that chrominance and luminance signals arrive at the CRT simultaneously.

The CRT has an electron gun for each of the three basic colors: red, green, and blue. The three picture grids will receive signals of the correct amplitude and phase to reproduce the color combinations viewed by the camera. Any color can be created if the three colors are present.

The combination of red and green produces yellow, green and blue make cyan, red and blue create magenta, etc.

Color *purity* is the capability of an electron beam to produce a field of pure color. This is accomplished when the individual beam strikes phosphor dots of *only one* color. Purity is the uniformity of the hue and the brightness of each color field as well as the entire area composed of red, green, and blue combinations.

Convergence is the ability of the three beams to converge at the same point in the shadow mask across the entire screen. All beams should pass through the same holes and strike the same phosphor dots.

COLOR BAR GENERATORS

The *color bar generator* provides rectangular pulses for application to the red, green, and blue input circuits of a colorplexer. This produces a color bar test pattern at the colorplexer output. The outputs of color bar generators usually provide a choice of six patterns for aligning, troubleshooting, and adjusting television receivers.

Types

There are two general types of color bar generators: the keyed rainbow and the National Television Systems Committee (NTSC) generator. The functions are basically the same, but the *NTSC generator* output produces exactly the same results as if the signals were transmitted from a television station. Both produce rainbow and other test patterns, and the NTSC generator also provides single color bars. A selection can be made of any one of six or more fully saturated colors.

Test Patterns

The six common test patterns are illustrated in Figure 5–11.

The *rainbow pattern* consists of ten color bars, and it is used for color alignment and color circuit troubleshooting. The color output of the receiver is adjusted to match a sample color pattern.

The *horizontal bar pattern* is a modulation display of 13 evenly spaced, parallel horizontal bars. It is used for picture centering and for adjusting the dynamic horizontal convergence. Usually only 11 of the 13 bars are visible on the screen.

The *vertical bar pattern* is a modulation display of nine evenly spaced, parallel vertical bars. Usually eight bars are visible. This pattern is used for centering and adjusting the dynamic vertical convergence.

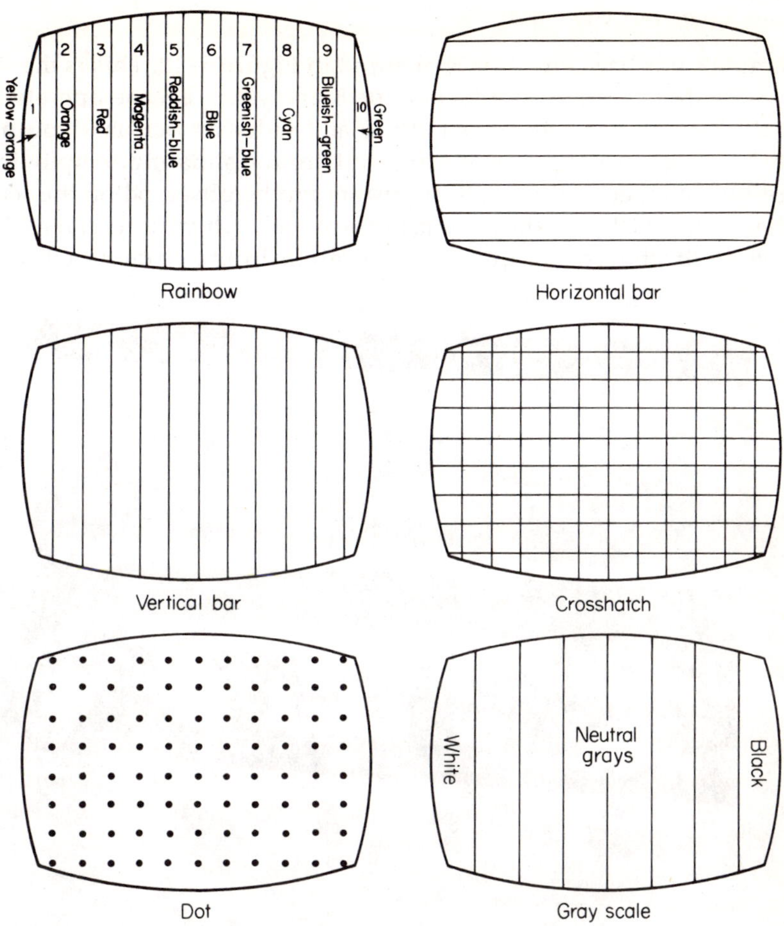

Figure 5–11 Television Test Patterns

The *crosshatch pattern* is a modulation display of the horizontal and vertical bars combined. This pattern is useful in adjusting centering, height, width, linearity, and dynamic convergence.

The *dot pattern* consists of about 120 small, evenly spaced dots. They are used for adjusting the static convergence.

The *gray-scale pattern* consists of six levels of gray, one of white, and one of black, displayed in wide vertical bars. This pattern is used to adjust the receiver for proper gray-scale tracking. These adjustments may be referred to as white balance or screen-temperature adjustments. A color receiver must be tuned for proper gray-scale tracking in order to faithfully reproduce a color picture.

Sample Color Bar Generator

A sample color bar generator is pictured in Figure 5–12. This is the RCA WR-515A, Master Chro-bar Generator/Signalyst. It can be operated on either Channel 3 or 4 and has both IF and video as well as RF outputs. It can be used for direct checking and troubleshooting, employing a stage-by-stage technique. Pattern results can be observed on the screen of the receiver CRT as they are processed through various stages. This is a very effective technique for application in the modular television chassis.

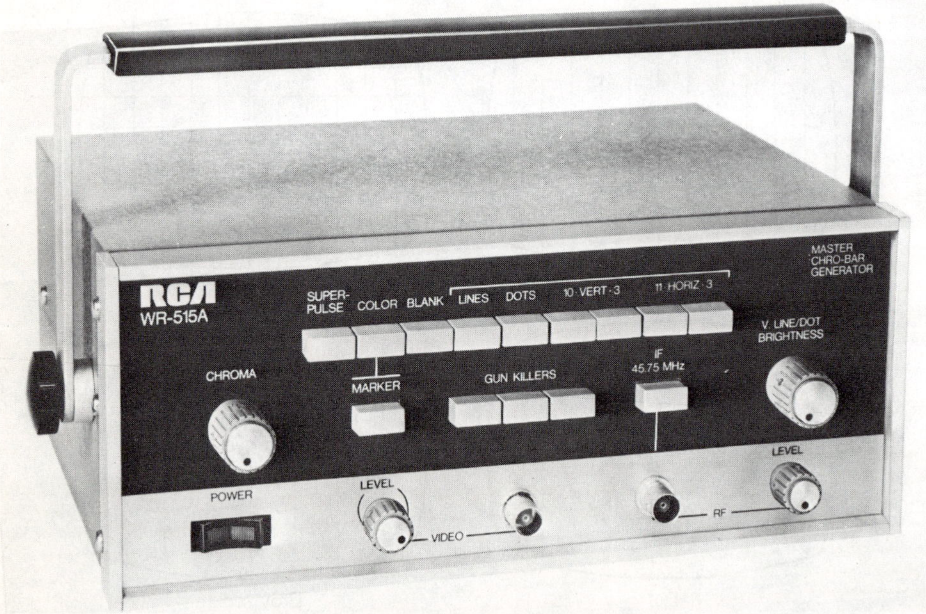

Figure 5–12 Color Bar Generator (Courtesy RCA)

All signals of the WR-515A are derived from crystal-controlled oscillators. They provide the test signals required for adjusting convergence, color phasing, gray-scale tracking, purity, and linearity of color in television receivers. In addition to the six test patterns shown in Figure 5–11, this generator has a *blank raster*, and a *super pulse*. The vertical and horizontal patterns can be selected in groups of three or ten lines. The output levels of RF, IF, and video are adjustable. The brightness of the vertical line and dot patterns is also adjustable. Figure 5–13 lists the adjustments and the recommended test pattern to use with each.

The rainbow pattern from the generator shown in Figure 5–13 has a quick identification feature. Pressing the mark pushbutton places a

Adjustment or Procedure	Recommended WR-515A Pattern
AGC	Lines; 10V, 11H
Video Peaking	Superpulse
H and V Centering	Lines; 3V, 3H
Purity	Blank Raster
Overscan	Lines; 3V, 3H
Linearity	Lines; 10V, 11H
Pincushion	Lines; 11H
Gray Scale Tracking	Superpulse
Convergence	
Static	Dot, Lines (3V, 3H) or Superpulse
Dynamic	Lines; 10V, 11H
AFPC (Tint)	Color Bars with "mark".
Troubleshooting	
Signal Injection	Superpulse, Color Bars, Lines
Signal Tracing	Superpulse, Color Bars

Figure 5–13 Adjustments and Patterns (Courtesy RCA)

bright line on the red, blue, and blueish-green color bars, as shown in Figure 5–14.

The blank-raster pattern applies sync modulation to the RF carrier. This generates a white raster without snow. This pattern is designed especially for purity adjustments.

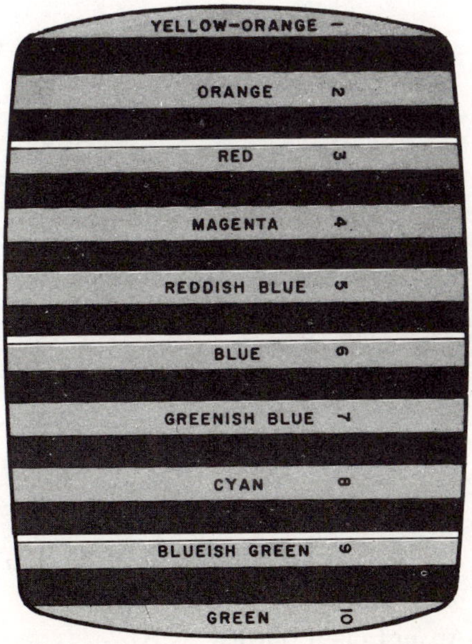

Figure 5–14 Color Bars with Markers (Courtesy RCA)

The super-pulse pattern is a white rectangular area in the center of the screen as shown in Figure 5–15. The super-pulse pattern is used for adjusting gray-scale tracking, purity, smear, ringing, video peaking, and static convergence. In troubleshooting it provides a simple, sharply defined waveform which is easy to follow on the oscilloscope.

Figure 5–16 describes the function of the WR-515A controls, switches, and terminals.

Figure 5–15 Super-Pulse Pattern (Courtesy RCA)

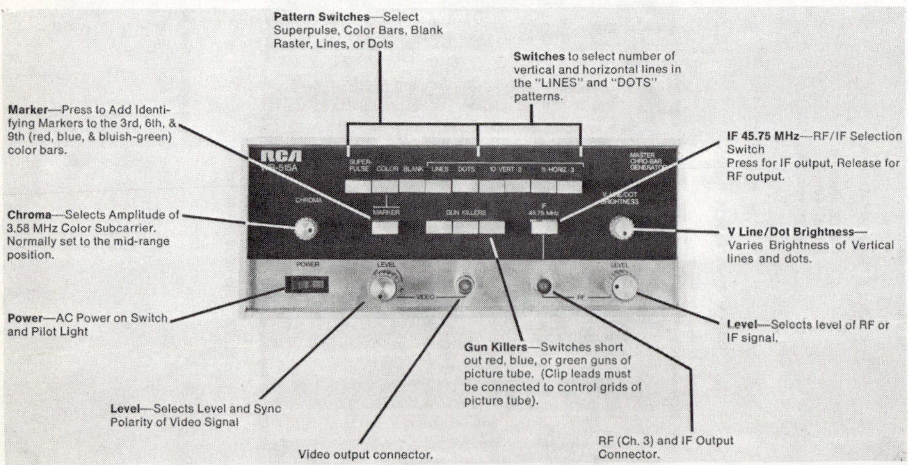

Figure 5–16 WR-515A Controls (Courtesy RCA)

Video Sweep Generator

The video sweep generator is specifically intended for use with television equipment, but its wide-band characteristics enable the use of other applications. It is used to inject a signal into a test circuit. The oscilloscope can then be used to measure the amplitude response in respect to the frequency characteristics. Figure 5–17 is a functional diagram of a video sweep generator.

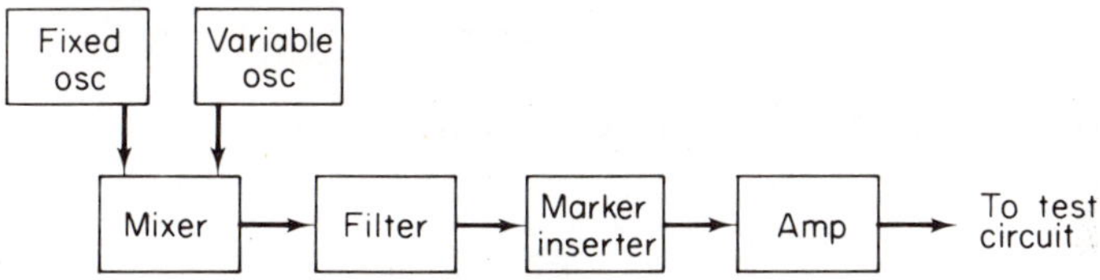

Figure 5–17 Video Sweep Generator

The fixed oscillator produces a continuous RF at approximately 70 MHz. This signal is heterodyned with the variable frequency signal from the variable oscillator. The output of the variable oscillator sweeps between 69 and 80 MHz. These signals are mixed, and when the difference is about 10 MHz, markers are inserted and the signal with markers is amplified and sent to the test circuit. The marker is a frequency amplitude notch on the 10 MHz signal.

A sweep generator of this type has limited accuracy—2 to 3% —but it still provides a signal for displaying complete sweep characteristics.

Other sweep generators incorporate a calibrated continuous wave (CW) oscillator for a choice of CW or sweep operation. These types usually have a tunable frequency range from 100 kHz to 10 MHz.

Multiburst Generators

The multiburst generator is an instrument used for complete operational checks of television broadcasting facilities. It has applications in checking individual circuits.

The multiburst generator provides an output burst composed of various frequencies. These frequencies are superimposed on a pedestal, and they are supported by standard blanking and synchronizing signals. All bursts must be of equal amplitude. A sample output burst is illustrated in Figure 5–18. This test signal simulates actual television signal frequencies. The operational response can be determined by feeding this signal into the test circuit and checking the output on an oscilloscope.

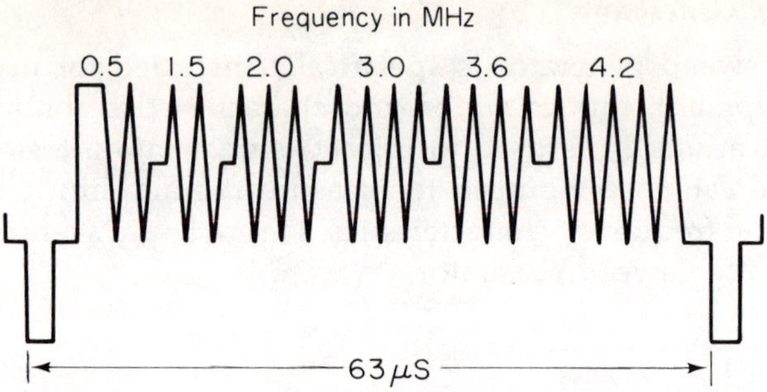

Figure 5–18 Output of Multiburst Generator

chapter six

ELECTROMAGNETIC WAVE MEASURING INSTRUMENTS

This chapter deals with instruments used to measure modulation, field strength, transmission line characteristics, standing-wave ratios, and antenna power output. Several instruments utilized in this area have been discussed in previous chapters and will not be repeated here. For instance, the oscilloscope and spectrum analyzer are both widely used in electromagnetic wave measurements, but they were covered in Chapter 4. This discussion is limited to the electromagnetic waves located in the vicinity of the transmitter and the modulation, propagation, and radiation of these waves.

MODULATION CHARACTERISTICS

Modulation is a process of impressing intelligence on a carrier wave by altering either its amplitude or frequency. In amplitude modulation, the degree is the percentage of amplitude change from the unmodulated value of the carrier. In frequency modulation, the frequency deviation is the change in frequency in one direction from the center frequency of the carrier. The degree of modulation (or the frequency deviation) is of primary concern in a transmitter, because it has a direct bearing on output power and bandwidth.

Amplitude Modulation

A carrier wave has 100% modulation when the total amplitude change from crest to trough is equal to twice the amplitude of the unmodulated carrier. Fifty-percent modulation is present when the amplitude of the trough is half the amplitude of the unmodulated carrier. When the trough of a wave reaches zero amplitude and remains there for a finite period of time, the wave is over 100% modulated. These three conditions are illustrated in Figure 6–1.

The percentage of modulation of a modulated carrier is calculated by dividing the change in amplitude by the amplitude of the unmodulated carrier and multiplying the result by 100. For example, in Figure 6–1(b), the percent of modulation is either $100\,b/a$ or $100c/a$. When the modulating wave is not symmetrical, b and c will be of different amplitudes. In this case, the largest amplitude is used.

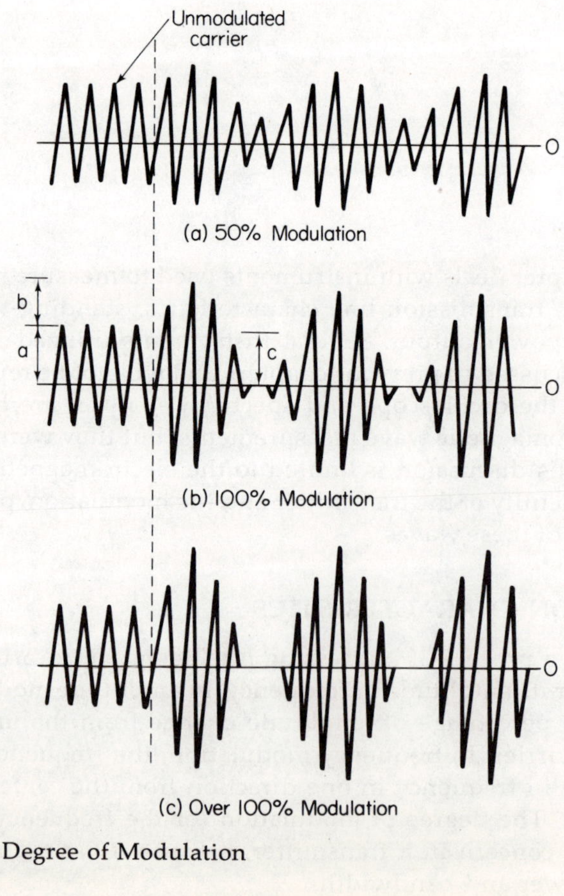

Figure 6–1 Degree of Modulation

Modulation of 100% is attained on an RF carrier when the sine wave AF modulating signal power is equal to half the carrier power. Under these conditions, the average power of the modulated carrier is equal to the sum of the powers of the two waveforms.

During peaks of 100% modulation, the amplitude of the carrier is doubled, and the peak power is equal to four times the peak power of the unmodulated carrier.

When the modulating wave is a voice signal, modulation is not symmetrical. Only the highest amplitude peaks should be permitted to modulate the carrier to a full 100%. Most speech components will then modulate much less than 100%. When the modulating signal power is 25% of the carrier power, the voice peaks usually cause 100% modulation.

Frequency Modulation

The change in frequency (frequency deviation) in a frequency-modulated carrier corresponds to the amplitude change in amplitude modulation. Deviation is expressed in Hz (hertz) and is equal to the difference between the unmodulated carrier frequency (center frequency) and the frequency reached at the maximum excursion. The maximum permissible deviation may be as large as desired so long as it remains within the assigned bandwidth.

The modulation index is the ratio between the maximum frequency deviation and the frequency of the modulating signal. For example, when a modulating signal of 15 kHz causes a frequency deviation of 75 kHz, the modulation index is five. The modulation index is a number that can be located in a modulation index table. A sample modulation index table is provided in Appendix E. The table provides the number of significant sidebands and other helpful information for any given modulation index.

Figure 6–2 illustrates the results of frequency modulation. Figure 6–2(a) illustrates modulation by two different amplitude sine waves of the same frequency. The amplitude of the modulating signal determines the degree (amount) of deviation from the center frequency of the carrier (f_0). Figure 6–2(b) illustrates modulation with two sine waves of the same amplitude but with different frequencies. The rate of deviation is determined by the frequency of the modulating signal.

INSTRUMENTS FOR MEASURING AMPLITUDE MODULATION

One method of measuring the degree of modulation makes use of a wave component separation and comparison test circuit. This circuit

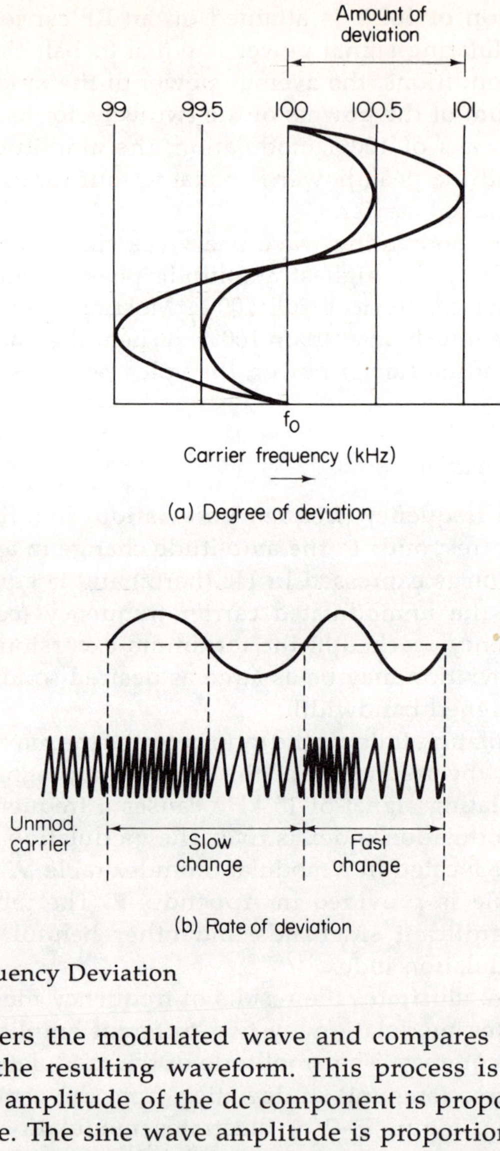

Figure 6–2 Frequency Deviation

rectifies and filters the modulated wave and compares the ac and dc components of the resulting waveform. This process is illustrated in Figure 6–3. The amplitude of the dc component is proportional to the carrier amplitude. The sine wave amplitude is proportional to the amplitude of the modulating signal.

A basic test circuit for measuring degree of modulation is illustrated in Figure 6–4. The modulated RF carrier is picked up on the antenna and rectified by the diode. The coils and capacitors filter out RF components and leave an audio signal riding the dc component. The meters indicate the values of the ac and dc components. M_1 is a dc meter, and its indication is proportional to the amplitude of the un-

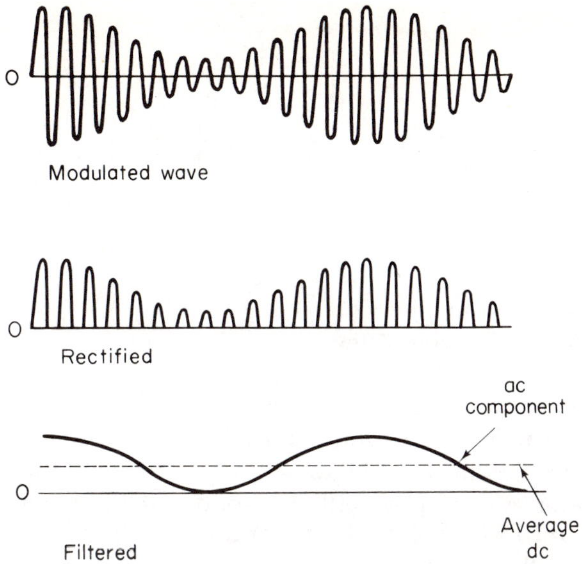

Figure 6–3 Component Comparison

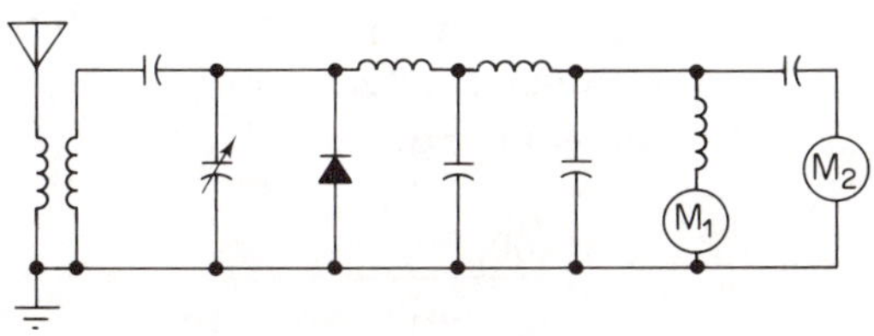

Figure 6–4 Single Rectifier Test Circuit

modulated carrier. M_2 is an ac meter, and its indication is proportional to the amplitude of the modulating signal.

Since the ac meter reads effective values, 100% modulation causes the reading on M_2 to be 0.707 of the reading on M_1. This shows that the modulating change in amplitude is equal to twice the amplitude of the unmodulated carrier. With 50% modulation, the carrier amplitude is equal to the modulating change in amplitude. In this case, the reading on M_1 is nearly three times the value read on M_2.

The circuit in Figure 6–4 can be modified to provide a continuous reading directly in percentage of modulation, as illustrated in Figure 6–5. This circuit rectifies the difference in voltage between points *A* and *C*. This causes a current through the meter proportional to the percentage of modulation. The meter is calibrated directly in modulation percentage.

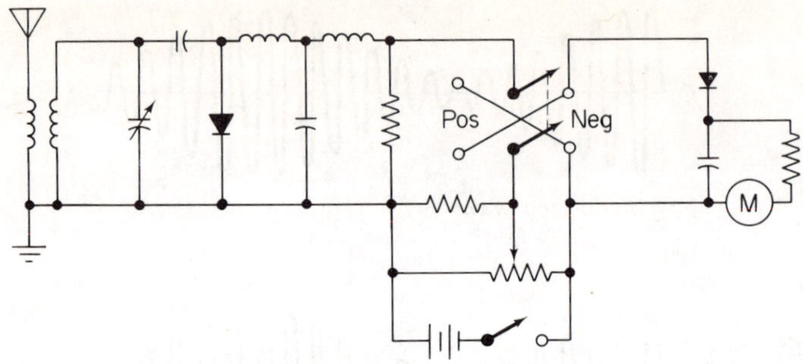

Figure 6–5 Modulation Meter

Another method of measuring modulation percentage uses the slide back vacuum-tube voltmeter (VTVM). The basic circuits of such a meter are illustrated in Figure 6–6. This is a common type of modulation meter and may be either a solid-state meter or a VTVM. R_1 is adjusted until M_1 indicates zero, and R_2 is adjusted until M_2 indicates zero, when there is no-signal input. The unmodulated carrier is then coupled in, and the tuned circuit is adjusted for a convenient deflection on M_2 (usually half-scale). R_1 is then reset for zero on M_2, and the reading on M_1 is recorded. The modulated carrier is now coupled to the tuned circuit, R_1 is readjusted for zero on M_2, and the new reading on M_1 is recorded. The percentage of modulation is

$$\frac{100(E_2 - E_1)}{E_1}$$

(E_1 and E_2 are first and second readings taken from M_1.)

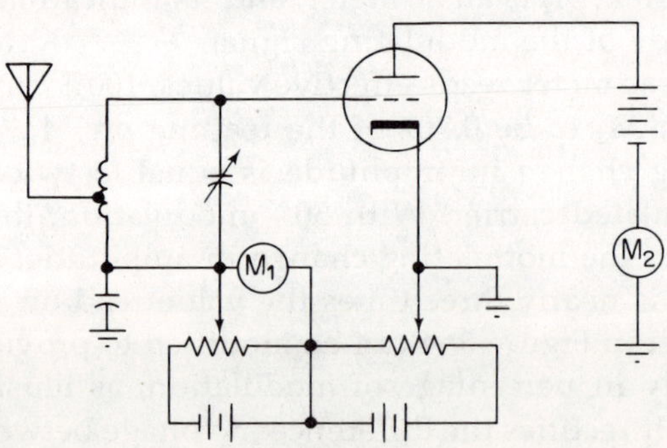

Figure 6–6 Circuits of Slide Back VTVM

INSTRUMENTS FOR MEASURING FREQUENCY MODULATION

The principal equipment for measuring frequency modulation consists of a radio receiver or a spectrum analyzer. The procedures are discussed in a later chapter.

TRANSMISSION LINES

There are a great many types of transmission lines associated with electricity and electronics. Some of these are telephone lines, power lines, coaxial lines, and waveguides. Since this section is concerned with electromagnetic waves, the discussion here is limited to transmission lines that carry these waves. That narrows the field to two-wire lines, coaxial lines, and waveguides.

Specifically these are the lines that carry the electromagnetic energy from the transmitter to the antenna and from the antenna to the receiver. Since, in each case, the antenna is technically an extension of the transmission line, the discussion will at times include antennas.

Properties

All transmission lines possess the three inherent properties of all electrical circuits. These are: resistance, inductance, and capacitance. These three properties always exist together. In transmission lines, the majority of the resistance is in the line itself, and this is consistent as a specified number of ohms per unit length. The inductance and capacitance are results of line characteristics. Imperfect insulation causes inductive effects. Capacitance results from the nearness of the lines to one another and to other conducting media. The quantity of inductance and capacitance for specific lines is also obtainable in terms of unit length. The impedance of a line is determined by the physical properties of the line, the length of the line, and the frequency of the electromagnetic waves that it carries.

Traveling Waves

When RF energy is coupled to the end of a transmission line, traveling waves of voltage and current are set up in the line. These waves travel along the line from the source to the destination. In the case of transmission the transmitter is the source, and the antenna is the destination. With reception, the antenna is the source, and the receiver is the destination. These traveling waves represent the propagation of energy from the source to its destination.

When a line is terminated in its characteristic impedance, all the energy is absorbed by the load. This represents perfect coupling and perfect impedance matching so there is no loss of energy in the line.

Standing Waves

Since the ideal situation is never quite accomplished, all transmission lines have some degree of mismatch. This means that a portion of the energy will be absorbed by the load, and the remainder will be reflected back toward the source. This is an energy loss, and it creates a reflected wave that travels back down the line in opposition to the traveling wave previously described.

The mismatched line then has two types of traveling waves: incident and reflected. The *incident wave* is traveling from source to load, and the *reflected wave* is traveling from load back to source. The incident and reflected waves combine to produce *standing waves*. This is an undesirable characteristic. Not only do these standing waves decrease the energy transfer, they also attenuate the signal and cause distortions.

Decreasing Standing Waves

Since standing waves are evidence of energy losses, it is prudent to keep these waves at an absolute minimum. This is done of course by obtaining the best possible match between the energy source and the transmission line. This is easily accomplished with a calibrated variable resistor. The line is terminated with this resistor, and a standing wave indicator is applied to the line. The resistor is adjusted until standing waves are at an absolute minimum. The resistance of the resistor is now very close to the characteristic impedance of the line. If the signal source is set to this value, the standing waves should remain at a minimum value.

TWO-WIRE AND COAXIAL-LINE MEASURING INSTRUMENTS

Current Indicator

Current indicators can be very useful instruments, but they need not be overly complicated. Such a device is illustrated in Figure 6–7. This instrument may be used to detect minimum and maximum current points in a transmission line. This is not a means of taking precise current measurements, but the relative values of minimum and maximum are indicated. The ammeter is coupled to the line indirectly through the

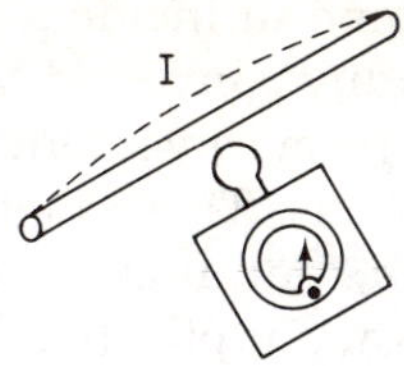

Figure 6–7 Current Indicator

coupling loop. A minimum current indication occurs at points of minimum current and maximum voltage.

Voltage Indicator

The voltage indicator may also be a very simple device, as illustrated in Figure 6–8. This indicator uses a dc voltmeter with a rectifier circuit. The meter is capacitively coupled to the RF line. The rectified dc produces a minimum reading at points of minimum voltage and maximum current.

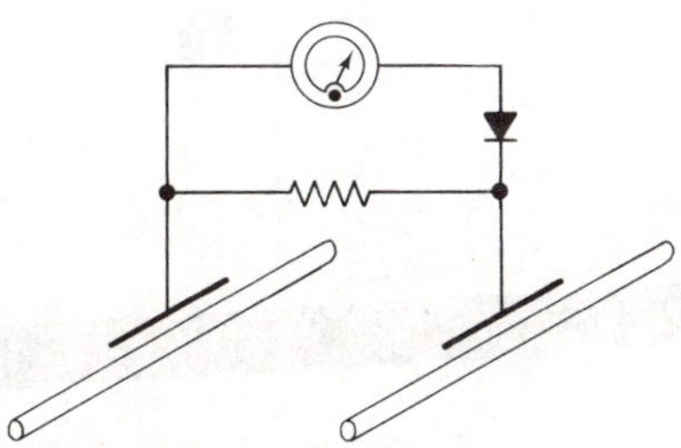

Figure 6–8 Voltage Indicator

Several other very simple indicators may be improvised. For instance, a neon bulb held near an RF line will glow. If this bulb is held at a uniform distance from the line and moved along the line, the voltage peaks (minimum current and maximum voltage) will be indicated by maximum brilliance of the light. Well-designed, accurate indicators of voltage and current may be obtained for more precise work.

Standing-Wave Ratio Meters

The *standing-wave ratio* (SWR) is E_{max}/E_{min}. An SWR of 1 to 1 indicates a perfect line and no mismatch. There are many SWR meters to choose from. Most of them use a directional coupler to measure incident voltage and reflected voltage. Some meters are calibrated in SWR, some in reflection coefficient, and some in incident and reflected voltage. Any of these readings can be readily converted to SWR.

Many meters, although primarily intended for SWR measurements, have several auxiliary functions. These meters may measure attenuation, frequency, wavelength, admittance, conductance, detect resonance, measure gain, and other applications. One example of a highly versatile SWR meter is shown in Figure 6–9.

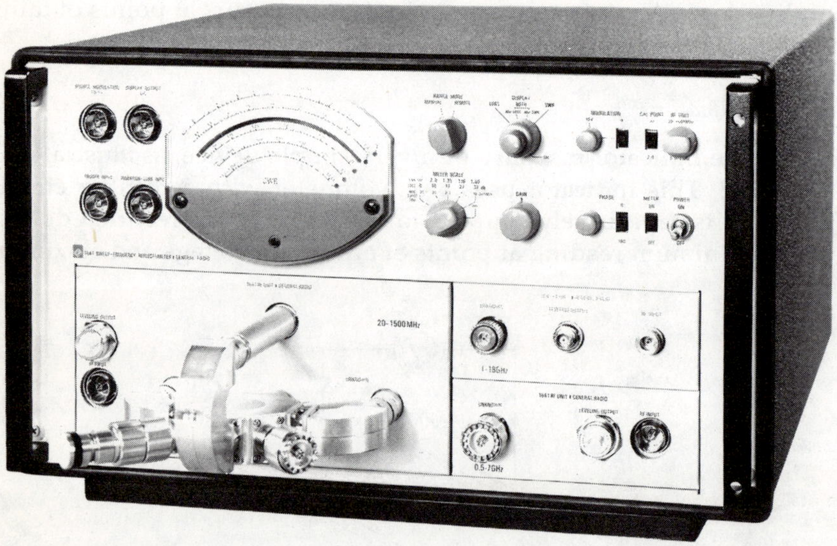

Figure 6–9 Sweep Frequency Reflectometer (Courtesy General Radio)

This General Radio 1641 Sweep Frequency Reflectometer contains all the required coaxial hardware, including directional couplers, detectors, and terminators. It provides a direct SWR reading and loss indication. It measures SWR from 1.005 to infinity and insertion loss from 0 to 5 dB. The frequency range is 20 MHz to 18 GHz. Many vital characteristics can be measured with this test instrument, including such items as admittance, Q, and impedance.

There are simpler instruments that can be used for measuring specific characteristics, for instance, a simple standing-wave ratio meter for only standing waves or for standing waves plus a few other applications. A simple reflectometer may be used to measure reflection coefficient. However, a good multiple instrument is generally more economical in the long run.

Null Impedance Meter

One type of impedance measuring instrument utilizes the Byrne bridge. It usually provides a direct reading in either admittance or impedance. This instrument is especially designed for measurements with coaxial transmission lines. The functional circuits are illustrated in Figure 6–10.

In this instrument the oscillator supplies RF energy through a coaxial transmission line to an unknown impedance. Two sampling elements are coupled to the line as close to the load as possible. One of these elements is responsive to the current of the load, and the other is responsive to the voltage of the load. The current element is an inductance pickup loop, and the voltage element is a capacitive probe. Loop and probe can be adjusted for more or less coupling by moving them in respect to the coaxial line. The loop and probe move together, so that one moves closer as the other moves farther away.

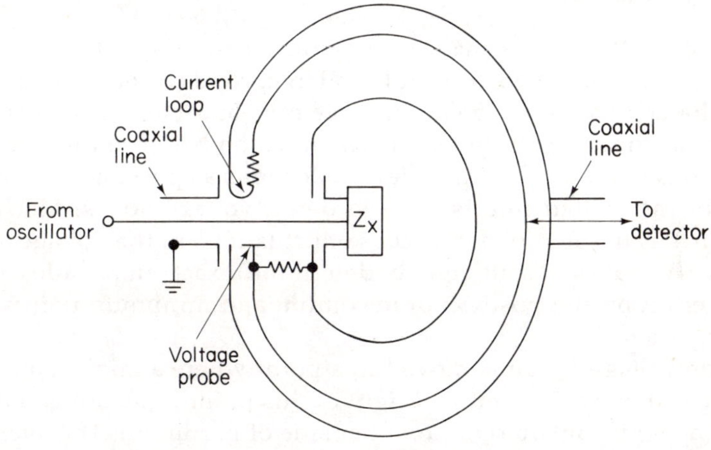

Figure 6–10 Byrne Bridge

Energy picked up by either loop or probe travels to the opposite end of the line where it is absorbed by the terminating resistor. The loop and probe are connected by a slotted auxiliary line. Energy from both loop and probe produces a traveling wave. The net voltage on the line is the resultant of superimposing the two signals traveling in opposite directions. The relative magnitudes of the two waves in the auxiliary line depend on the vector relationship of current and voltage in the unknown impedance as well as on the relative positions of the sampling loop and probe.

The output is coupled to a movable probe. This probe is set to a null position to begin a measurement; this is the point on the line where the two signals are equal in amplitude and 180° out of phase with each other. The dial of the detector should now indicate the impedance of the load. If the load is pure resistance, the movable probe is at the midposition on the line. If not, the phase angle of the impedance being measured is proportional to the displacement of the moveable probe in respect to the midposition. An indicator that shows the position of this probe may be calibrated in phase angle at some convenient reference frequency.

The Byrne bridge is equivalent to an ac bridge. It can be calibrated to measure both magnitude and phase angle of an unknown impedance. It is manufactured with very critical tolerances to provide reliable accuracy over a wide frequency range. Two advantages over an SWR detector are: ease of operation and lower cost.

Coaxial SWR Detector

A simple SWR detector may be constructed from a section of coaxial line. The detector is constructed from large, rigid coaxial line for measuring characteristics of smaller, flexible coaxial cable. Such an arrangement is illustrated in Figure 6–11. The junction between the lines must be well designed to contain as few deflections as possible. The voltage probe in the slotted line is used to detect voltage nodes. Mechanical irregularities in the probe may cause fluctuations in the voltage magnitude. This makes it difficult to determine exact amplitudes, but it has no effect on the position of maximum and minimum points of the standing wave.

The voltage probe is moved to a point where a minimum voltage is located, then shifted left and right of this point to obtain equal readings of a significant amount to either side of minimum. The average of

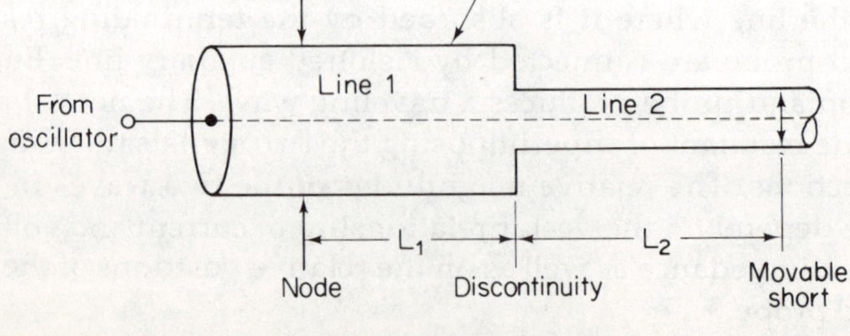

Figure 6–11 Coaxial SWR Detector

these two positions is the correct location of the node. One plus the total variation in the quantity is equivalent to the standing-wave ratio. This method of detecting SWR is primarily used with coaxial lines, but it can be adapted for application to two-wire lines and waveguides.

Impedance Bridge

The transmission line impedance bridge is a special bridge constructed from sections of coaxial line. It is useful for measuring impedance at frequencies too high for ordinary bridges and null networks. The circuit is illustrated in Figure 6–12. The signals into lines *A* and *B* are equal in amplitude but opposite in phase. When the arrangement is perfectly symmetrical electrically, the known impedance, Z_s, is exactly equal to the unknown impedance, Z_x. The output to the detector, in any case, will indicate the difference between Z_s and Z_x.

WAVEGUIDES

Waveguide transmission lines become increasingly necessary at higher and higher frequencies. When wavelengths are reduced to fractions of

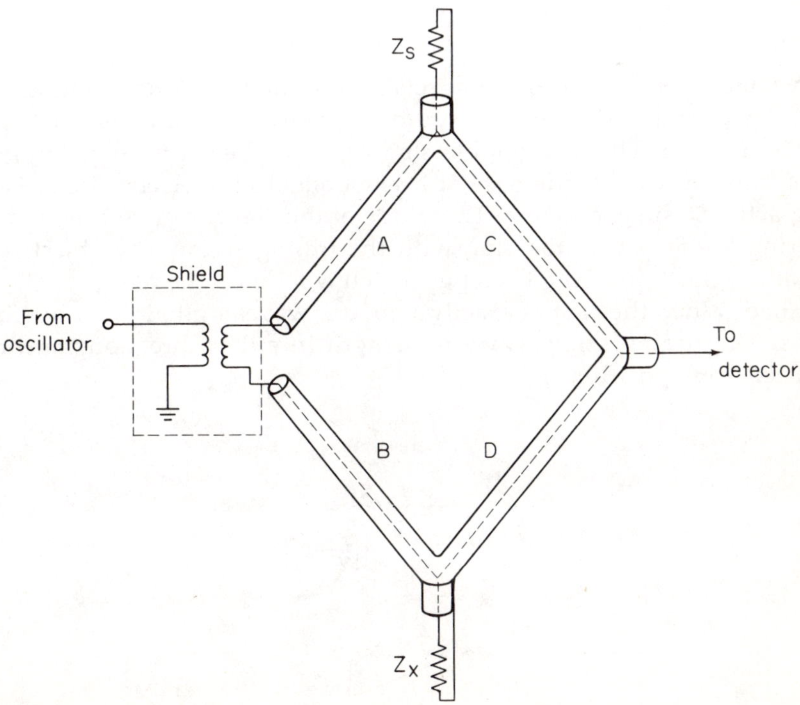

Figure 6–12 Coaxial Bridge

an inch, as in the SHF band, the waveguide is the only efficient transmission line.

A waveguide is equivalent to a coaxial cable with the center removed. This leaves a hollow pipe that may be either cylindrical or rectangular in shape. It has the form of a two-wire line supported by continuous metal insulators. The insulators are shorted quarter-wave stubs, and they form the walls of the waveguide. The characteristics of the rectangular guide are determined by the height and width of the walls. The narrow wall is considered as the height even when it is oriented on the horizontal plane. The characteristics of the cylindrical guide are determined by the diameter. The dimensions for both types are internal measurements.

Impedance

The *impedance* of a waveguide at any point is the ratio between the current and voltage at that point. Another way of saying the same thing is that the impedance is the ratio between the transverse magnetic fields at that point. In either case, the phase relation of voltage and current must be considered when computing the ratio.

Tuning

Waveguide *tuning devices* are specially designed sections of plumbing that can be attached to the waveguide. A type of slug tuner is illustrated in Figure 6–13. This is a double-slug tuner, and each tuning slug projects into the waveguide so that it is parallel to the electric field. Each slug acts as a small antenna. The phase of the signal can be changed by altering the length of the slug with the tuning screw. The object is to produce reflections from the slugs which will cancel the reflections from the load. When the slug is exactly a quarter wavelength, it has no reflections. The tuning screws vary the length in either direction from the quarter-wave position.

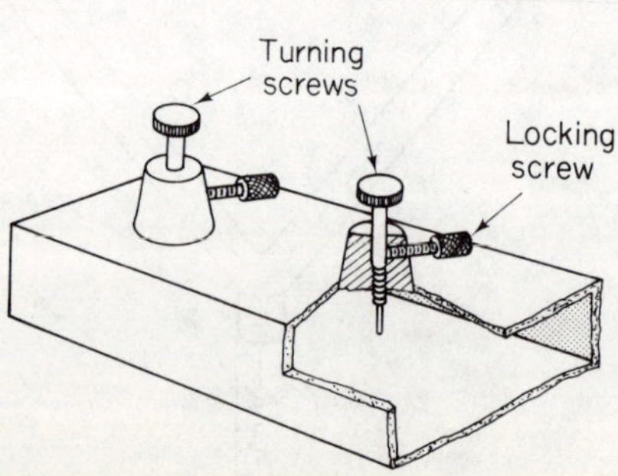

Figure 6–13 Slug Tuner

Shorted-stub tuners are also used. Such a device is illustrated in Figure 6–14. The shorted-stub tuner may be added to either wall of a waveguide. It consists of a section of waveguide with an adjustable plunger. The impedance at the groove is the infinite impedance of a shorted quarter-wave section. The impedance at point *A*, the forward end of the plunger, is nearly zero.

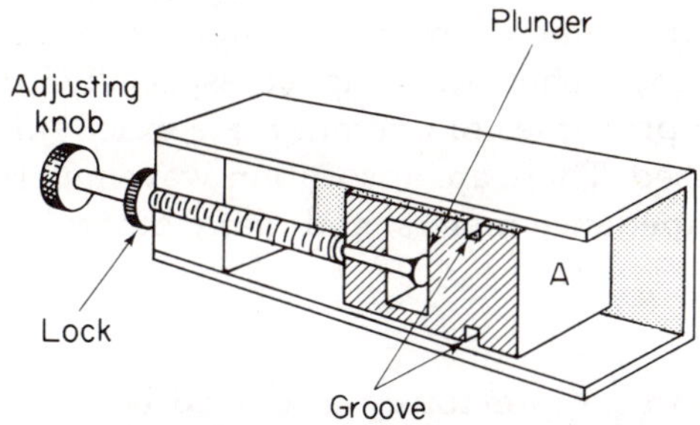

Figure 6–14 Shorted-Stub Tuner

When mounted on the wide wall, perpendicular to the *E* field, the stub impedance is in series with the impedance of the main waveguide. When mounted on the narrow wall, perpendicular to the *H* field, the stub impedance is parallel with that of the main line.

Other Tuners

A waveguide can be tuned by any device which introduces the required reflection into the line. Small conducting plates inserted into the waveguide are sometimes used. These are called window tuners. They are seldom adjustable, but they are useful when they can be preset and left alone.

Attenuators

It is sometimes necessary to adjust the *power* in a waveguide. Any of the various tuning devices may be used for this purpose. The tuning device can be adjusted to cause a mismatch of impedance that will lower the output power. In addition to these devices, there are several types of waveguide attenuators. The attenuators are designed to attenuate the signal without introducing reflections.

The attenuator usually employs the principle of a shutter or dissipative element which is introduced into the waveguide. The shutter limits the amount of energy allowed to pass by, closing-off part of the main waveguide. The dissipative element absorbs part of the energy from the waveguide and produces the same limiting effect.

MICROWAVE MEASURING INSTRUMENTS

Generally speaking, the same type of instrument is used to measure microwaves that is used to measure electromagnetic waves at lower frequencies. These instruments measure impedance, admittance, SWR, power, and frequency. They are designed especially for microwave frequencies, but the principles of operation are exactly the same as those previously discussed. There are absorption wavemeters, power meters, reflectometers, impedance meters, and many others.

Echo Box

One test instrument not previously discussed is the *echo box*. This is a device which is designed to provide testing for the overall performance of a radar facility. It consists of a tunable resonant cavity. Energy is generally coupled into the cavity from a small antenna. One type of echo box is illustrated in Figure 6–15. Energy from the transmitter is picked up by the antenna and coupled into the cavity resonator. The output may be fed through a crystal rectifier to a metering circuit. It may also be coupled to one of several types of CRT for display and analysis.

The micrometer is used to tune the cavity to resonate at the received frequency. Rise and decay of the signal in the resonant chamber are related to the Q of the cavity. It rises to 99% of maximum in 1.5 Q

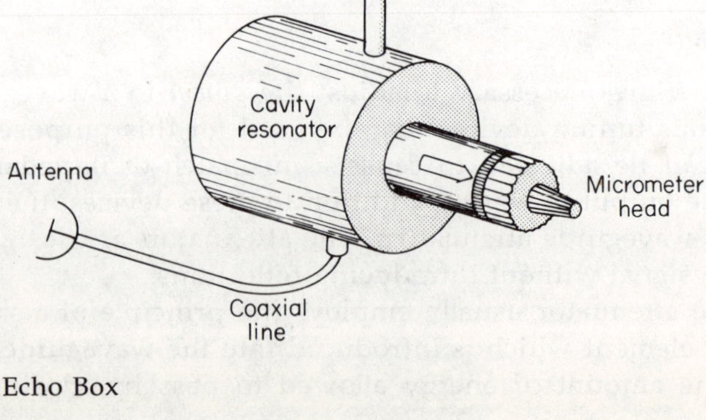

Figure 6–15 Echo Box

cycles and decays to 1% of maximum in the same time period. When the transmitter pulse terminates, oscillations in the cavity continue for a period of time. The space of time between termination of the transmitter pulse and termination of the oscillations in the cavity is called ringing time.

The ringing time of the echo box can be used to adjust transmitter power to peak efficiency. The adjustment is simple, and the objective is to obtain maximum ringing time on the echo box. The same characteristic can be useful in adjusting the receiver sensitivity. In this case, the output of the echo box is radiated back into the radar receiver and is processed as a target return. Again, the adjustments are made for maximum ringing time. The percentage of improvement in radar range against a selected type of target is about half of the increase in the ratio of the echo box signal to the receiver noise signal.

ANTENNAS

As previously stated, antennas are extensions of transmission lines. Several instruments already discussed for transmission-line applications can be used just as well for antenna measurements. Among these are the echo box, the reflectometer, frequency meters, power meters, and SWR detectors.

Some important characteristics of antennas, however, are different from transmission lines. Antennas therefore require different measuring instruments. For instance, the object of a transmission line is to keep radiation to a minimum, while with the antenna the object is maximum or optimum radiation. Some of the important antenna measurements are: SWR, impedance, reactance, resonant frequency, gain, power, directivity, and field strength.

Absolute Field Strength

The absolute electromagnetic field strength at any given point is expressed in volts per meter. This is also known as the field intensity; in this discussion the terms are interchangeable. The volts-per-meter strength is obtained from the magnitude of voltage that the field induces into a standard wire antenna, one meter in length, when placed at that particular point in the radiation field. This antenna must have a specified diameter as well as a specified length. For instance, 30 μV induced on such an antenna, when it is placed at a specified point from the transmitter, means that the electromagnetic-field intensity at that point is 30 μV/m.

This requirement does not imply that only one-meter antennas can be used for field-strength measurements. It does mean that any readings taken must be referenced to this one-meter antenna. This provides a field strength which is independent of the type of antenna used for the reception. Any antenna may be used, but it must be calibrated against the standard antenna in order to obtain an absolute field-strength measurement.

Relative Field Strength

Instead of measuring absolute field strength, it may be sufficient to obtain only relative indications of the field intensity. It may be desirable to know when the signal at a given point is growing stronger or weaker. Or the radiation pattern may need to be charted in terms of strong and weak field strength. Relative field strength is just as useful for these applications as absolute field strength.

INSTRUMENTS FOR MEASURING FIELD STRENGTH

The instruments for measuring field strength are highly specialized. The equipment for measuring absolute field strength must have a high degree of accuracy and is a complex arrangement. The older types are rather bulky and difficult to manage. The relative field-strength measurements require only one simple instrument: a field-strength meter.

Field-Strength Meter

Field-strength meters are available to cover any part of the frequency spectrum from low RF through the microwave band. They consist of a simple radio receiver with a meter to indicate relative signal strength. Figure 6–16 is a schematic of such an instrument. A field-strength meter of this general type may be used to indicate relative field strength for

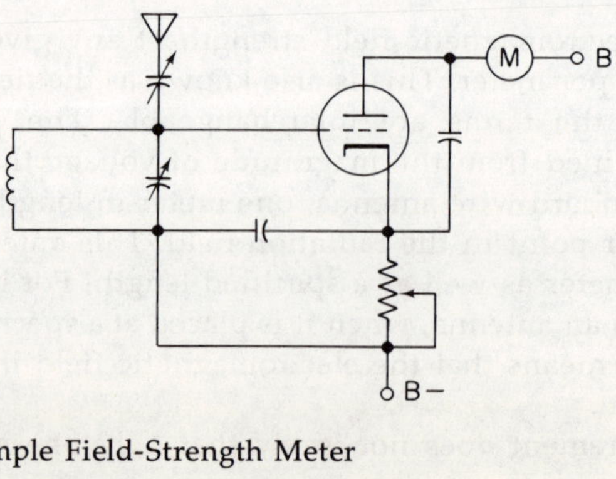

Figure 6–16 Simple Field-Strength Meter

frequencies up to 150 MHz. For higher frequencies, additional stages are added. This arrangement uses a combination amplifier and detector. The placement of the meter provides a high degree of sensitivity and enables indications of very weak fields.

The antenna used with field-strength meters is not a calibrated antenna. Its only function is to pick up a small quantity of the passing electromagnetic wave. For this reason, it is called a pickup antenna. It is mounted as close as possible to the circuitry of the amplifier to minimize noise and other spurious signals. The entire circuitry is well shielded for the same reason. The meter is generally a logarithmic type with indications directly proportional to decibels. In fact, the scale is usually calibrated in decibels.

Measuring Absolute Field Strength

The test equipment for measuring absolute field strength must contain a calibrated antenna, a well-shielded communications receiver, an indicating meter, and a calibrated RF signal generator. The relationship of these elements is illustrated in Figure 6–17.

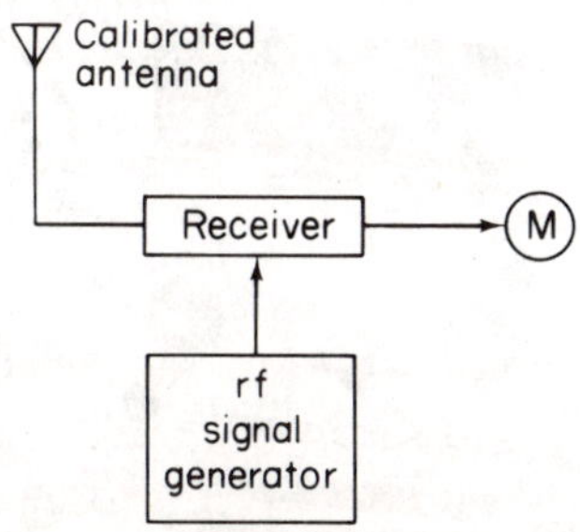

Figure 6–17 Test Equipment for Measuring Absolute Field Strength

The detected signal on the calibrated antenna is fed into the receiver and compared with the output of the calibrated RF signal generator. The meter is directly calibrated to the antenna. When the antenna actually is a one-meter standard antenna, the meter is calibrated in μV/m. If the antenna is any other type, the calibration factor must be applied to convert the meter indication to μV/m. For instance, a loop antenna may detect three times as much energy as the standard antenna. In this case, the calibration factor is three. When the meter reads 150 μV, this figure is divided by 3 to obtain 50 μV/m.

Most receivers cannot maintain a constant level of performance with a wide range of signal strengths. For this reason, field-strength instruments incorporate a calibrated attenuator between the antenna and the RF stage of the receiver. Signals are then attenuated by a factor proportional to the strength at the antenna. This provides the receiver with signals of a relatively constant value. In this case, the meter reading is multiplied by the attenuation factor to obtain actual field strength. For example, with an attenuation factor of 3 and a meter reading of 50 μV, the field strength is 3×50; 150 μV/m.

Figure 6–18 is a picture of a field-strength meter. This is the Sencore FS-134. It is a versatile, portable, accurate, and reliable instrument for measuring field strength. This instrument covers all television and frequency-modulation channels. It has a 30 μV sensitivity and uses calibrated attenuators on the input.

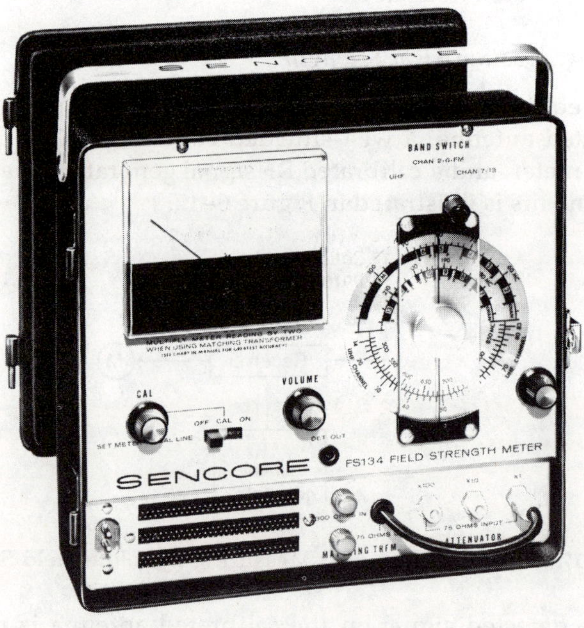

Figure 6–18 Field-Strength Meter (Courtesy Sencore)

ANTENNA GAIN

The gain of a transmitting antenna is a comparison of the radiated power of a directional antenna to that of a nondirectional antenna which is fed with the same amount of power. The directional antenna does not increase the power; it simply uses the available power more effectively by concentrating it in the proper direction. The field-strength ratio of the two antennas may be used to determine the antenna gain.

chapter seven

INSTRUMENTS FOR TESTING ACTIVE COMPONENTS

Instruments for measuring the passive circuit components, such as resistance, capacitance, inductance, reactance, and impedance, have been discussed in previous chapters. However, very few circuits are complete without some active component like the electron tube, the solid-state diode, a transistor, or some integrated circuit elements. These active components must be tested also. Many common types of general-purpose instruments such as voltmeters, ammeters, and ohmmeters are used to make certain tests on active components, but a better job can be done with the proper tools. In this case, the proper tools are instruments designed to test active components.

The characteristic curve tracers which were discussed in the chapter concerning waveform measuring instruments are one type of instrument designed especially for testing this type of component. This chapter examines some additional instruments used for testing active components.

INSTRUMENTS FOR TESTING ELECTRON TUBES

A number of instrument types have been developed for testing electron tubes. Bridge instruments are used for laboratory measurements of the parameters of grid-controlled tubes. These parameters are amplification factor (μ), plate resistance (r_p), and transconductance (g_m). Diode characteristics can be *adequately* checked by measuring plate current with specified amounts of plate voltage.

General-purpose work is not as demanding as laboratory measurements for accuracy, but it is more demanding from the standpoint of time and efficiency. Field-test instruments use the same principles as laboratory instruments, but the methods are generally simplified and streamlined. These general-purpose instruments are valuable troubleshooting tools, but their applications are limited. They can be used for testing the most often encountered tubes, but special tubes such as reflex klystrons, infrared converters, and electrometer tubes must be tested in a different fashion. These tubes must be tested either with special test instruments or in the equipment where they are used.

There are two classes of general-purpose test instruments for use in the field: emission testers and transconductance testers. In addition to their main function, most test instruments in either class can be used for testing such common performances as filament activity, noise, gas, cathode leakage, and shorted elements. These functions are generally added features to increase the usefulness of the instrument.

Emission Testers

The primary purpose of an *emission tester* is to measure the condition of the emitting surface of the cathode. When the electron emissivity starts to reduce, the useful life of the electron tube is almost over. This is usually the result of the gradual aging process. It renders the cathode incapable of supplying the quantity of electrons required for proper tube operation.

Figure 7–1 illustrates the circuit for an emission tester. This is a simple instrument; it applies operational voltages and measures the plate current of the tube under test. Notice that the grid (or grids) is returned to the plate. The tester is checking only the quantity of emission from the cathode.

The emission test is essential, but it is not a precise measurement. The meter indicates only good, weak, and bad tubes. Tube manufacturers do not specify a definite amount of emission that can be used as a reference. The emission test can prove the tube is bad, but it cannot prove conclusively that a tube is good.

Very high emission can be caused by a hot spot on the emitting surface, which will lead to complete tube failure in a very short time. The tube is not operated under the same conditions as in the actual equipment, so a good indication here may not be good under the work-

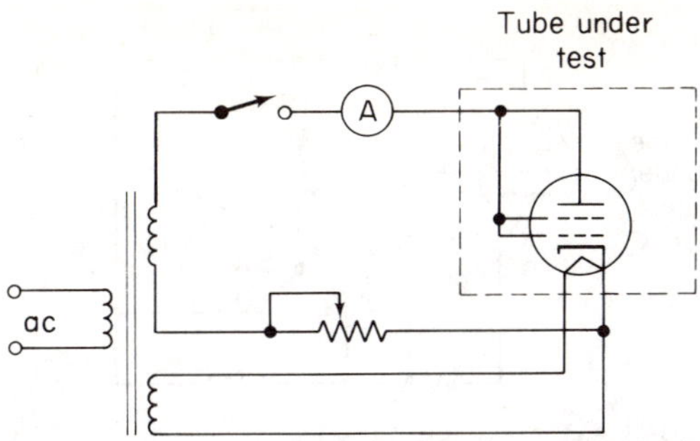

Figure 7–1 Emission Tester

ing conditions. The alternating voltage may cause gas to be released inside the tube, which causes erroneous readings. In normal operation, the grid potential controls the plate current. The grid has no control in this tester. This fact sometimes causes an invalid indication.

Transconductance Testers

The *transconductance* test instrument provides a more reliable indication of grid-controlled tube conditions than can be obtained from the emission tester. It evaluates the amplification ability of the tube under both static and simulated circuit conditions. Also manufacturers issue standard tube tables that specify proper transconductance for each type of grid-controlled tube. A tube is defective when the transconductance decreases to 70% of the value specified in the standard table.

Transconductance (g_m), sometimes called mutual conductance, indicates the effects of a grid voltage in controlling plate current. It is the ratio of change in plate current to a small change in grid voltage as shown below:

$$g_m = \frac{\Delta I_p}{\Delta E_g}$$

Transconductance is expressed in micromhos.

In the static test for g_m, the tester supplies the tube with direct voltage. The grid voltage is then changed by a specified amount, and the change in plate current is noted. The changes are used to calculate the transconductance. This circuit is illustrated in Figure 7–2.

The tester also supplies filament voltage, which is not shown on this diagram. With the switch set as indicated, the tube is functional with a definite bias level. Grid voltage and plate current are recorded. The switch is then placed to Position 2. This changes E_g by a specified

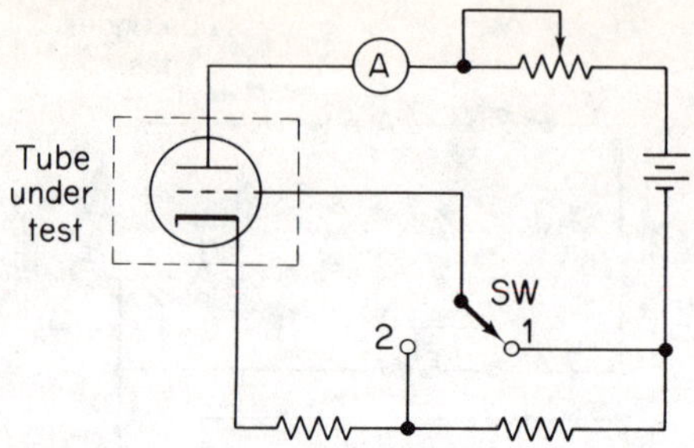

Figure 7–2 Tester Circuit for Static g_m Test

number of volts in a less negative direction. The I_p increases; the change in I_p is noted. These values are then substituted into the formula to calculate g_m:

$$g_m = \frac{\Delta I_p}{\Delta E_g}$$

During the dynamic test of g_m, the tester supplies an ac signal and a dc bias to the control grid. The simplified tester circuit for this test is illustrated in Figure 7–3. Without the ac signal to the grid of the tube under test, the two halves of the rectifier conduct equally. The meter, which reacts to the difference between the currents from the two plates, would remain at zero.

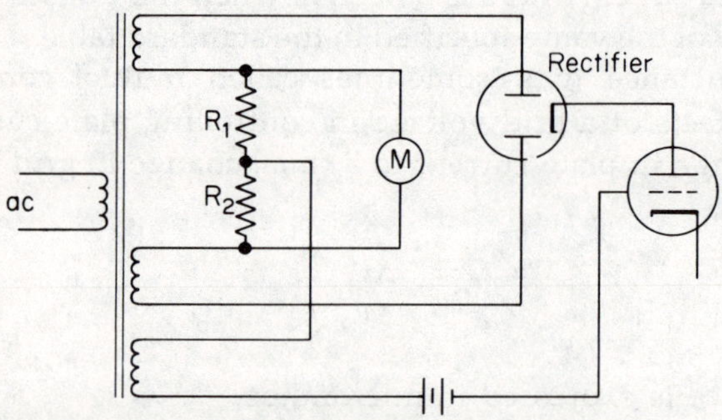

Figure 7–3 Tester Circuit for Dynamic g_m Test

The ac signal alternately adds to and subtracts from the dc bias and causes the tube under test to conduct more on one alternation than it does on the other. This produces an unbalanced condition with more current through R_1 than through R_2. The meter deflection is proportional to the difference in current between R_1 and R_2.

The meter may be calibrated as simply good, fair, or poor, but in most cases it is calibrated directly in micromhos of transconductance.

As previously stated, either emission or transconduction measuring instruments may be used for several auxiliary tests. The switches are arranged so that the internal circuitry is altered to fit the desired test.

Short-Circuit Test

If the quality of a tube is in doubt, the short-circuit test should be made first. If it proves that elements are shorted, time is saved by the fact that no further testing is necessary. The circuitry for this test is illustrated in Figure 7–4. With the switch positions as shown, the plate is connected to one side of the neon indicator, and all other elements are connected to the other side of the indicator through the transformer. If any element is shorted to the plate, the neon lamp circuit is completed, and the lamp glows to indicate that a short is present. Lightly tapping the tube during this test will aid in detecting loose elements that can cause intermittent shorts. Each element can be checked for shorts to all other elements by rearranging the switch positions.

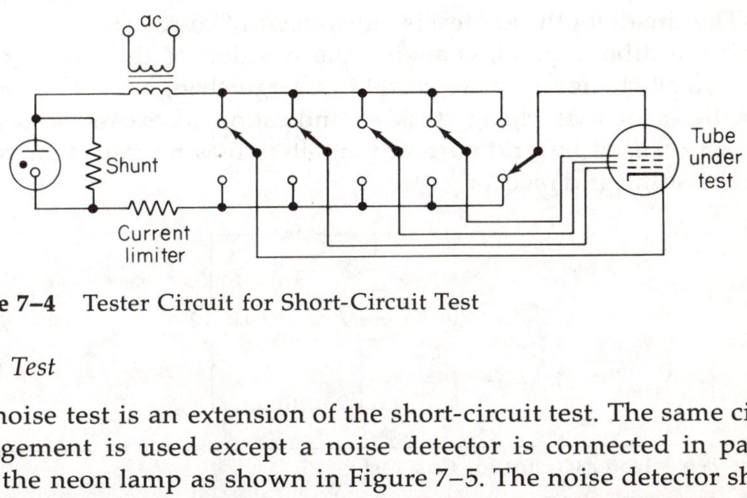

Figure 7–4 Tester Circuit for Short-Circuit Test

Noise Test

The noise test is an extension of the short-circuit test. The same circuit arrangement is used except a noise detector is connected in parallel with the neon lamp as shown in Figure 7–5. The noise detector shown here is a pair of headphones; this is adequate for most work. For a highly sensitive noise check, the headphones may be replaced by an external amplifier and speaker. The ordinary radio receiver makes an excellent noise detector. If a receiver is used, the antenna terminal is connected to one side of the neon lamp, and the ground terminal is connected to the other side. Noise detectors are not normally supplied with the test equipment.

Once the noise detector is in place, the procedure is the same as for checking shorts. If the tube has a tendency toward excessive noise,

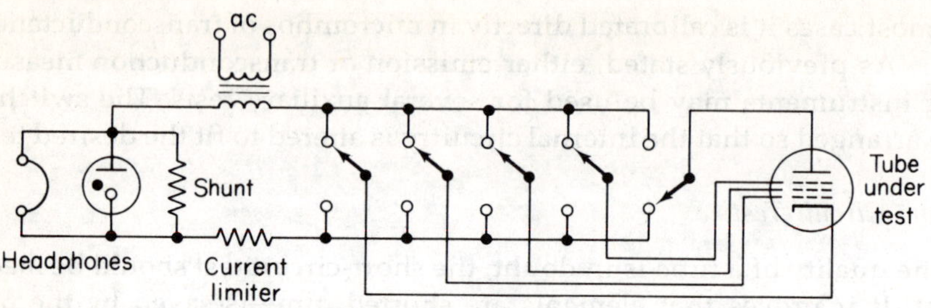

Figure 7–5 Tester Circuit for Noise Test

tapping the tube will produce loud crashes of noise over the normal noise level.

Gas Test

Except for tubes that are deliberately filled with gas, excessive gas inside the tube is an undesirable condition. The usual result is a cloud of ions that clusters about the control grid and causes excessive grid current. The circuit for the gas test is illustrated in Figure 7–6.

If the tube is good, changing the position of the switch causes only a small change in plate current. A large change in plate current, when the switch is changed, is an indication of excessive gas. The change is caused by grid current that alters bias by causing a voltage drop across the grid resistor.

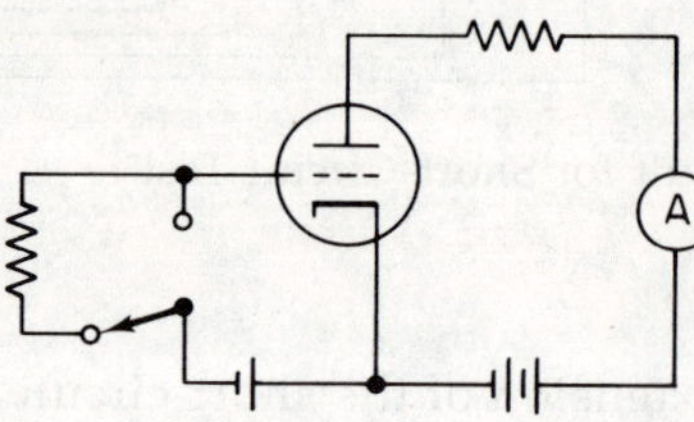

Figure 7–6 Tester Circuit for Gas Test

Cathode-Leakage Test

Cathode leakage refers to a breakdown of the insulation between the heater and the cathode. This is usually a slow process as the insulation deteriorates from age and constant heat. As the insulation becomes brittle and cracks, high-resistance paths are opened up between the grid and cathode. In extreme conditions, there may be a direct short between these elements. The circuit for the cathode-leakage test is illustrated in Figure 7–7.

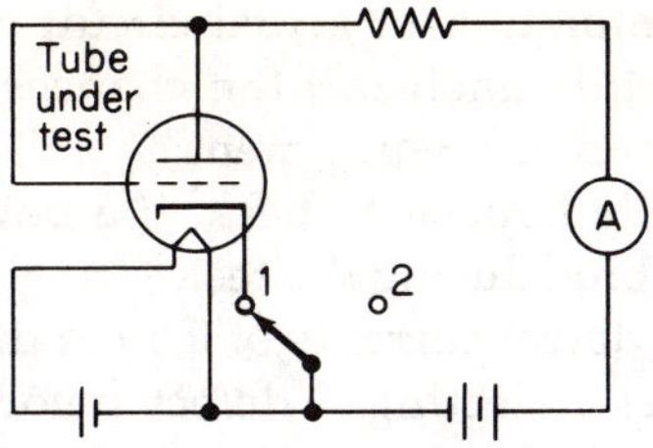

Figure 7–7 Tester Circuit for Cathode-Leakage Test

With the switch in position 1, the circuit is complete and the meter indicates a given amount of plate current. When the switch is changed to position 2, the cathode is open. If the tube is good, the plate current drops to zero because of the floating cathode. However, if there is leakage between heater and cathode, a small amount of plate current will still be registered on the meter.

Filament Activity Test

A tube is designed in a manner that allows it to operate with near normal plate current when the heater (filament) voltage is reduced by 10%. The tube checker has a two-position filament activity test switch. In the normal position, normal heater potential is applied to the tube under test. The switch remains in the normal position for all tests except the filament activity test. When it is placed in the test position, it reduces the heater voltage by 10%.

A reasonable time is allowed for the cathode to cool; then the meter is checked for a change in plate current. If the tube is good, no appreciable change in current will occur. If the tube is nearing the end of its useful life, the meter will register a noticeable decrease in plate current as the cathode cools.

Diode Tests

Diodes are tested in the same manner as the emission test previously described. A known value of ac is applied, and the meter indicates the rectified plate current. With double diodes, each section is checked as a separate diode.

Electron-Tube Analyzer

The *electron-tube analyzer* incorporates both the emission and transconductance testing capabilities, and it can be used to make all the tests described here. Such an analyzer is pictured in Figure 7–8. This is the Sencore MU-150, and it has many advanced features. One important item is the simulation of actual circuit conditions. If the tube performs properly in this analyzer, the chances are very good that it will perform properly in the equipment. A full load is placed on the tube during the cathode-emission check. If a power tube turns blue in the circuit, it will turn blue during the test.

The MU-150 can detect currents of 0.5 μA during the grid-leakage (gas) test. This enables detecting a defect before it is bad enough to cause serious problems.

Instead of applying an ac sine wave to the tube, this analyzer uses a 5 kHz square wave. This provides a g_m check with a signal that is rich in harmonics and enables interference-free checks at higher frequency levels. Actual operating potentials are applied, and the tube is checked for its ability to amplify under operating conditions.

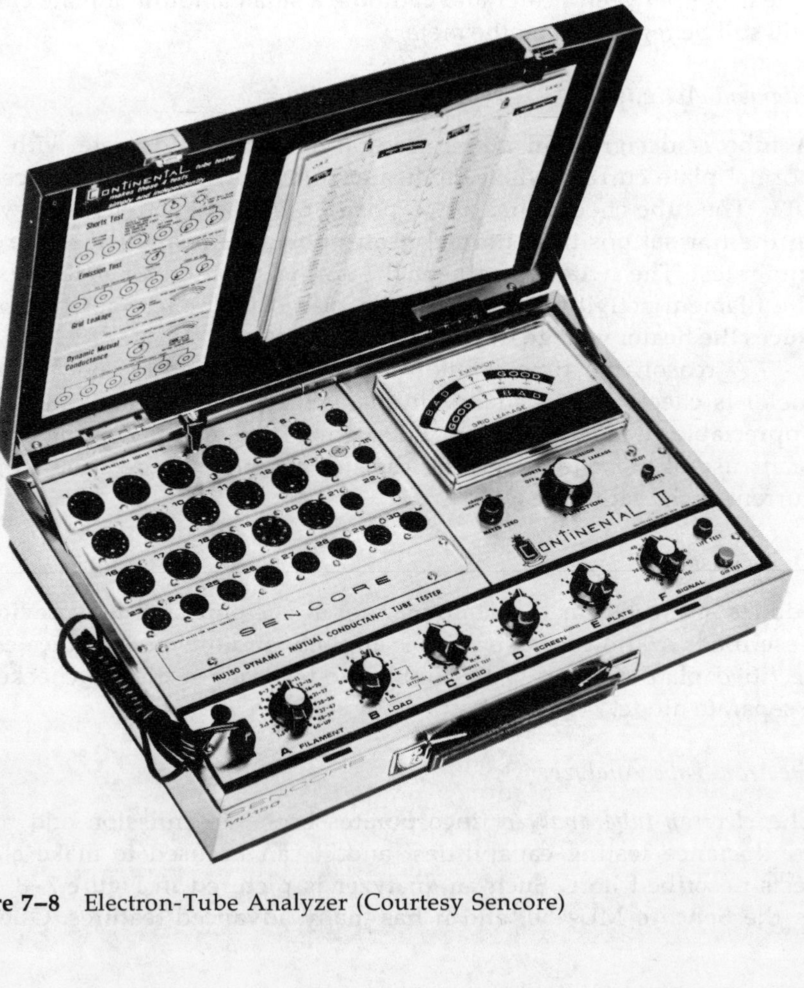

Figure 7–8 Electron-Tube Analyzer (Courtesy Sencore)

CRT Tester

Cathode-ray tubes, especially the picture tubes used in television, require special types of tests and specialized testing instruments. The CRT auto-tracker featured in Figure 7–9(a) is such an instrument. It is not good enough to check emission alone on a CRT, and this CR-161 improves on that old method. To avoid a change in background color, the three guns of a color CRT must look alike from the time the brightness control is minimum until it is all the way up. This is called tracking, but tracking has in the past been very difficult to check.

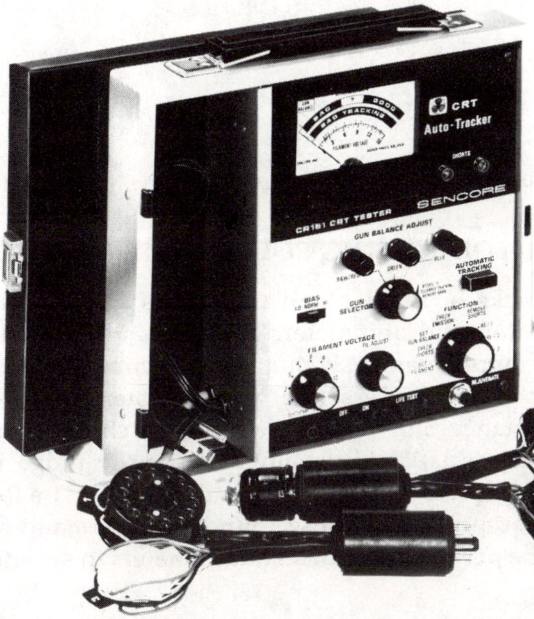

Figure 7–9(a) CRT Auto-Tracker (Courtesy Sencore)

The Sencore auto-tracker simplifies the procedure and supplies highly reliable results. It charges a memory circuit during the emission test and switches the potentials to field-effect transistors and comparison diodes. If there is a difference of more than a 1.5-to-1 ratio (the recommended tolerance) between the guns, the diodes conduct and cause the meter to indicate a faulty condition. This memory circuit is illustrated in Figure 7–9(b).

The charges on the memory circuit will remain undisturbed for a prolonged period of time. This enables the test to be made in a leisurely fashion with no necessity for recharging.

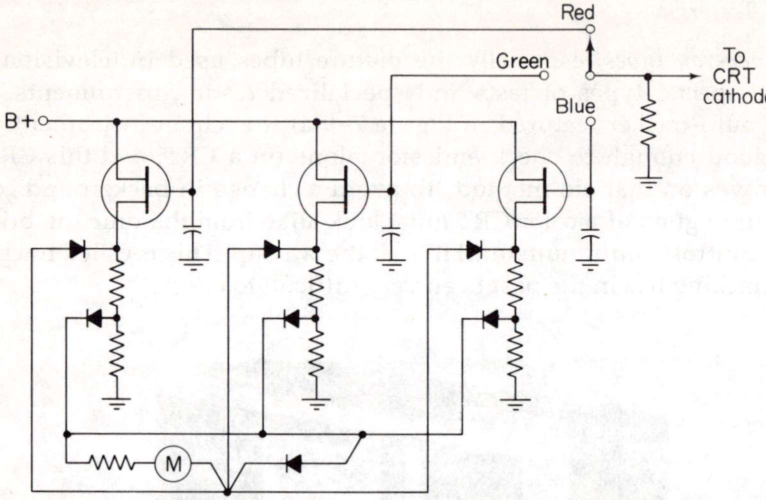

Figure 7–9(b) Auto-Tracker Memory Circuit (Courtesy Sencore)

INSTRUMENTS FOR TESTING SOLID-STATE DEVICES

Many useful checks on diodes and transistors can be made with the ammeter, voltmeter, and ohmmeter. However, specialized equipment is essential for complete analysis under actual circuit conditions. Complete parameter checks are essential in experimental work but not in ordinary maintenance and repair activities. A check of two or three performance characteristics is usually sufficient to judge whether the device should be replaced or returned to the circuit. The following test circuits are representative of those found in testers and analyzers. A similar test can be performed with standard meters in several cases.

Tests of Rectifier Diode

Many crystal diodes are primarily used as rectifying elements. A check of the front-to-back resistance and current is an adequate check on the rectifying qualities. A simplified circuit for this test is illustrated in Figure 7–10. This is a simple series circuit with an ammeter to measure diode current. With the switch in the positions indicated, forward current is being registered on the meter. Changing the switch positions reverses the current through the diode. The meter will then indicate the reverse current. The variable resistor enables the checks to be performed with different levels of voltage.

The results of this test are evaluated by comparing with typical values. The typical values may be furnished with the test instrument or

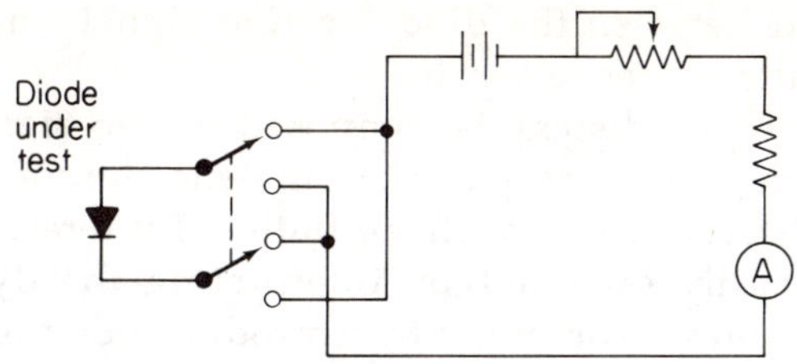

Figure 7–10 Tester Circuit for Rectifier Diode Test

obtained from the manufacturer's data sheets. Generally, a good rectifier is indicated when the forward-to-reverse current ratio is at least 10 to 1.

Dynamic Tests of Diodes

General-purpose diodes and switching diodes for computer applications are most effectively tested under dynamic conditions. This means that it is necessary to measure forward-to-reverse ratios with a variety of voltage levels. A dynamic diode tester is used for this purpose, and for some tests, the tester output is fed to an oscilloscope. A functional diagram for such an instrument is illustrated in Figure 7–11.

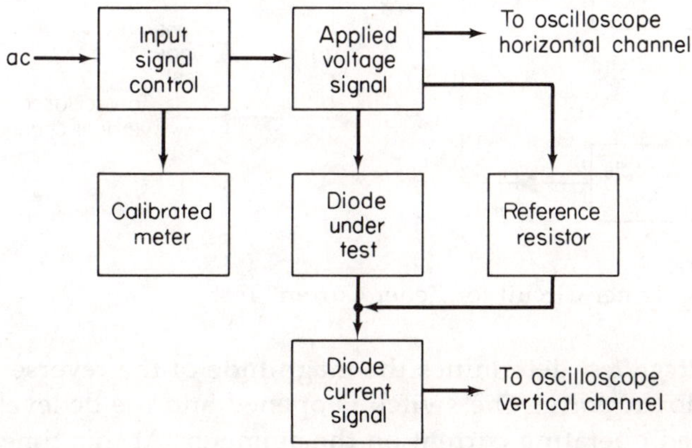

Figure 7–11 Tester Functional Diagram for Dynamic Diode Tests

In this diagram the applied voltage is adjusted to the desired amplitude of test voltage by the input controls. A portion of this test voltage is connected to the horizontal channel of an external oscilloscope. The diode current signal is actually a voltage which is produced by diode current. The signal to the vertical channel of the oscilloscope can be switched between the diode current signal, and the signal developed across the reference resistor.

The oscilloscope display is composed of two signals. The values of voltage and current correspond to the deflection of the oscilloscope trace. Diode characteristics such as flutter, hysteresis, and negative resistance are easily obtained by interpreting the dynamic display. For instance, a comparison of the trace produced by the diode with the trace produced by the reference resistor voltage reveals the reverse current characteristics of the diode.

Zener Diode Tests

Zener diodes can also be tested in a dynamic circuit with the output going to an oscilloscope. A simplified tester circuit for the current measuring test is illustrated in Figure 7–12. Slight alterations in this circuit enables measurement of voltage and impedance as well as current. A switching arrangement, not shown on the diagram, switches the output to represent either the voltage applied to the Zener or the current through it.

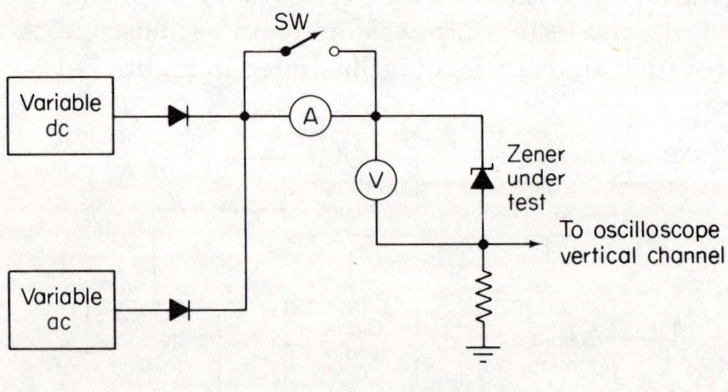

Figure 7–12 Tester Circuit for Zener Current Test

A *voltage test* determines the magnitude of the reverse voltage at the breakdown point. The switch is opened and the dc level is set for the specified operating current on the ammeter. At this time, the voltmeter must indicate a voltage which conforms to the recommended Zener operating voltage.

A dynamic impedance test will check the Zener's regulatory ability. This test measures the impedance of the diode in the *Zener region.* The dc input is adjusted so that the diode reverse current is at the recommended value. This places the diode in the Zener region. The ac input is then adjusted until the oscilloscope display indicates the current specified in the impedance test procedure. This value is designed to keep the diode in the Zener region. The voltage across the diode is then read from the voltmeter. The direct voltage value is taken from the voltmeter, and the alternating voltage value is obtained from the oscilloscope. The dynamic impedance of the Zener is then calculated from these values of voltage and current as follows:

$$Z = \frac{E_{ac} - E_{dc}}{I_{ac} - I_{dc}}$$

Transistor Tests

Five basic types of transistor tests are covered in this section. They are: (1) collector-leakage current, (2) direct-current gain, (3) voltage punch-through, (4) alternating-current gain, and (5) the four-terminal parameter test. The transistors shown are standard bipolar three-terminal PNP, but the same tests apply to the NPN. For testing NPN transistors all the bias voltages are reversed, but otherwise the test procedures are the same.

Collector-Leakage Current Test

Collector-leakage current is a function of temperature and the resistivity of the transistor materials. Excessive leakage generally occurs when the conductor surface is contaminated. Other cause of this condition is overheating or other types of damage. This is a reverse current from collector to base with the emitter open. It is abbreviated I_{cbo}. An excessive I_{cbo} indicates a faulty transistor. The tester circuit for this test is illustrated in Figure 7–13. The collector-base junction is reverse biased, and the emitter is open. The ammeter (in the micro range) indicates the reverse current.

Current from collector to emitter with an open base is termed I_{ceo}. Figure 7–14 illustrates the circuit for this test. I_{ceo} should be expected to be much larger than I_{cbo}.

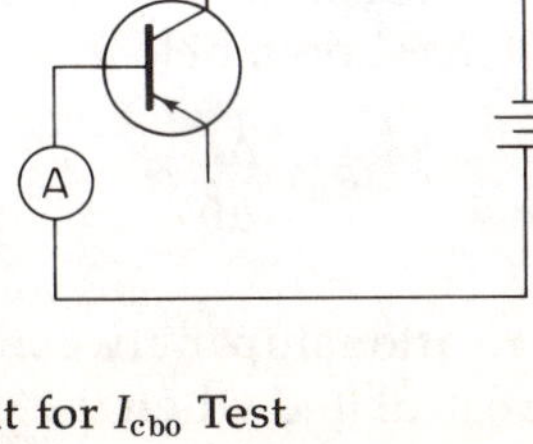

Figure 7–13 Tester Circuit for I_{cbo} Test

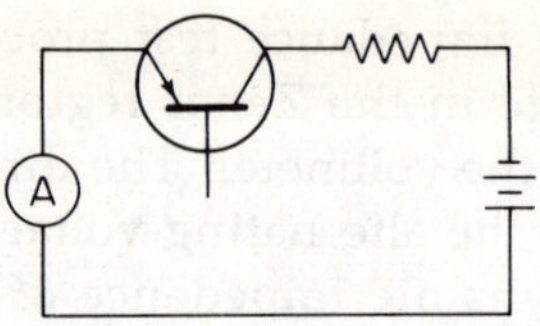

Figure 7–14 Tester Circuit for I_{ceo} Test

Direct-Current Gain Test

Direct-current gain is a measure of the effectiveness of the base in controlling collector current. This is a useful parameter for transistors used in low-frequency power, amplification, switching, and control circuits. This test is useful only if the transistor is used in a common-emitter configuration.

Figure 7–15 illustrates the tester circuit during this test. R_1 is adjusted for a null on the voltmeter, and at this point, dc gain $= I_c/I_b$. Also under these conditions, dc gain $= R_x/R_2$. R_1 usually has a calibrated dial to provide a direct reading of dc gain.

As a transistor ages, there is a tendency for the dc gain to decrease. This causes less amplification, which leads to distortion of the signal.

Alternating-Current Gain Test

Alternating-current gain is expressed in two different factors depending on the type of configuration the transistor is used in. For a common-base configuration, the amplification factor is *alpha,* and the symbol is the Greek letter α. Alpha is the ratio of the change in collector current to a change in emitter current with the collector voltage held constant. The collector has an *ac* short to the base during this test.

$$\alpha = \frac{\Delta I_c}{\Delta I_b}$$

If the transistor is in a common-emitter configuration, the amplification factor is *beta,* and it is symbolized by the Greek letter β. Beta is the ratio of collector-current change to a base-current change when the collector voltage is held constant. The collector has an ac short to the emitter during this test, as shown below:

$$\beta = \frac{\Delta I_c}{\Delta I_e}$$

There is a definite relationship between alpha and beta so that the value of either may be calculated when the other is known:

$$\alpha = \frac{\beta}{1+\beta} \qquad \text{and} \qquad \beta = \frac{\alpha}{1-\alpha}$$

The test circuit for measuring beta is similar to that for measuring dc gain; see Figure 7–15. The primary difference is that the beta measurement requires an ac signal at the transistor base.

The proper values of either beta or alpha will appear on the manufacturer's data sheet. Measured and calculated values should match these fairly closely if the transistor is good.

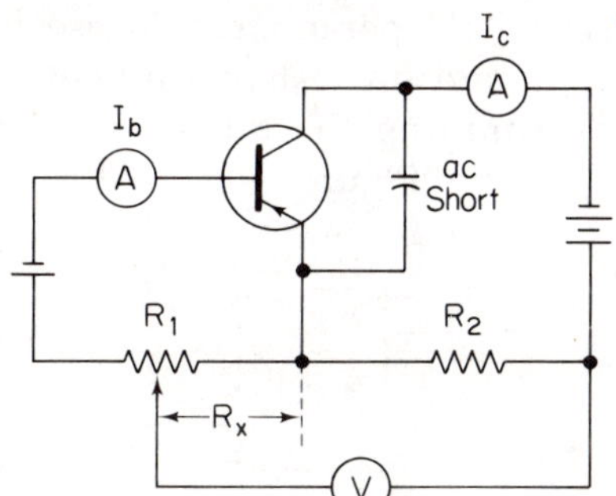

Figure 7–15 Tester Circuit for Direct-Current Test

Punch-Through Voltage Test

The punch-through voltage test is a dc test to determine the quality of a transistor for use in switching circuits. The punch-through voltage is the potential between collector and base which will cause a low-resistance path of contact between collector and emitter. This is the point where the base region disappears and transistor action stops. This test, handled properly, does not damage the transistor. It will resume normal characteristics when the collector-to-base potential is lowered. The tester circuit for this test is illustrated in Figure 7–16.

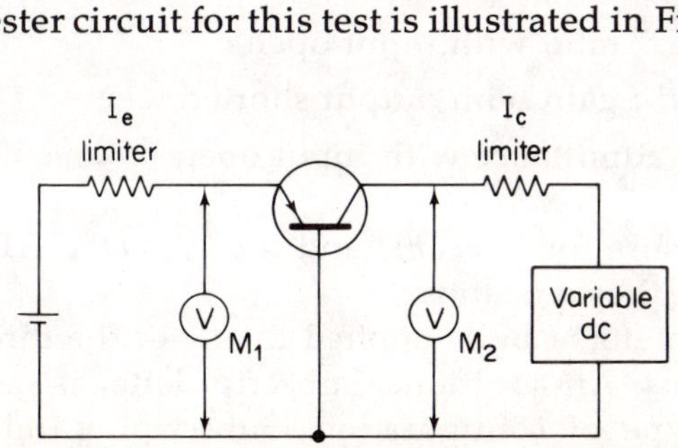

Figure 7–16 Tester Circuit for Punch-Through Voltage Test

The variable dc is adjusted until punch-through voltage is obtained. This is indicated when the reading on M_1 reaches a specified reference value. At this time, the punch-through voltage can be read from M_2. The two current-limiting resistors keep the punch-through current to a safe level and prevent transistor breakdown.

Four-Terminal Parameter Test

A transistor may be considered as a *four-terminal network* in order to determine the relationships of inputs and outputs. These relationships are referred to as hybrid (*h*) parameters. These *h* parameters are referred to in data sheets and on test instruments. Hybrid parameters are very useful in determining the quality of a transistor. The four-terminal network is shown in Figure 7–17.

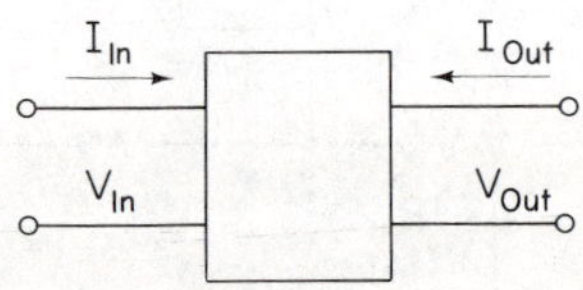

Figure 7–17 Four-Terminal Network

With this arrangement, there are two currents and two voltages to consider. If the two currents are considered as dependent variables, the resulting parameters are short-circuit parameters, and they are measured in mhos. When the two voltages are considered as dependent variables, the resulting parameters are open-circuit parameters, and they are measured in ohms. Hybrid (*h*) parameters are obtained by using one current and one voltage as dependent variables. These are the designations for the four *h* parameters:

h_i = input Z with output shorted
h_r = reverse *V* ratio with input open
h_f = forward *I* gain with output shorted
h_o = output admittance with input open

The unit of measure for h_i is *ohm* and for h_o is *mho*. There is no unit for h_r and h_f since they are ratios.

The *h* parameters can be applied to any of the three basic amplifier configurations. An additional subscript letter is generally used to designate the type of configuration. Subscript *b* indicates common base, *e* designates common emitter, and *c* is used for common collector.

Here are the h-parameter designations for a common emitter:

h_{ie}

h_{re}

h_{fe}

h_{oe}

Alpha in a common-base circuit is equal to h_{fb}. *Beta* in a common-emitter circuit is equal to h_{fe}.

The tester circuit for obtaining h parameters is illustrated in Figure 7–18. G_1 is a calibrated current generator, and G_2 is a calibrated voltage generator. The ac meter is used to indirectly measure current. The switches are four-ganged sections of a five-position rotary switch.

Position 1 of the switch connects the calibrated current generator to the emitter of the transistor under test. The ac meter indicates $1 + h_f$, which is the ratio of ac base current (i_b) to ac emitter current (i_e).

Position 2 of the switch connects the calibrated voltage generator to the collector, and the meter indicates output admittance, h_o. The emitter is open during this measurement.

Position 3 measures h_r; position 4, h_i; and position 5, alpha.

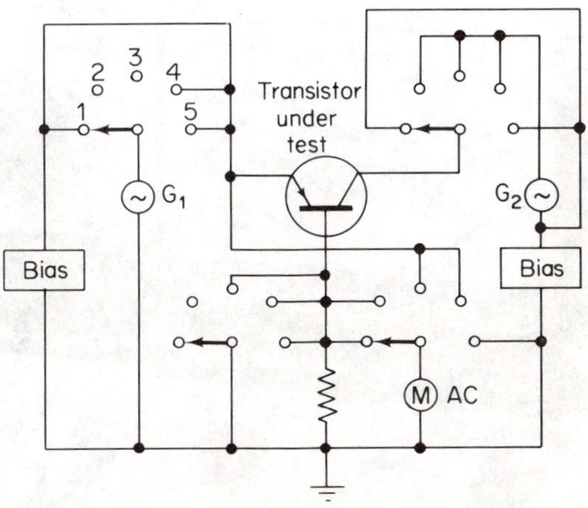

Figure 7–18 Tester Circuit for Measuring Hybrid Parameters

Sample Transistor Tester

Most transistor testers may be used for all these tests and for others that the instrument manufacturer considers pertinent. One mark of an advanced transistor tester is its ability to measure current leakage and voltage gain of field-effect transistors. The field-effect transistor (FET) has become so popular that a tester without FET measuring capabilities is likely to fall short of doing a complete job.

Figure 7–19 is a picture of an advanced transistor tester. This is the Sencore TF-17. It is capable of measuring the operational qualities of nearly 14,000 transistors, including FETs. Gain and leakage tests for regular and field-effect transistors can be performed either in or out of circuit.

The gain of regular transistors is the increase in current from input to output. This is either alpha or beta depending on the configuration used. With the FET, the gain is transconductance (g_m) because the FET is a voltage-sensitive device. In fact, the parameters of an FET resemble those of an electron tube more than those of a transistor. Beta of regular transistors and transconductance of FETs—both are made directly in the circuit where they are used.

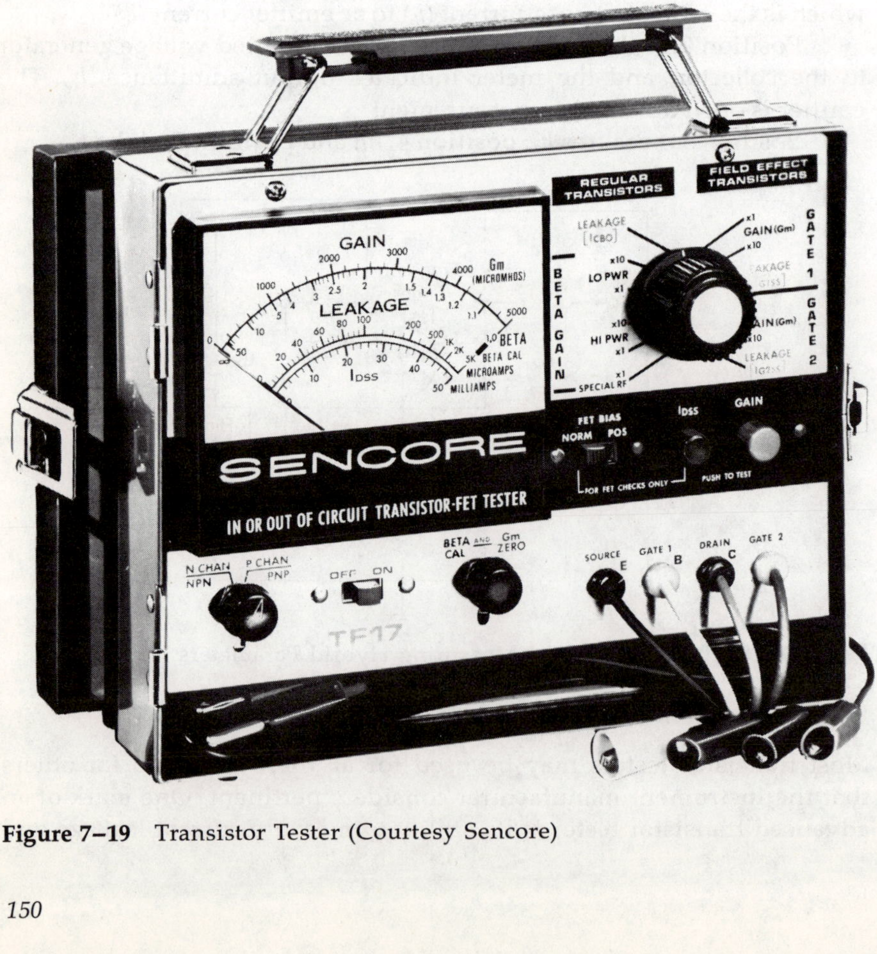

Figure 7–19 Transistor Tester (Courtesy Sencore)

Special tests can be performed on RF and high-power transistors with the TF-17. A reference book of functional data for over 12,000 transistors accompanies this instrument. The book is not needed in most cases to check for good or bad transistors; but when it is important to know exact information, the book is very handy.

chapter eight

INSTRUMENT ACCESSORIES

Nearly all instruments require one or more accessory devices. Even a simple dc meter requires a probe for obtaining a proper input. The manufacturers build a great many accessory circuits into the test instruments, but that still leaves a great variety of accessory devices to be selected by the user as attachments to the equipment. There are probes, delay lines, sensors, electronic switches, dummy loads, couplers, and attenuators—just to mention a few.

Selecting the proper accessories for electronic instruments is just as important as selecting the instruments themselves. A gun will not fire without matching ammunition, and test instruments will not function without matching accessories. Not only must the accessory match the instrument, it must match the measuring *job* the instrument is expected to perform.

ATTENUATORS

An *attenuator* is a sampling device used in a coupling network to reduce the signal amplitude or to pick off a reduced sampling of the signal. For the lower frequencies, one attenuator is usually a resistive network designed to introduce a known loss between its input and output terminals. Attenuators at microwave frequencies take the form of a *waveguide* section. The loss introduced by an attenuator is intended to reduce voltage, current, or power in controllable amounts for testing purposes.

Impedance Matching

Most attenuators are designed to operate between specific impedances. This is a critical consideration because a definite signal ratio between input and output must be maintained. A mismatch of impedance at either end of the attenuator will cause a loss of energy and result in an upset of the signal ratio.

Some attenuators are designed for operation between different impedances or between unknown impedances. This complicates the networks, but these devices have a wider application. For instance, an impedance network that operates between impedances of 500 ohms and 200 ohms is illustrated in Figure 8–1. In this case, an impedance-matching transformer is used on the input to the attenuator. Sometimes it may be necessary to use such a transformer on both input and output. In the case of unknown impedances, tunable transformers may be used.

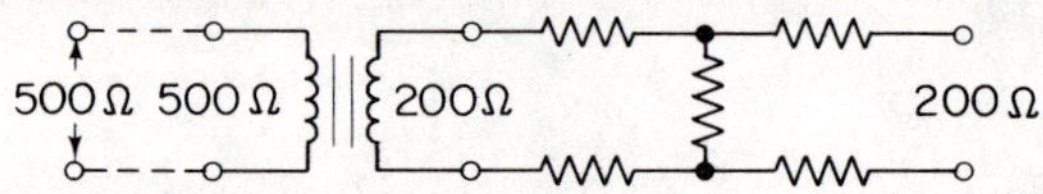

Figure 8–1 Attenuator between Different Impedances

Balanced Pad

A resistive attenuator is commonly called a *pad*. A *balanced pad* has equal corresponding impedances on both sides of the line. That is, the impedance from *both* input terminals to ground are the same, and the impedance from both output terminals to ground are the same. Such a pad is illustrated in Figure 8–2.

The input impedance does not necessarily equal the output impedance, but like-numbered resistors have the same resistance. If R_1 and R_3 were of equal values, then input and output impedance would be the same. This latter case is called a symmetrical balanced pad.

The pad in Figure 8–2 is an *H* network. The name is descriptive of the shape of the schematic diagram. Other popular shapes for balanced pads are *L* and *O*.

The balanced pad undergoes a major reaction change as frequencies increase. This is caused by the capacitance-to-ground effect. Another disadvantage is the fact that this pad *cannot* be used with a trans-

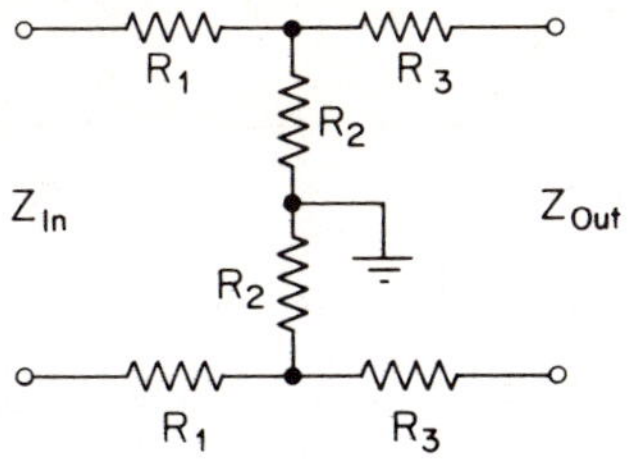

Figure 8–2 Balanced Pad

mission line that has one side grounded, without adding a transformer between the pad and the line.

Unbalanced Pad

The unbalanced pad has one side grounded. It has less capacitive effect at higher frequencies and works quite well with a transmission line with one grounded side. Such a pad is illustrated in Figure 8–3.

Again Z_{in} and Z_{out} may or may not be equal. When R_1 is equal to R_3, the two impedances are the same, and the pad is symmetrical. The pad in Figure 8–3 is a T-network. Other popular shapes for the unbalanced pad are L and π.

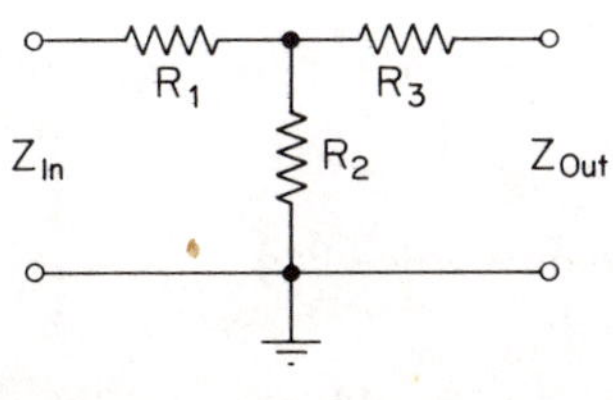

Figure 8–3 Unbalanced Pad

Extended Attenuator Networks

Any compatible attenuator pads may be connected end-to-end to form additional attenuation. The two most popular designs are ladder and lattice.

A *ladder attenuator* is composed of either T or π sections connected in cascade with a half section at each end for matching purposes. Such an attenuator is illustrated in Figure 8–4.

This ladder chain is composed of three π-attenuator pads with a terminating resistor on each end. The terminating resistor should be equivalent to the image impedance in each case. The image impedance on the input is the source impedance. The image impedance on the output is the load impedance. These terms are derived from the fact that

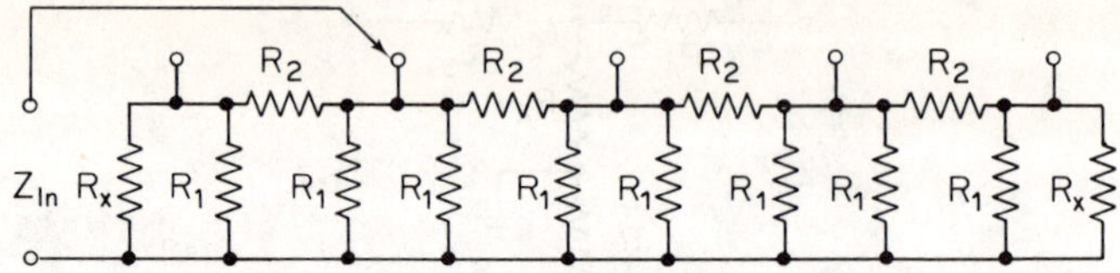

Figure 8–4 Ladder Attenuator

when attenuators are properly matched, the terminal resistor is a reflection or image of the input or output impedance.

Any *T*- or π-shaped attenuator pad may be resolved to an equivalent *lattice* section. Figure 8–5 illustrates two equivalent pads. In each section, resistors and impedances of like numbers are of identical values. Since the lattice pad always contains two identical horizontal and two identical diagonal sections, most schematics show only one horizontal and one diagonal section and represent the others with broken lines.

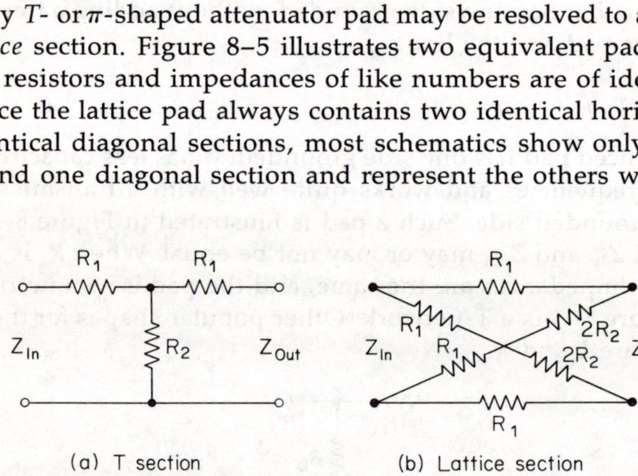

Figure 8–5 *T* and Equivalent Lattice Pad

The lattice attenuator can be used to physically represent any four-terminal network. The image impedance of a lattice attenuator depends on the lattice resistance product. The image transfer constant depends only on the branch impedance ratio. These characteristics enable the lattice attenuator to control transmission and impedance independently. This makes the lattice network the most popular design for attenuators.

Microwave Attenuators

There are two general classes of microwave attenuators: dissipative and nondissipative. The dissipative type utilizes a movable, absorptive strip mounted within the waveguide section. The nondissipative type reduces the dimensions of the waveguide section to reduce its effectiveness. These attenuators were adequately discussed in Chapter 6 under Waveguides.

DECADE RESISTORS

Resistive decade boxes are enclosed units containing fixed, conveniently selectable values of precision resistance. They are used for resistance standards in a wide range of instrument applications. They may be used to calibrate ohmmeters, to form ratio arms on low-frequency bridges, or to fill any other need requiring the use of precision resistors.

Structure

Improper use of a decade box can result in damage to the resistors or to the equipment it is used with. The maximum current for each decade switch is indicated. These limits must not be exceeded.

The front of a decade box is illustrated in Figure 8–6. Each switch controls a bank of ten resistors. All resistors are precision-type, and all resistors of a bank are the same value. A schematic is illustrated in Figure 8–7. Notice that the rotary switch in each case can select up to ten resistors. When a switch is on position zero, that bank of resistors is bypassed. In any other position, the switch selects a number of resistors corresponding to the number the switch is set to. For example, the units switch is set to position 9 to allow current through nine one-ohm resistors.

The entire network is a series circuit. This greatly simplifies the problem of avoiding overload. Total resistance is a simple summation of all resistors selected. Starting from the upper left on the schematic, the total resistance, with the indicated switch settings, is obtained like this: $2000\ \Omega + 0 + 0.6\ \Omega + 9\ \Omega + 300\ \Omega + 80{,}000\ \Omega = 82{,}309.6\ \Omega$.

The maximum current in any case must not exceed the maximum specified for the largest resistance bank in use. In Figure 8–7, the 10,000 ohm bank is in use, and this limits the maximum safe current to 7 mA. The planned voltage to be applied, divided by the total selected resistance, will quickly reveal whether or not excessive current will result. In this example, the network can withstand voltages up to 576.1672 volts without exceeding the 7 mA maximum.

Design

Precision is the key work in the design of decade resistors. The resistors are wire-wound, but many other circuit characteristics are also considered. Elimination of inductance and capacitance is impossible, but these factors must be kept at an absolute minimum. This consideration, of course, takes on increasing importance as frequencies go higher.

One technique uses two identical wires to wind a resistor. The

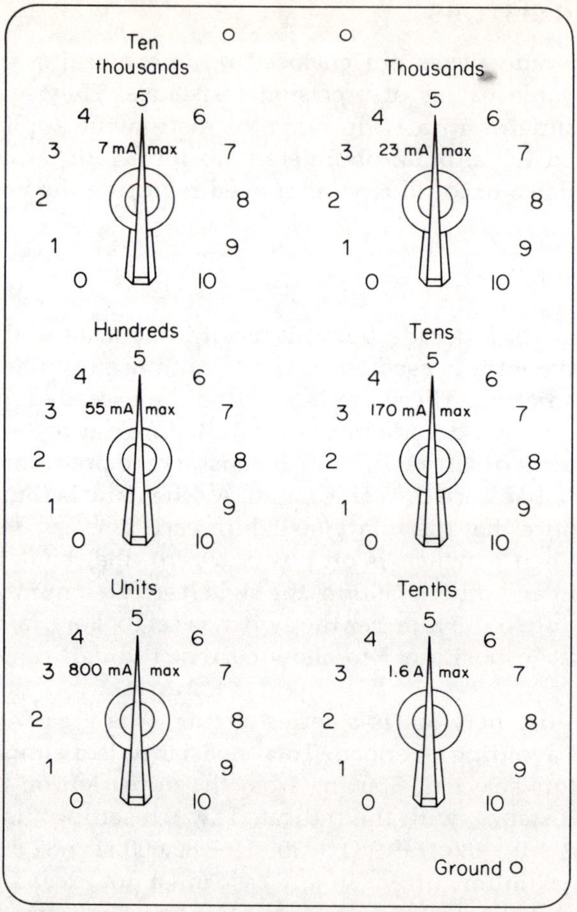

Figure 8–6 Resistive Decade Box

wires are wound in opposite directions. This causes currents in opposite directions, which result in a cancellation of any induced magnetic field. This is called the Ayrton-Perry winding method. It is commonly used for resistors of 100 ohms or less, and its purpose is to minimize the residual inductance. Other specialized techniques are employed to minimize capacitance.

A decade resistance must retain its precision over a wide temperature range. This is accomplished by using manganin alloy wire. This alloy has a low-positive temperature coefficient which holds fairly constant even through drastic temperature changes.

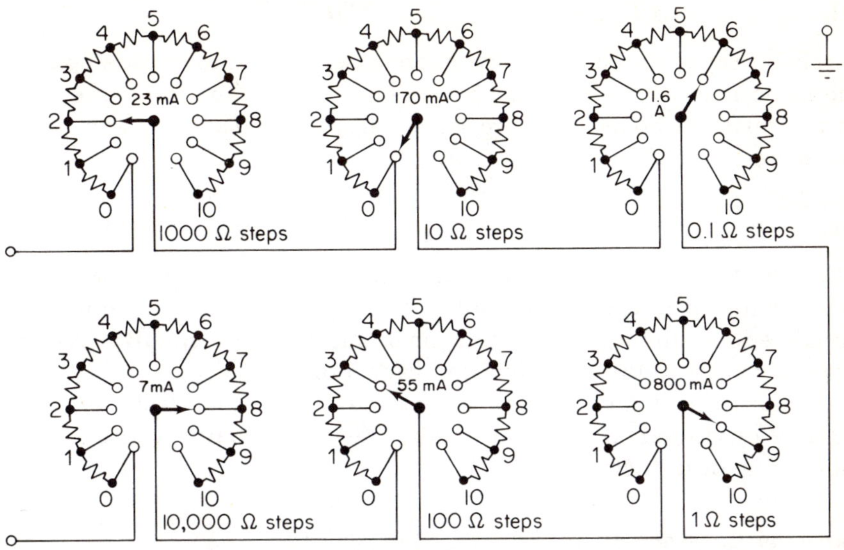

Figure 8–7 Resistive Decade Schematic

Sample Decade Resistor

A sample decade resistor box is pictured in Figure 8–8. This is the General Radio 1434-B decade resistance box. It features six decades in resistance of one ohm per step and a total resistance of 1,111,110 ohms. This instrument has solid silver alloy contacts, and the resistors are wound from low-positive temperature coefficient alloy wire. The precision on most resistors is within ±0.02%. This decade resistor box is a laboratory-quality instrument designed for use in a variety of testing and measuring applications. The accuracy is adequate except in situations calling for the greatest degree of exactness.

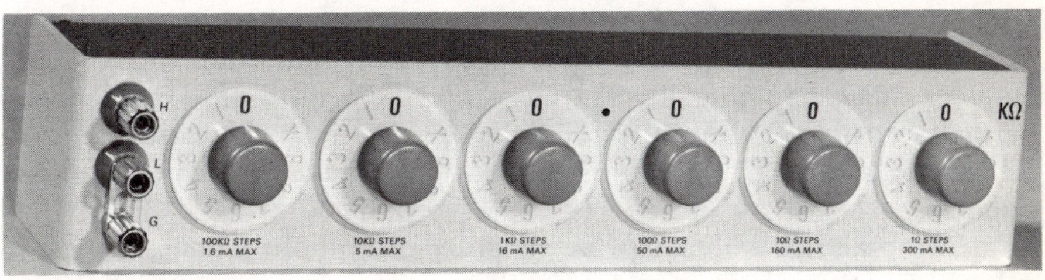

Figure 8–8 Sample Decade Box (Courtesy General Radio)

DECADE CAPACITORS AND INDUCTORS

Precision capacitors and inductors are assembled into decade boxes in the same manner as resistors. These boxes have the same appearance as the resistive boxes, and precision values of capacitance and inductance are selected in the same way.

Uses

The capacitor decade boxes are handy for substituting precision values of capacitance into tuned circuits, wave filters, oscillators, amplifiers, equalizers, frequency multipliers, or in short, in any circuit where test or repair work requires a precision capacitor.

The *inductor* decade boxes are used in the same fashion as previously described for capacitors when tests or repair require a precision value of inductance.

Capacitor Construction

There are a variety of materials employed for constructing capacitors. The intended use of the decade box governs the material that is selected for the capacitor. If the dissipation factor is not important, a paper dielectric may be used. However, silver mica is a better insulator and has a lower dissipation factor than paper. Polystyrene is generally found where a constant dissipation factor is required with constant capacitance values. In fact, polystyrene is the best dielectric for applications involving frequency changes or high-ambient temperatures. Dielectric absorption is an important consideration, and polystyrene keeps it at a minimum value. Every effort is made in the design of the capacitors and the box assemblies to keep the resistance and inductance to a negligible value.

Inductor Construction

The inductors for an inductive decade box are generally *toroids* to minimize any external field. The purpose here is to avoid mutual inductance between adjacent inductors. The cores may be formed of molybdenum permalloy or some similar ferromagnetic alloy dust. The coils are designed for a high *Q*. This provides a high storage factor at low frequencies.

The practical coil cannot be constructed without having inherent values of both resistance and capacitance. The object is to keep these undesirable values to a minimum. A well-designed coil should have about 45 Ω/H.

DUMMY LOADS

A *dummy load* is a passive device designed to absorb the transmitted power that normally goes to the antenna. For this reason a dummy load may be referred to as a *dummy antenna.* Dummy loads are designed in many forms and are composed of a variety of materials. Any device capable of absorbing the total transmitter power is a dummy load when it is fitted with the proper coupling devices. A dummy load may consist of light bulbs, sand, space-wound resistors, mica cards, or water-cooled resistors.

Uses

Dummy loads are indispensable in most transmitter tuning operations. For many reasons it is undesirable to saturate the airwaves with off-frequency transmissions of any type. Since most transmitter tuning requires a live transmitter, the only way to avoid undesirable transmission is to connect a dummy load to the transmitter output instead of the antenna. The load should be designed to handle the maximum power from the transmitter it is to be used with.

Coaxial Dummy Load

A dummy load capable of dissipating up to 1000 watts of power may be constructed from a section of coaxial line. Such a load is illustrated in Figure 8–9. The coaxial line section is composed of a brass tube 1.75 in. in diameter. A portion of the dielectric around the center conductor has been replaced by a mixture of graphite and a special cement. This forms a conical-shaped resistive element. The RF power is converted to heat, which is dissipated by the cooling fins. A fan forces air over the fins to carry off the heat.

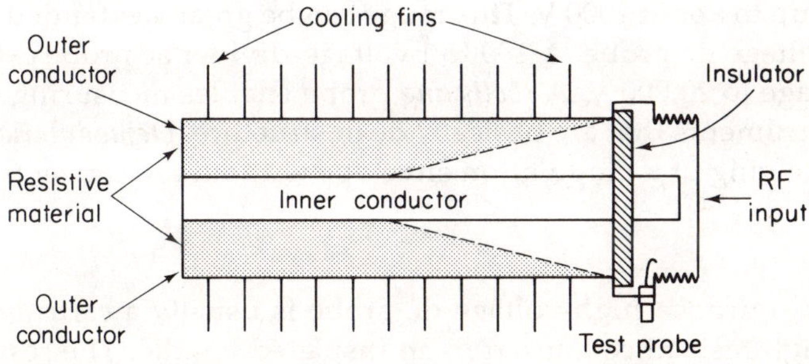

Figure 8–9 Coaxial Dummy Load

Waveguide Dummy Load

A section of waveguide filled with sand is a common type of dummy load for microwave frequencies. Such a device is illustrated in Figure 8–10. A tapered-end waveguide section feeds the RF power into a resistive load. The load consists of a mixture of sand and aquadag, and it is held in place by a piece of clear plexiglass. The cooling fins provide an increased heat dissipating ability.

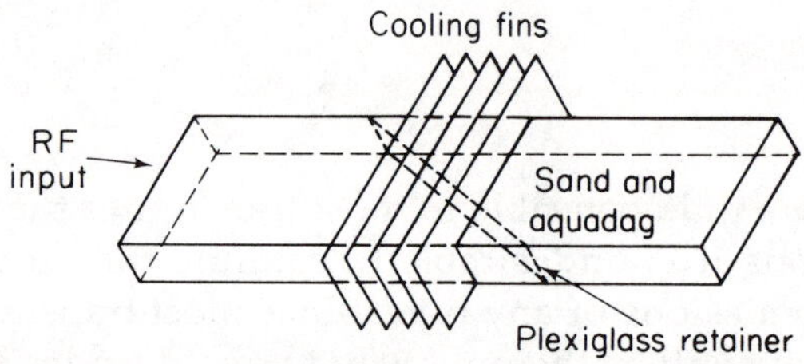

Figure 8–10 Waveguide Dummy Load

Electrical Dummy Load

Dummy loads may be constructed from electrical components such as resistors and capacitors. These components can be arranged in a fashion that will match the impedance of the transmitter they are made for. These loads may be constructed to handle any reasonable level of transmission power. The resistive networks are generally enclosed in a metal case with forced air cooling.

INSTRUMENT PROBES

The effective range of many instruments largely depends upon the type of probes available. An electronic voltmeter is normally used to measure voltages up to about 2000 V. This range can be greatly extended by using a high-voltage dc probe. A 100 to 1 voltage-divider ac probe extends the 200 V range to 20,000 V. A *rectifying* probe enables measuring values of ac on instruments that are basically dc in structure. *Demodulation* probes enable viewing of a modulation envelope without the carrier frequency.

dc Probes

A general-purpose high-voltage dc probe is usually a cartridge design with a high resistance housed in an insulated handle. The resistor may be a spiral film-type resistor printed on a ceramic core. The resistor is housed in a handle of insulation composed of a high-dielectric strength and a low-leakage factor. Such a probe is illustrated in Figure 8–11.

The probe tip may be either pointed for easy contact or clip-on for holding the connection. The cable, extending from handle to instrument plug, is a high-voltage shielded cable which is grounded through the instrument. Care must be exercised to make sure that there is a good physical ground between the instrument and the equipment under test.

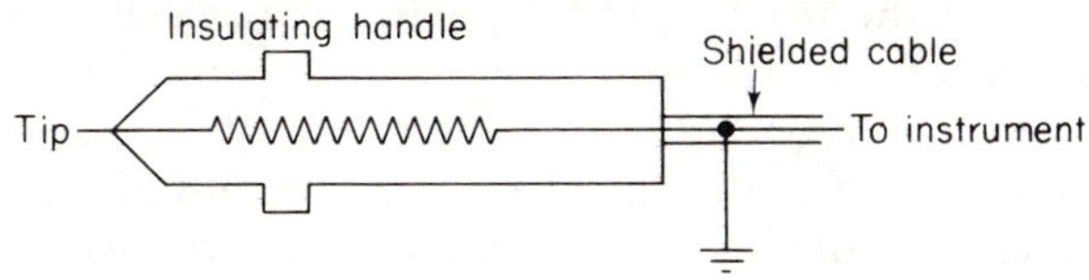

Figure 8–11 High-Voltage dc Probe

ac Probes

The two most common types of ac probes are voltage-divider and isolation. The voltage-divider probe uses capacitors to reduce the input voltage to a manageable level. This type of probe is illustrated in Figure 8–12. The input voltage is divided between C_1 and C_2 in direct proportion to the respective reactances. Most of the voltage is normally dropped across C_1, so it must have a much lower capacitance (higher reactance) than C_2. The remaining voltage across C_2 is small enough to apply to a measuring instrument.

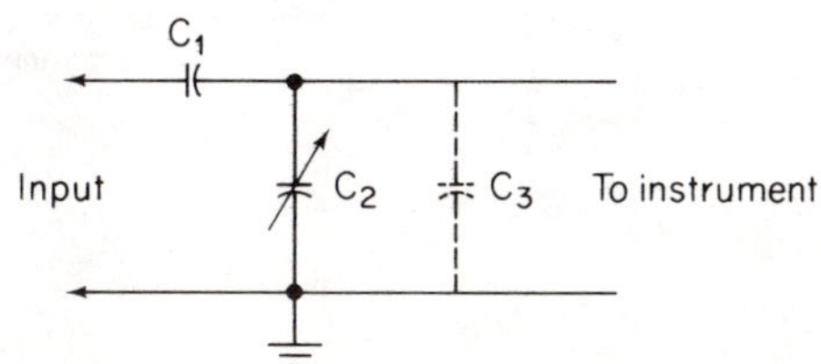

Figure 8–12 Voltage-Divider ac Probe

This probe must have a high-input impedance to prevent loading of the circuit under test. This may be done by a high-voltage rectifying element used as C_1. This is an inexpensive capacitor with a high breakdown voltage. C_2 is a trimmer capacitor which can be varied to provide the proper voltage ratio; this ratio is usually 100 to 1. C_3 is the distributive capacitance between the inner and outer conductors.

Isolation probes are essential in many cases to prevent the test instrument from loading the circuit under test. A low-capacitance, high-impedance probe is effective for isolation purposes. Such a probe is illustrated in Figure 8–13.

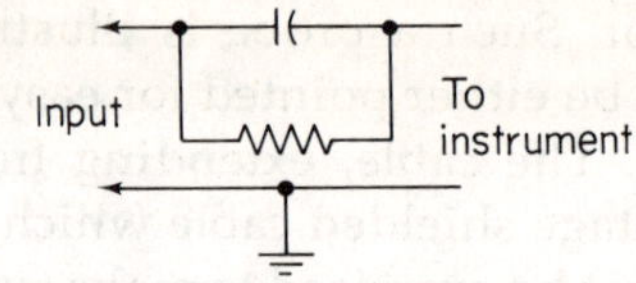

Figure 8–13 Isolation Probe

The values of the resistor and capacitor are selected to match the input impedance of the instrument which the probe is intended for. These values are also intended to provide the proper voltage ratio at all frequencies likely to be measured. This arrangement effectively shunts the input impedance of the test instrument and reduces the loading effect.

Rectifying Probes

The rectifier probe contains a rectifying element and a filter circuit built into the probe handle. This probe is illustrated in Figure 8–14. The rectifier is sealed in a metal shield to minimize the effects of stray inductance and capacitance. A particular probe is designed to handle a specified range of signal frequencies. This factor must be taken into consideration when selecting a probe.

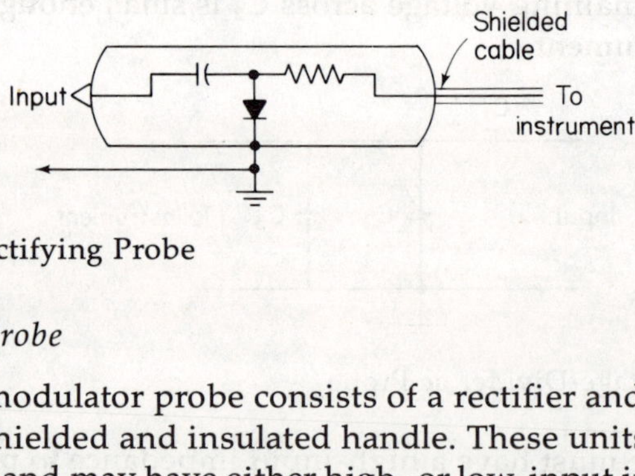

Figure 8–14 Rectifying Probe

Demodulation Probe

Basically, a demodulator probe consists of a rectifier and filter enclosed in a properly shielded and insulated handle. These units may be either shunt or series and may have either high- or low-input impedance. Figure 8–15 illustrates a series type with high-input impedance. The series blocking capacitor blocks the dc, and the bleeder resistor provides a discharge path for this capacitor. The signal is rectified by the diode, and the RF component is removed by the filter. The filter also serves as a shunt to minimize hum interference.

DELAY LINES

Delay lines have many applications in tests and measurements. Apart from those that are built into the test instruments, there are a variety

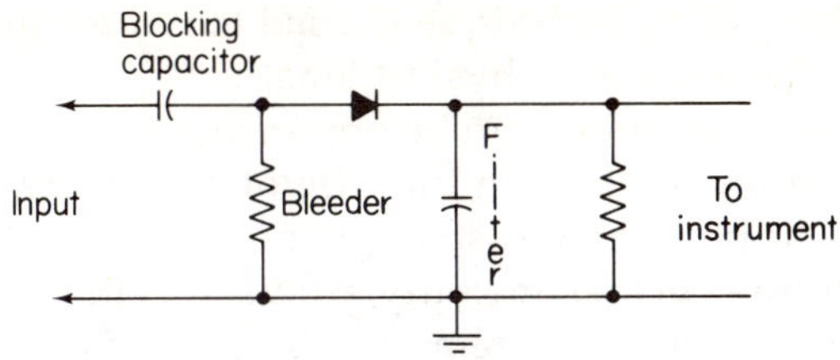

Figure 8–15 Demodulation Probe

of delay lines available as accessories. Delay lines are used in pulse networks for calibration of voltage, current, and timing. A special application uses a delay line to simulate altitude in calibrating an electronic altimeter.

Types of Delay Lines

There are nearly as many types of delay lines as there are uses. All are classed roughly as either electromagnetic or ultrasonic. Electromagnetic delay lines can provide delays from a few nanoseconds (ns) to several hundred microseconds (μs). Ultrasonic delay lines can produce delays on the order of milliseconds (ms). Both electromagnetic and ultrasonic lines may have either a variable or a fixed delay.

Electromagnetic Delay Lines

Probably the most common type of electromagnetic delay line is the artificial transmission line. This type uses physical inductors and capacitors lumped into convenient quantities to obtain specific sections. Each section is designed for an exact delay time.

The simplest form of delay line consists of a section of actual two-wire transmission line. One type of coaxial cable constructed for RF applications has a delay on the order of 1.5 ns/ft. Useful delay lines have been constructed by coiling a specified length of such a cable into a box. However, ordinary RF coaxial cable is not satisfactory for this application because the delay is only about 1 μs/600 ft of cable.

Ultrasonic Delay Lines

The ultrasonic delay line is based on the relatively slow velocity of propagation in liquid or solid media. Mercury, fused quartz, epoxy resin, and silica are some of the transmitting media. A crystal transducer is the usual method of coupling signals into or out of an ultrasonic delay line. A folded mercury delay line is illustrated in Figure 8–16. Propagation through liquid mercury is approximately 4.76 ft/ms at a temperature of 20°C. The folding technique is used to reduce the physical bulk while retaining the required physical length.

Different delay intervals can be obtained from a single line by providing more than one output. In the case of a mercury line, additional transducers would be placed at strategic points along the line.

In a liquid line such as mercury, provisions must be made for expansion and contraction due to temperature changes. A diaphragm consisting of a gas-filled rubber bag will take care of the expansion problem when the line is not in use. When being used, the line should be at a constant temperature since propagation rates vary with temperature. This is generally accomplished by a thermostatically controlled heater.

TRANSDUCERS

A *transducer* is a sensing device that accepts an input of one type of energy and provides an output in another type of energy. For instance, a piezoelectric transducer changes electrical energy to mechanical energy at the input to a delay line. The same type of transducer changes mechanical energy to electrical energy at the output of a delay line.

Transducers are common input devices to electronic instruments, and there are many varieties of transducers. A very large and important group of transducers converts mechanical energy or physical displacement into electrical signals. There are also types of transducers designed to convert any type of energy to electrical energy.

General Classification

There are two basic classes of transducers: passive and self-generating. A passive transducer must have an external voltage source. The sensed energy changes the physical aspect of some component and alters the current or voltage in the circuit. For example, a capacitor diaphragm has voltage applied across a capacitor. Sound pressure on the diaphragm varies the capacitance and alters the current. These variations represent speech, music, and noise.

The self-generating transducer requires no external voltage source. It converts the detected energy directly into electrical signals. For example, a thermocouple is a self-generating transducer. It generates an electric current directly proportional to the intensity of the heat at the junction of the two dissimilar metals.

In these two general classes of transducers, there are many types, and each type is designed for specific applications. Space does not permit a detailed treatment of all types, even if it were desirable to do so. A few types of general interest are covered here in order to develop the general principles of transducers.

Piezoelectric Transducer

Two types of *piezoelectric transducers* were mentioned previously. One type changes mechanical energy to electrical signals; the other changes an electrical signal to mechanical energy. Both of these were illustrated in Figure 8–16.

Certain types of crystals, quartz for instance, contain a quality known as the *piezoelectric effect*. When these crystals are subjected to a mechanical stress, they vibrate at their own resonant frequency and generate an electrical voltage at this frequency. Conversely, when a voltage is applied across the crystal, it generates a mechanical pressure.

The transducer at the delay line input in Figure 8–16 has a signal applied to the crystal. This electrical signal causes mechanical vibrations which are transmitted from the crystal into the mercury. These vibrations travel through the mercury to the opposite end where another transducer is encountered.

The transducer on the output is set into vibration by the mechanical force. These vibrations produce electrical signals on the output line.

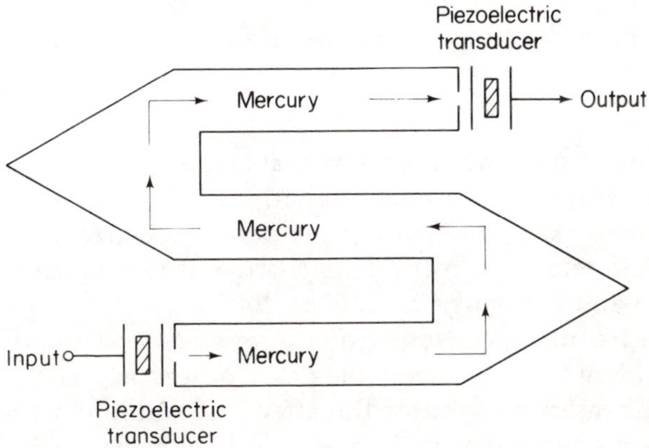

Figure 8–16 Ultrasonic Delay Line

Linear-Motion Transducers

A linear-motion transducer senses an analog quantity of movement and provides a signal with amplitude in proportion to the movement. The sensor may be a potentiometer, a coil, a capacitor, or a transformer. Such a transducer is illustrated in Figure 8–17. The sliding arm on the transducer rheostat is controlled by the linear motion to be measured. The external circuit applies the activating voltage and a meter to measure the results. The position of the movable arm determines the resistance and, consequently, the current in the circuit.

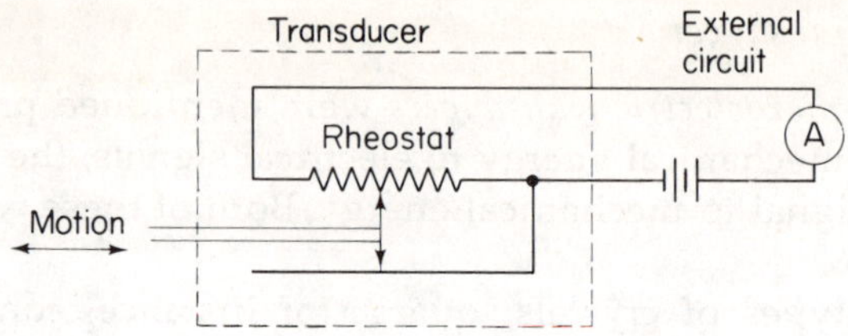

Figure 8–17 Linear-Motion Transducer

Instead of a rheostat, the same principle can be used for a coil with a movable core, a capacitor with a movable plate, or a transformer with a movable core.

Angular-Motion Transducers

The angular-motion transducer is based on the same principle as the linear-motion transducers just described. The only difference is application. Here a shaft's angular position is the controlling factor. This position can be sensed by several types of rotary devices, such as rheostats, potentiometers, capacitors, and transformers.

Temperature Transducers

One of the most popular sensors of heat is the *thermister*. The thermister is a bead of semiconductor material with two metal leads. The resistance of the thermister drops when the temperature increases, and increases when the temperature drops. This change in resistance is fairly linear over a temperature change from 0° to 100°C.

There are many varieties of thermisters and many methods of packaging. Some are encased in glass; others are enclosed in metal boxes with heaters to counter the effects of ambient air temperatures. All thermisters are very small, and this is one reason for their popularity. They can be used in areas too small for other types of transducers.

Selecting a Transducer

Whatever the measuring job, there is a transducer to fit the need. Selecting the proper transducer is the first step to obtaining reliable results. The physical quantity to be measured, the degree of accuracy required, and the calibration features of the equipment are some of the factors to be considered.

part two

TESTING PRACTICES

This part, consisting of Chapters 9 through 15, illustrates and discusses *practical* tests and measurements using the instruments described in Part I.

chapter nine

BASIC MEASUREMENT TECHNIQUES

This chapter discusses common measurements that are accomplished by use of the basic instruments described in Chapter 1. These measurements include resistance, direct- and alternating-current, and voltage. Uses of basic instruments for checking capacitors and power are also discussed.

SAFETY PRECAUTIONS

Since the majority of measurements are made on live circuits, safety practices are essential to the prevention of damage to both persons and equipment. The following guidelines are provided for a safe electronic environment. Live circuits should not be feared, but they must be respected.

1. Remove all hand and arm jewelry, such as rings, watches, and bracelets. Even a small current through a piece of jewelry can do irreparable damage to both the jewelry and the wearer.

2. Stand on rubber mats or wear rubber-soled shoes when testing live circuits. These precautions won't prevent shock, but they will lessen the severity when shock occurs.

3. Make sure that test equipment and equipment being tested have a common ground.

4. Remove all power while connecting and removing probes when it is practical to do so. Otherwise, use only one hand. Keep the other hand in a pocket or behind the back. This will prevent current from finding a path through the vital organs in the event that a hot contact is made.

5. Use insulated probes and make sure the fingers do not touch the metal tips.

6. Avoid use of test leads with frayed or broken insulation.

7. Discharge all storage capacitors with a shorting bar before touching the circuit. Even with the power off, these capacitors can cause a severe shock.

8. Remember that all shocks are dangerous. Even a mild shock may cause the recipient to react violently and further injure himself against hard or sharp objects.

9. Respect the test instruments. They are both expensive and delicate.

10. Be familiar with the capabilities and limitations of the instruments to be used. Study the instruction manuals before attempting to use a new instrument.

11. Be familiar with the equipment being tested. Know approximately what is expected, and match the instrument to the test circuit.

MEASURING CURRENT

The word *current* will mean *electron current* throughout this book. Electron current is the flow of electrons through a conductor. One *ampere* of current is present when one *coulomb* of electrons flows past a point in one second. A coulomb is 6.24×10^{18} electrons.

This should cause no confusion to the person who thinks in terms of conventional current (positive to negative). The procedures are the same, and the meter readings are the same.

Measuring Direct Current

Observe the test circuit to determine the direction of current and the approximate quantity of current in the circuit. Make sure that the ammeter leads are inserted firmly into the proper jacks. Select a current range at least twice as high as the estimated current value. If the estimated current is fairly close to the actual current, the measurement will be an accurate midscale reading. If the estimated value is wrong, the meter has the benefit of the doubt and is less likely to be damaged.

Power should be removed from the circuit while connecting the meter. Do this by opening a switch or disconnecting a lead. Now break the circuit at the point where the ammeter is to be inserted, as shown in Figure 9–1.

The ammeter is connected to fill the gap in the circuit. This places it in series with the load and the power supply. Check to make sure the negative lead is connected to the point nearest the negative terminal of the power supply.

Observe the meter as the switch is closed. If the needle deflects too far or in the wrong direction, immediately open the switch. Readjust the current range or the meter lead connections. A reaction of this type shows an error in judgment. It should happen only rarely or not at all. If it does happen, prompt action on the switch may save the meter from serious damage. When the meter is connected properly and the switch is closed, electrons flow through the ammeter as shown in Figure 9–2.

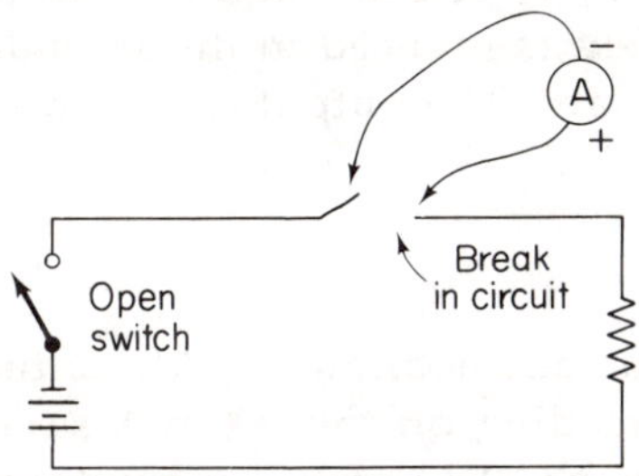

Figure 9–1 Preparing to Measure Current

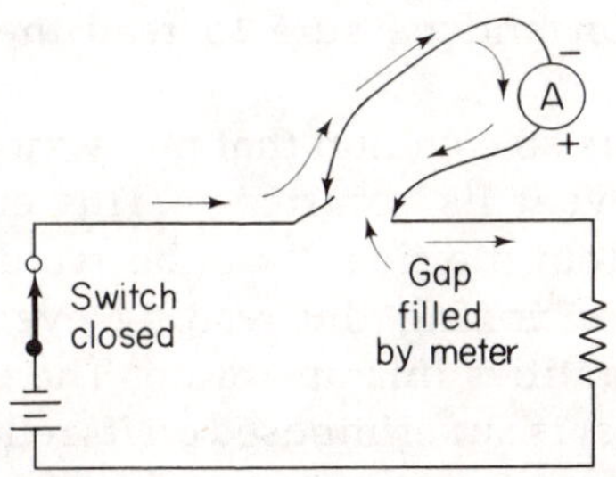

Figure 9–2 Ammeter Measuring Total Circuit Current

Measuring Alternating Current

Use of the ac ammeter is very little different from the use of the dc ammeter. Since current reverses direction with each alternation of the sine wave, there is no polarity to observe. However, the same technique for hook-up and reading is recommended. The meter is still connected in series with the load and the power supply.

When reading the regular ac scale, the indicated value is effective (rms) value. However, some meters, as illustrated in Chapter 1, have a peak-to-peak scale. If peak value is desired, it can be readily calculated by the following formula:

$$\text{Peak} = 1.414 \times \text{effective}$$

Using the Clamp-on Probe

It is frequently inconvenient to break a circuit and insert the ammeter for a current reading. The electromagnetic properties of ac enable measurement of current through transformer action. The *clamp-on probe* illustrated in Figure 1–13 can be used for this purpose.

When this probe is clamped around a conductor in the test circuit, the conductor is the primary of a transformer. The secondary is a winding in the probe. Current is induced in the secondary in proportion to the current in the primary. The output can be used as an input to an ac ammeter.

Ammeter Accuracy

A good ammeter is generally accurate to $\pm 1\%$ of the full-scale reading. This means that any reading on the 300 mA scale is accurate within ± 3 mA. Any reading on the 30 μA scale is accurate within ± 0.3 μA.

Accurate meters are often blamed for operator errors. One of the most common errors is reading the wrong scale. Select a scale for a near midscale indication and be sure to read the scale that *matches* the range.

The *parallax error* is so common that many meters have antiparallax mirrors to help prevent its occurrence. This error is caused from viewing the indicator from the side. It can be avoided by reading with one eye closed and positioning the reading eye directly above the pointer. The mirror simplifies this operation. The eye is in the correct position when the pointer is superimposed on its reflection in the mirror.

For an accurate reading, ammeters and voltmeters must indicate zero when there is no input. This should be checked frequently. If necessary, adjust the plastic screw at the bottom center of the meter to zero the pointer. The leads have no bearing on this adjustment, and it must not be confused with zeroing of the ohmmeter. The zero-adjust screw is clearly visible on the meter in Figure 9–3.

Static charges sometimes affect meter readings, especially on low ranges. Meters are treated with a coating of antistatic chemical to pre-

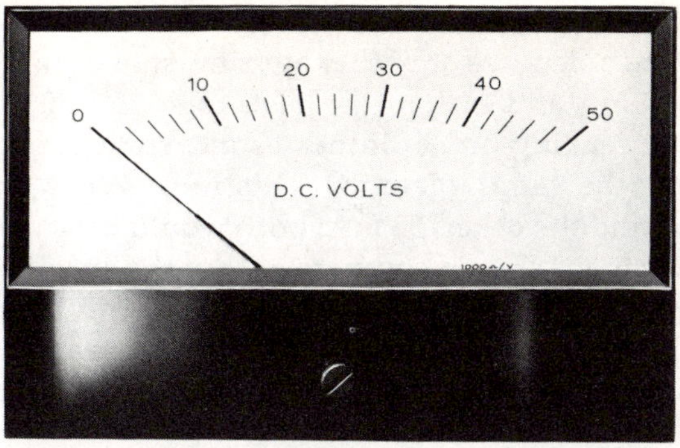

Figure 9–3 Voltmeter, Showing Zero-Adjust Screw (Courtesy Simpson Electric Company)

vent this problem. If the pointer fluctuates when a dry cloth is lightly rubbed on the glass face, the protective coating is defective. This coating can be renewed by spraying with an antistatic substance. This substance is available at most electric supply stores.

Checking Ammeter Accuracy

Ammeters should be calibrated periodically, and the easiest way to check an ammeter for accuracy is to check it against another meter that is known to be accurate. The test circuit for dc should contain a rheostat or potentiometer and sufficient voltage for near full-scale deflection on the highest current range. The meters are connected in series, so the same current passes through both meters. The test circuit is illustrated in Figure 9–4. The variable resistor permits comparison of the standard meter (M_1) with the test meter (M_2) for any number of readings on all dc ranges.

A very important item for consideration in this type of test is meter tolerance. A very accurate meter probably has a tolerance of

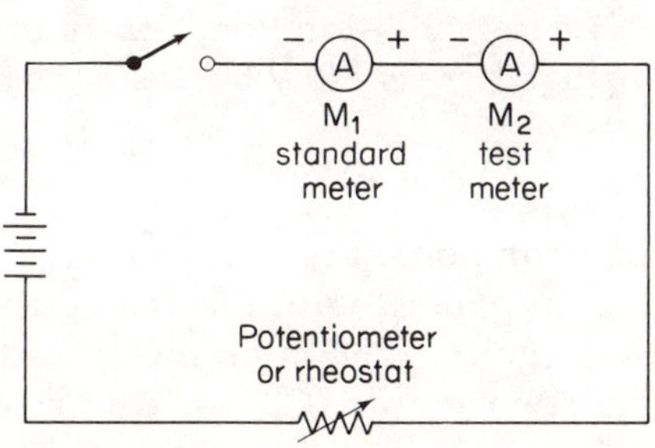

Figure 9–4 Checking Accuracy of dc Ammeter

±1% of full-scale current. The critical point that many people overlook is the fact that the tolerance of both meters must be considered. Suppose the range is 300 mA and each meter has a tolerance of ±1% of full-scale value. A 50 mA reading on M_1 indicates a current between 47 mA and 53 mA. M_2 has the same tolerance. This means that one meter could indicate 47 mA and the other, 53 mA; both would be accurate. The meter readings on all ranges should compare within the specified tolerance.

When checking an ac ammeter accuracy, the battery in the test circuit is replaced by an ac voltage source. The variable resistor will still serve, but if the ac supply is through a variac, the variable resistor can be replaced with a fixed resistor.

The circuit arrangement and procedure are the same as those described for dc meters, but polarity is of no importance.

Sensitive Circuits

Very sensitive circuits are subject to upset by almost any type of current reading. This is especially true of resonant circuits. The coil in the meter movement is a reactive component that will cause a drastic change in a resonant circuit. In some cases, the ill effects of adding an ammeter to a circuit may be overcome by using the clamp-on probe. Some circuits are too sensitive even for the clamp-on probe. When it is necessary to ascertain the current, and a measurement may affect the circuit, the safest procedure is to measure the voltage and calculate the current.

MEASURING VOLTAGE

Voltage is electrical pressure between two points, electromotive force, or potential difference. Voltage exists only when two points are being compared. When only one point is specified, the other point is understood to be ground. The statement, "500 V at test point *A*," implies 500 V between test point *A* and ground. Single-phase ac voltage is supplied by a two-wire transmission line. One of these lines is ground. Thus 120 V ac means that there is 120 V between the two lines.

Use of Voltmeter for Direct Voltage

Voltage drop across a component is measured by connecting the voltmeter into the circuit in parallel with the component. In dc circuits, it is important that the negative (common) meter lead be attached to the most negative of the two points. Circuit connections are illustrated in Figure 9–5.

The voltmeter in Figure 9–5(a) is measuring the terminal voltage

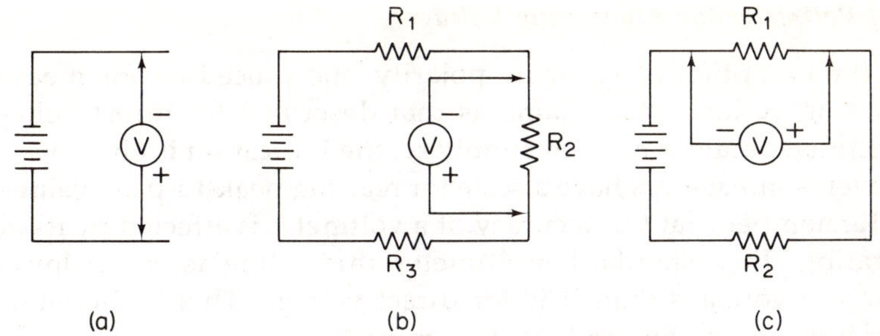

Figure 9–5 Measuring Direct Voltage

of the battery; in 9–5(b), the voltmeter is measuring the voltage drop across R_2; and in 9–5(c), the voltage across R_1 is being measured. Notice in each case that the negative lead is on the most negative of the two relative points.

The test circuit should be studied and voltage estimates made before connecting the meter. Select a voltage range twice as high as the estimated voltage. If no approximation can be made, start with a voltage range capable of handling the highest possible voltage. The range can be worked down until an accurate reading is obtained at approximately half-scale.

It is good practice to remove all voltage from the circuit while connecting and disconnecting the meter leads. This is especially important in high-voltage circuits. However, if safety rules are not practiced in all circuits, they are likely to be forgotten when they are vitally important.

In complex circuits there may be more than one voltage source. In this case, the voltmeter measures the effective voltage drop across a component. The effective voltage drop is the algebraic sum of all voltage drops. This is illustrated in Figure 9–6.

The voltmeter in Figure 9–6(a) indicates 20 V. This is the algebraic sum of +40 V and −20 V. In 9–6(b), the voltmeter indicates 50 V; the algebraic sum of +40 V, +20 V, and −10 V.

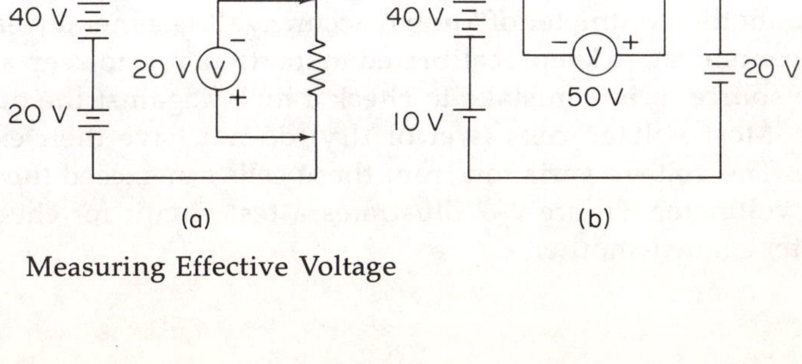

Figure 9–6 Measuring Effective Voltage

Use of Voltmeter for Alternating Voltage

With the exception of ignoring polarity, the procedure for measuring alternating voltage is the same as that described for direct voltage. If the voltmeter scale is marked simply ac, the indication is effective value. However, some meters have a scale for reading peak-to-peak values.

Remember that the accuracy of a voltmeter is affected by its ohms/volt rating. In a standard multimeter this rating is much lower for alternating voltages than it is for direct voltage. This is shown on the face of the meter in Figure 1–10 (see page 12).

Voltages at frequencies below 15 Hz cannot be measured on ordinary ac voltmeters. This is seldom a problem since most power supply frequencies are 60 Hz, and 15 Hz is the bottom audio frequency.

The average meter encounters another frequency problem at frequencies in excess of 30 kHz. The meter coil and coupling capacitor may be resonant to the frequency of the circuit voltage. This will cause abnormally high readings. At frequencies in excess of high audio (20 kHz), a radio-frequency probe is necessary.

Voltage Accuracy

Most multimeters, including the electronic types, have a voltage accuracy of ±3% of full-scale value. A reading on the 50 V scale can vary by ±1.5 V; on the 500 V scale this variation may be ±15 V.

Voltage accuracy is affected by the relative size of the meter resistance and the resistance it is connected parallel to. A meter rated at 10 kΩ/V has a resistance of 100 kΩ on the 10 V scale. This meter will cause drastic circuit loading when placed across a resistance of 100 kΩ or higher.

Circuit loading is not a problem with electronic meters unless voltage is measured across a resistance in excess of 10 MΩ.

When measuring voltages in excess of 600 V, special leads and a high-voltage probe should be used.

Checking Voltmeter Accuracy

A voltmeter may be checked against a standard voltage source or against another voltmeter of known accuracy. It is generally easier, and more economical, to keep a calibrated meter than it is to keep a standard voltage source. It is a mistake to check a meter against the output of a battery. Most voltage cells (wet or dry) do not have their exact rated voltage. The voltage variation from these cells can exceed the tolerance on the voltmeter. Figure 9–7 illustrates a test circuit for checking one voltmeter against another.

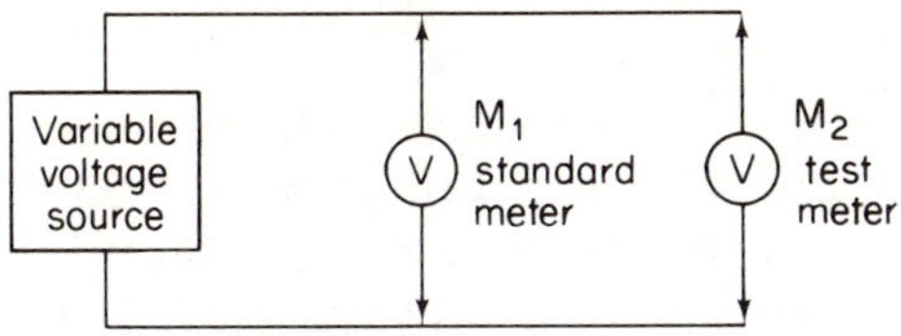

Figure 9–7 Checking Voltmeter Accuracy

The voltage source should have enough range for several checks on all ranges. The procedure is the same for both alternating voltage and direct voltage meters. As the voltage source is varied over all ranges, the indication on the test meter should always match that of the standard meter. Of course, the tolerance on each range must be taken into consideration. If the difference between the indications is within the meter tolerance, the meter under test is good.

MEASURING RESISTANCE

Ohmmeters are designed to measure the resistance connected between the external leads. An internal battery furnishes the current for the meter reading.

Use of the Ohmmeter

Before taking a reading on any range, it is good practice to zero the meter. This is literally true with a series-type ohmmeter; the leads are shorted together and the zero-adjust control is turned until the meter indicates *zero* (full-scale deflection). With the shunt ohmmeter, zeroing is just a figure of speech. The meter leads are not shorted together, but the zero-adjust is set for full-scale deflection. However, since zero is on the left of the shunt ohmmeter scale, the adjustment is literally a full-scale setting—not zero.

Readings should be taken on the low half of the ohmmeter scale. When an indication occurs on the high end of the scale, select a higher range and try again.

When taking a reading with an ohmmeter, care must be exercised to have a firm electrical connection between the meter leads and the wire terminals of the resistor being measured. Sometimes it is necessary to scrape away the insulating varnish from the resistor terminals. Make sure the fingers are not in contact with the metal when taking the reading. If the fingers touch the metal of both leads, the meter indicates the equivalent resistance of the operator's body in parallel with the test resistor.

Resistive Circuits

When measuring a resistor in a circuit, all power must be removed from the circuit, and the resistor must be isolated from the rest of the circuit. The only way to truly isolate a resistor is to completely disconnect one end. As a rule, however, this is an undesirable practice unless there are indications that the resistor is faulty.

The reason for resistor isolation before measuring becomes apparent when a relatively complex circuit is examined. There is an infinite number of resistor arrangements possible, but all resistive circuits can be reduced to one of those shown in Figure 9–8.

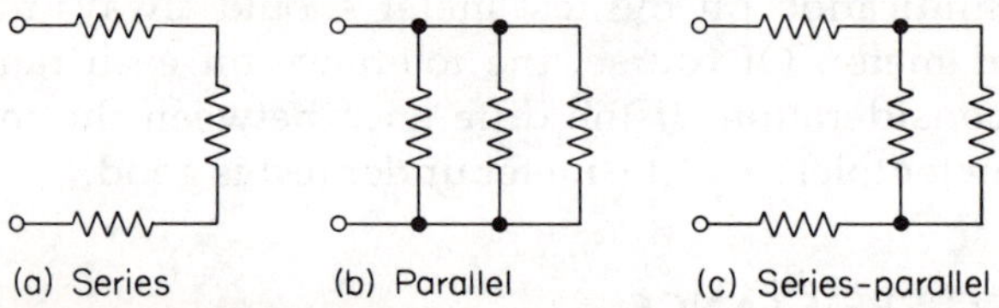

Figure 9–8 Resistive Circuits

In practice, there are very few pure series or pure parallel circuits. Most of them fall into the series-parallel category. Figure 9–9 illustrates possible electron paths when the test resistor is not isolated.

The ohmmeter in Figure 9–9(a) indicates a short, but there is no way of telling where the short may be located. The same indication would be obtained by trying to measure either of the resistors. In Figure 9–9(b), there is an open resistor, but the meter is indicating a definite value of resistance. The resistance actually measured in this case is the equivalent resistance of all good branches.

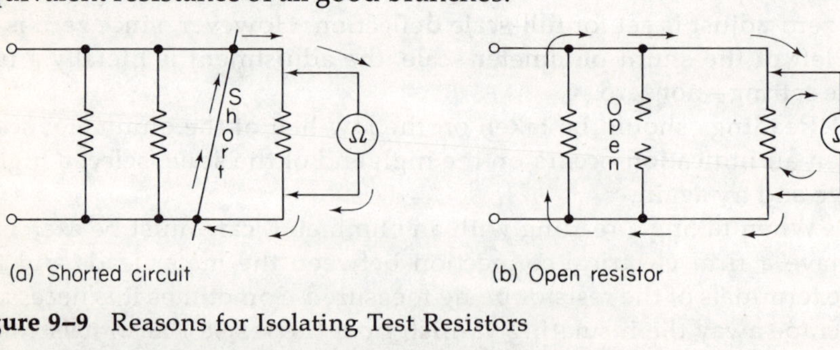

Figure 9–9 Reasons for Isolating Test Resistors

Point-to-Point Resistance Charts

Many manufacturers make available charts (or diagrams) that indicate the resistance from test point to test point or from test points to ground. These charts enable the operator to verify a good circuit or isolate a faulty section of a circuit without the need of disconnecting each resistor.

The fallacy in using resistance charts is that the testing ohmmeter is very likely to be different from that used to construct the chart. This makes it necessary to evaluate the test results. This is easily done by taking several readings and comparing the test meter reading with chart values. A correction factor can then be applied to make test indications compare favorably with chart values. For example; three chart values are: 800Ω, 1700Ω, and 2400Ω. Readings between indicated points produce indications of 824Ω, 1751Ω, and 2472Ω. In each case, the test reading exceeds the chart value by 3%. As long as this meter range is used, the 3% correction factor should be added to the value specified on the chart.

Resistor Tolerance

Common resistors have a wide tolerance on their ohmic value. The most common tolerances are ±10% or ±20%. The rated ohmic value of molded composition resistors and the percent of tolerance are indicated by color code. See Appendix B for standard color code. When measuring the resistance, the resistor is good if the measured value falls within the indicated tolerance.

Checking Ohmmeter Accuracy

Precision resistors can be used to check the accuracy of an ohmmeter. For a complete check, several readings should be made on each range. The average ohmmeter has an accuracy tolerance of ±2% or ±3%. The rated tolerance of the test resistors should be ±1% or better. The meter is accurate if indicated values match rated resistor values within the tolerance of resistor plus meter. For example; a test resistor is rated at 100Ω ±1%, and the meter tolerance is ±3%. Meter tolerance + resistor tolerance = ±4%. The measured value of the resistor is acceptable for any indication between 96Ω and 104Ω.

CHECKING CAPACITORS

Special equipment is needed for measuring capacitance and capacitive reactance, and these measurements are covered in a later chapter. However, certain checks can be made, with an ordinary multimeter, on the condition of the capacitor.

Capacitor Checks Using Ohmmeter

Three capacitor defects are readily detectable with an ohmmeter: short, open, and partial short. If the capacitor is in a circuit, it must be isolated by disconnecting one lead. The shorted capacitor has had a dielectric breakdown. When this condition is present, the ohmmeter will show a steady-zero reading. All capacitor checks should be made on the highest resistance range of the ohmmeter. Make all checks with current in both directions. Connect the leads in a selected manner for one check; reverse the leads and check again. This double check will prevent erroneous indications from partially charged capacitors.

The most common reason for an open capacitor is a broken internal connection. When checked with an ohmmeter, the open capacitor will cause a steady reading of infinity. Remember that very small capacitors charge so rapidly that some ohmmeter needles will not have time to react.

The partial short is a condition of excessive leakage. This condition will cause a definite resistance reading on the ohmmeter. Remember that the electrolytic capacitor is supposed to allow some leakage and will normally cause a resistance indication. Also remember that current through an electrolytic capacitor in the wrong direction will destroy the capacitor. When checking the electrolytic capacitor, observe polarity and do not reverse the ohmmeter leads.

When a good capacitor is placed between the leads of an ohmmeter, the needle will swing from zero to infinity as the capacitor charges. The speed of needle movement will be determined by the size of the capacitor. After the capacitor is charged, the reading will hold steadily on infinity.

Checking Capacitors with a Voltmeter

Opens, shorts, and partial shorts in capacitors can also be detected by using a voltmeter.

The out-of-circuit capacitor is placed across a direct voltage source and allowed time to charge (allow about 10 sec.). The applied voltage should approach, but never exceed, the working voltage of the capacitor. Remove the voltage source and check the charged capacitor with a voltmeter. This procedure is illustrated in Figure 9–10.

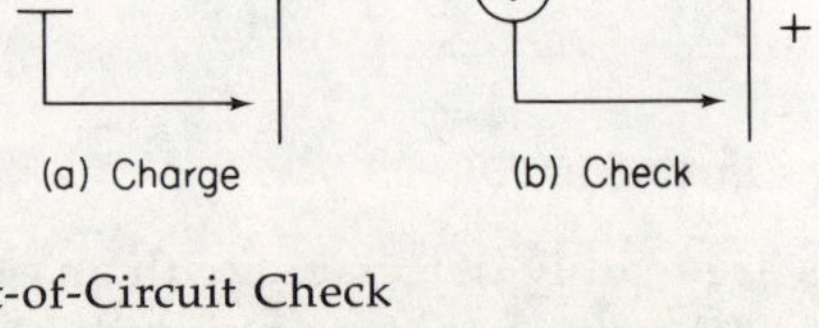

Figure 9–10 Out-of-Circuit Check

If the capacitor is good, the initial voltage reading will be approximately the same as the source voltage. An open or shorted capacitor will cause a reading of zero volts. The partial short will show a low voltage.

If the capacitor is in a live circuit, it may be checked there. Disconnect the ground side of the capacitor, and *bridge* the gap with a voltmeter as shown in Figure 9–11.

A good capacitor or an open capacitor should show zero voltage. However, the initial reading may show some value of voltage due to capacitor charge. If the voltage indication remains, the capacitor is leaking.

If this check produces a zero-voltage indication, power should be removed from the circuit and the ohmmeter should be used to differentiate between a good capacitor and an open capacitor. Again this check may not provide satisfactory results when dealing with very small capacitors.

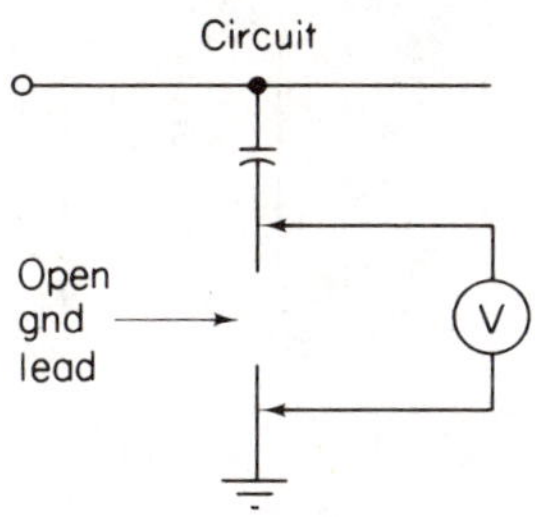

Figure 9–11 In-Circuit Check

CHECKING POWER RATIOS

Power gain is the ratio of power in to power out (P_{in}/P_{out}). When this ratio is a positive quantity, it is called a positive power gain. A negative ratio is a negative power gain; sometimes inaccurately referred to as power loss. One of the popular units for measuring power ratios is the decibel meter (dBM). Most multimeters have a dBM scale.

Decibel Meter

The dBM unit is in respect to 1 mW across a 600 ohm load. This is the load and power when the dBM scale indicates zero. A dBM reading may be converted to actual power by this calculation:

$$\text{dBM} = 10 \log \left(\frac{P_{out}}{1\text{ mW}}\right)$$

Actual conversion is not necessary in most cases because the power ratio is the item of interest. The dBM reading however is accurate only when the load is a 600 ohm resistor and the voltage is a pure sine wave. The decibel meter is not used on complex waveforms, and correction factors must be used when input and output loads are different. The chart in Figure 9–12 is a sample list of correction factors.

These correction factors were calculated by the following formula:

$$\text{Correction Factor} = 10 \log \left(\frac{R_2}{R_1}\right)$$

where R_2 is 600 ohms and R_1 is the actual load resistance.

Actual load (Ω)	Add dBM
500	0.8
300	3.0
250	3.8
150	6.0
50	10.8
15	16.0
8	18.8

Figure 9–12 Correction Factors

Ratio Measurements

First take a dBM reading at the input to the stage and compare it to a dBM reading at the output of that stage. The *ratio* is true for any value of load so long as input load and output load are identical. For a quick approximation of power gain, the following rule of thumb may be applied:

For each change of +3 dBM,
power is doubled.
For each change of −3 dBM,
power is halved.

Appendix C is a decibel conversion table for more accurate values; it also shows voltage and current ratios.

chapter ten

MEASURING REACTIVE COMPONENTS AND POWER

The reactive components considered here are capacitors and inductors. Any technician can perform simple ohmmeter tests to determine if these components are good or bad, but this does not measure the capacitance or inductance. This chapter treats some of these tests, as well as precision measurements of capacitance, capacitive reactance, inductance, inductive reactance, and impedance. Incidental to these measurements are quality (Q) and dissipation (D), which are also considered. Finally, power measurements are thoroughly examined.

CAPACITOR TESTS

Capacitors may be arranged in an infinite number of combinations, but all arrangements fit one of three circuit types: series, parallel, or series-parallel.

Capacitors in Series

Figure 10–1 illustrates a series capacitive circuit.

The formula for total circuit capacitance (C_t) is a reminder that total capacitance is less than the value of the capacitance of the smallest

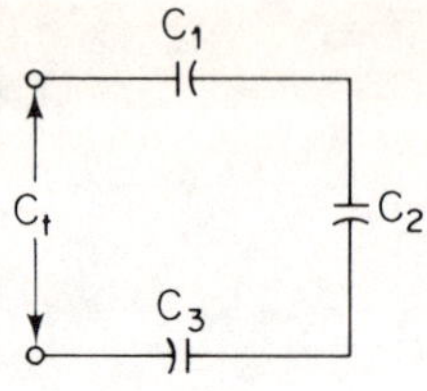

$$C_t = \frac{1}{\frac{1}{C_1} + \frac{1}{C_2} + \frac{1}{C_3}}$$

$$X_{C_t} = X_{C_1} + X_{C_2} + X_{C_3}$$

Figure 10–1 Capacitors in Series

capacitor in the circuit. This is the reciprocal formula, but the product over the sum is just as effective. The total reactance (X_{Ct}), of course, is a summation of the individual reactances.

Capacitors in Parallel

Figure 10–2 illustrates a parallel capacitive circuit. The total capacitance is now the sum of the individual capacitances, and the reciprocal (or product over the sum) formula is used to calculate the total reactance.

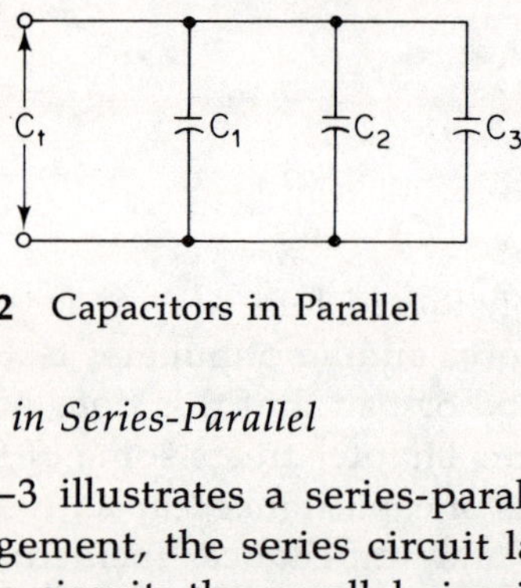

$$C_t = C_1 + C_2 + C_3$$

$$X_{C_t} = \frac{1}{\frac{1}{X_{C_1}} + \frac{1}{X_{C_2}} + \frac{1}{X_{C_3}}}$$

Figure 10–2 Capacitors in Parallel

Capacitors in Series-Parallel

Figure 10–3 illustrates a series-parallel arrangement of capacitors. In this arrangement, the series circuit laws are applied to the series portion of the circuit; the parallel circuit laws are applied to the parallel portion; and the two results are added to obtain the circuit totals.

Ohmmeter Test for Leaks and Opens

This test is similar to measuring a high value of resistance.

1. Isolate the capacitor by disconnecting one lead.
2. Set the ohmmeter to its highest resistance range.
3. Connect the meter leads across the capacitor terminals.
4. Observe the pointer.

A good capacitor will show some value of resistance when the leads are first connected. This will increase to infinity, and remain

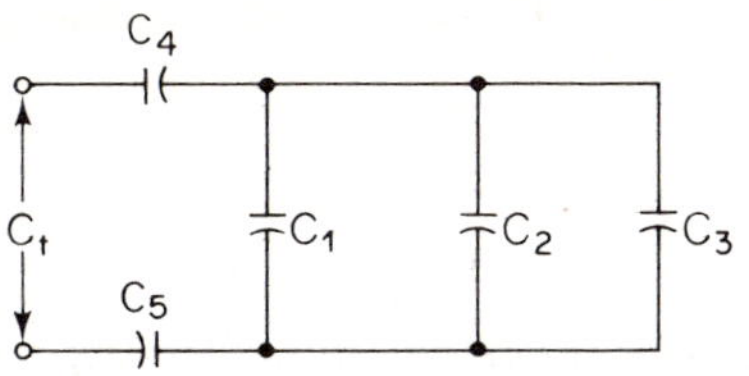

$$C_t = C_1 + C_2 + C_3 + \frac{C_4 C_5}{C_4 + C_5}$$

$$X_{C_t} = X_{C_4} + X_{C_5} + \frac{1}{\frac{1}{X_{C_1}} + \frac{1}{X_{C_2}} + \frac{1}{X_{C_3}}}$$

Figure 10–3 Capacitors in Series-Parallel

there, as the capacitor charges. If resistance becomes steady below 1000 MΩ, the capacitor is leaking. If the indicator goes instantly to infinity and remains there, the capacitor is open. For all capacitors except electrolytics it is a good idea to reverse the connections and double check the results. With electrolytics, polarities must be observed to avoid damage to the capacitor. Some capacitors have very low voltage ratings; make sure that the ohmmeter voltage does not exceed this rating. Also, low-voltage capacitors charge very rapidly; close observation is essential in order to avoid an erroneous indication.

Voltmeter Test for Leaks and Opens

1. Isolate the capacitor by disconnecting one lead.
2. Charge the capacitor to a known value of voltage.
3. Remove the voltage source, and measure the capacitor voltage with a voltmeter.

The voltage source should be near, but never greater than, the working voltage of the capacitor. A good capacitor will produce a voltage indication approximately the same as the source voltage. A leaky capacitor will show a low, gradually decreasing voltage indication. A zero-voltage reading indicates that the capacitor is either open or shorted. The ohmmeter check will differentiate between the open and short, but either way, the capacitor is bad.

Care must be exercised when conducting the voltmeter test on low-value capacitors. The multimeter or vacuum-tube voltmeter may discharge the capacitor too fast for a reliable reading to be taken. An electronic voltmeter is likely to produce better results because of its higher input resistance.

Capacitor Color Code

There are several methods of color coding capacitors. These codes provide not only capacitance but tolerance characteristics, working voltage, and operating temperature. Three sample color codes are presented in Appendix D.

CAPACITOR MEASUREMENTS

Three basic types of instruments are used to make capacitor measurements, involving both quality and value. The bridge type is the most accurate; the reactance type is generally an added feature to increase the usefulness of a testing device; and the Q meter is the type that makes use of the resonance technique.

Capacitance Bridges

The basic ac bridge shown in Figure 10–4 is used to measure capacitance. R_1 and R_3 are fixed precision resistors of equal ohmic value. It is important that these resistors be noninductive. The balancing resistor, R_2, is a variable wirewound resistor. The legs of the bridge are completed by C_s, a standard capacitor, and C_x, the capacitor being measured.

R_2 is adjusted until the bridge is balanced. This condition is indicated by zero current as shown on the meter. A dial on R_2 is calibrated in values of capacitance and may be read directly. This reading is equivalent to the capacitance of C_x.

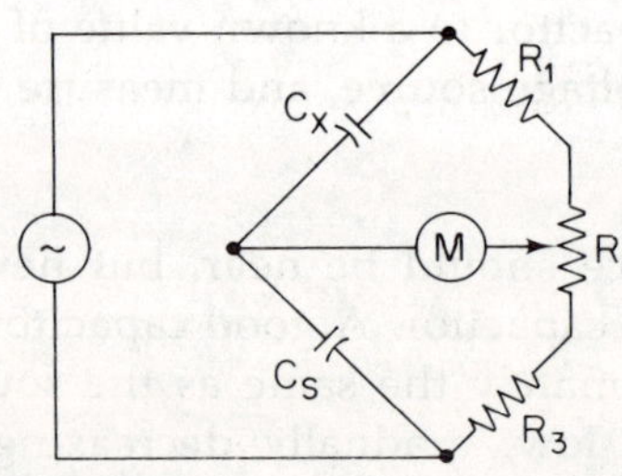

Figure 10–4 Capacitance Bridge

Another version of the capacitive bridge is illustrated in Figure 10–5. The balance formula in Figure 10–5 indicates that capacitive reactance is being balanced, rather than the capacitance. All capacitors have some loss, and this loss has resistive characteristics. Here the loss is represented by a simple series resistor, P_x. The loss is balanced by the adjustable resistor, P_s. This resistance (P_s) is calibrated in terms of dissipation factor, *D*. Control *A* is used to complete the balance, and its dial indicates the capacitance of C_x.

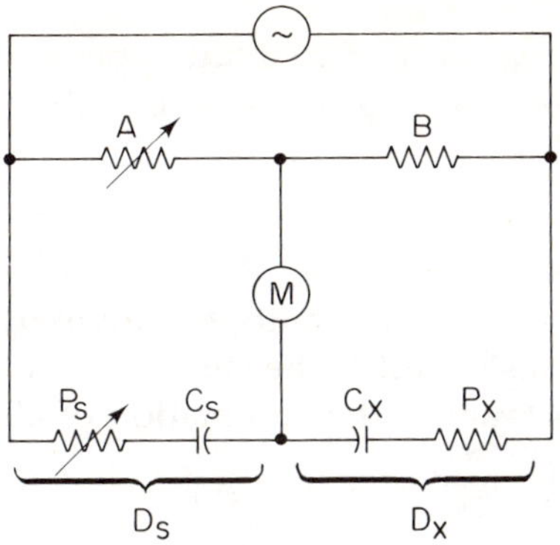

$$\frac{A}{B} = \frac{X_S}{X_X} = \frac{\frac{1}{2\pi f C_S}}{\frac{1}{2\pi f C_X}} = \frac{C_X}{C_S} \text{ also } 2\pi f C_S P_S = 2\pi f C_X P_X \text{, or } D_S = D_X$$

where $X_S = \frac{1}{2\pi f C_S}$ (the capacitive reactance)

$$X_X = \frac{1}{2\pi f C_X}$$

Figure 10–5 Variation of the Balanced Bridge

Capacitor Measurement through Reactance

Reactance-type capacitance-measuring equipment is based on the principle that the voltage drop across the reactance is *inversely proportional* to the capacitance. A small value of alternating voltage is applied across a capacitor and a resistor in series. The voltage drop across the reactance of the capacitor is used to actuate a meter which is calibrated in values.

The measurement technique just described is used where speed and portability are more important than accuracy. At best it indicates only the approximate values of capacitance. As capacitor losses increase, the indications become even more inaccurate. The losses effectively place a resistance in series with the capacitor. Voltage drops across this effective resistance are interpreted as voltage drops across the capacitor, which is in addition to the voltage actually dropped across the reactance. This causes the indicated capacitance to be lower than the actual capacitance.

INDUCTOR TESTS

The inductance and inductive reactance of a circuit depend not only on the type and number of inductors but also on the type of circuit arrangement.

Inductors in Series

Figure 10–6 illustrates a series inductive circuit. As shown by the equations, total inductance in a series circuit is the summation of all inductances; also total inductive reactance is the summation of all inductive reactances.

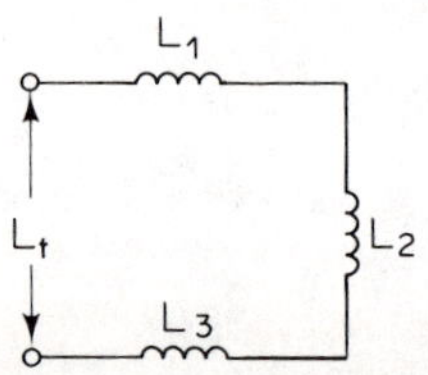

$$L_t = L_1 + L_2 + L_3$$

$$X_{L_t} = X_{L_1} + X_{L_2} + X_{L_3}$$

Figure 10–6 Inductors in Series

Inductors in Parallel

Figure 10–7 illustrates a parallel inductive circuit. In the parallel circuit arrangement, the reciprocal formula is used for both total inductance and total inductive reactance. The product over the sum formula can also be used.

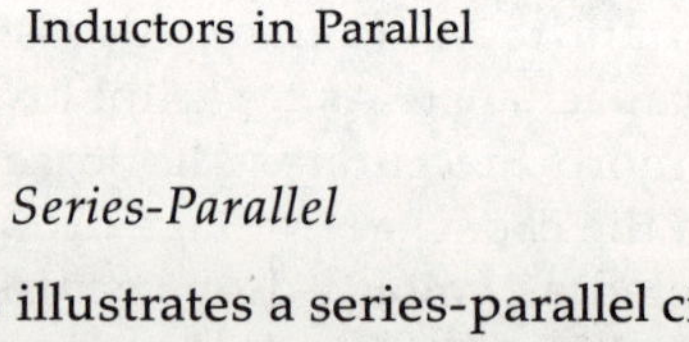

$$L_t = \frac{1}{\frac{1}{L_1} + \frac{1}{L_2} + \frac{1}{L_3}}$$

$$X_{L_t} = \frac{1}{\frac{1}{X_{L_1}} + \frac{1}{X_{L_2}} + \frac{1}{X_{L_3}}}$$

Figure 10–7 Inductors in Parallel

Inductors in Series-Parallel

Figure 10–8 illustrates a series-parallel circuit arrangement of inductors. The equations used in series-parallel circuits, for both L_t and X_{Lt1} show that series laws are applied to the series portion of the circuit, and parallel laws are applied to the parallel portions. The totals are then obtained by adding the two results.

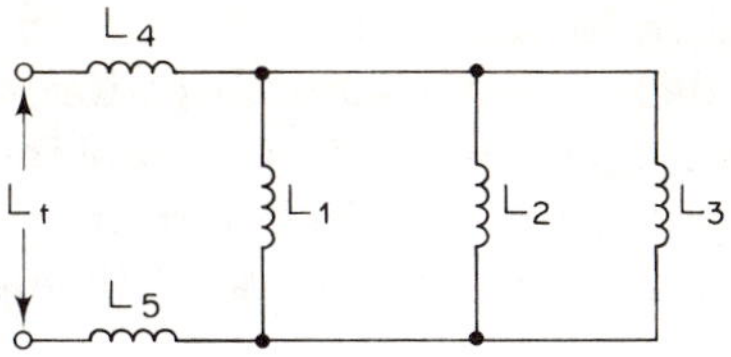

$$L_t = L_4 + L_5 + \frac{1}{\frac{1}{L_1} + \frac{1}{L_2} + \frac{1}{L_3}}$$

$$X_{L_t} = X_{L_4} + X_{L_5} + \frac{1}{\frac{1}{X_{L_1}} + \frac{1}{X_{L_2}} + \frac{1}{X_{L_3}}}$$

Figure 10–8 Inductors in Series-Parallel

Mutual Inductance

It has been assumed that the inductors in Figures 10–6, 10–7, and 10–8 have sufficient physical separation to prevent interaction of the magnetic fields of the individual inductors. When there is interaction between two fields, mutual inductance becomes an important factor.

Mutual inductance is calculated by

$$M = k\sqrt{L_1 \times L_2}$$

The constant, k, is the coefficient of coupling, and it is determined by the percentage of overlap of the magnetic fields. If all lines of force from L_1 cut across all the turns of L_2 and vice versa, $k = 100$. In practice, a coefficient of coupling of 100 is impossible to obtain, but some transformers may approach this magic figure. In fact, some transformers have a k in excess of 99.

Inductors that carry the same current are in series. When coils are in series and mutually coupled, they can be either series-aiding or series-opposing. When the current has the same direction through both coils, the fields are series-aiding. When the current is in opposite directions through the coils, the fields are series-opposing. In either case, mutual inductance is an important factor in determining the effective inductance of the circuit. The two conditions are illustrated in Figure 10–9.

Series aiding
$L = L_1 + L_2 + 2M$

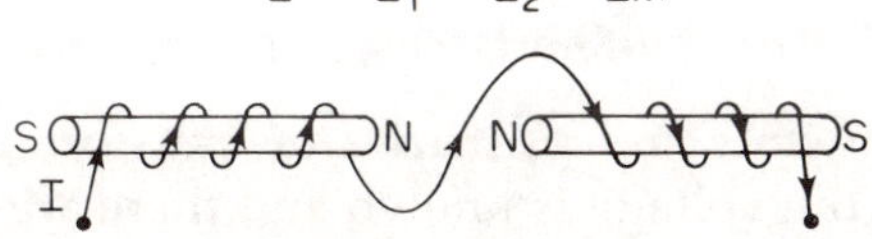

Series opposing
$L = L_1 + L_2 - 2M$

Figure 10–9 Mutually Coupled Series Inductors

Overlapping fields from parallel inductors also affect the inductance of the circuit. When the two fields are in the same direction, they are aiding. When they are in opposite directions, they are opposing. In either case, mutual inductance is an important factor in determining the effective inductance. Both of these conditions are illustrated in Figure 10–10.

$$L = \frac{1}{\frac{1}{L_1 + M} + \frac{1}{L_2 + M}}$$

Parallel aiding

$$L = \frac{1}{\frac{1}{L_1 - M} + \frac{1}{L_2 - M}}$$

Parallel opposing

Figure 10–10 Mutually Coupled Parallel Inductors

Ohmmeter Tests for Opens and Shorts

Each inductor has a specific ohmic value of resistance which enables simple ohmmeter tests to be performed. Isolate the inductor by disconnecting one lead, and measure the resistance. A zero reading on the lowest resistance range indicates a shorted inductor. A reading of infinity on the highest resistance range indicates that the inductor is open.

The partially shorted inductor is far more likely to occur, and it is more difficult to detect. A few turns of the coil may be shorted together. This may cause the inductor to perform unsatisfactorily and still give a definite resistance reading when checked with an ohmmeter. Measure the resistance of an identical coil which is known to be good and compare this reading with the reading from the coil that is suspected of being bad. If the suspect reading is several ohms less than the one used as a standard, the coil is partially shorted.

Voltmeter Tests for Opens and Shorts

Single-winding inductors and primaries of transformers have voltage applied to them. If the voltage is known and there are no parallel components, the inductor can be checked with an electronic voltmeter. A partial short will cause a lower than normal reading. A complete short will cause a zero reading.

The open coil is more elusive. An open will cause an indication equal to the total applied voltage. In many cases, this is the proper voltage. Such a reading requires further testing. If there is a resistor in series with the coil, the problem is simple—measure the voltage across the resistor. A normal voltage across the resistor confirms that the coil is good. A zero reading across the resistor confirms that the coil is open. If there is no series resistor, conduct the ohmmeter check on the coil.

INDUCTOR MEASUREMENTS

Precision measurements of inductors require the use of an ac bridge. This may be the Hay, Maxwell, Owen, Campbell, Carey-Foster, Felici, or Universal. A basic bridge is illustrated in Figure 10–11.

The Hay bridge measures the inductance of a coil by comparing it with a capacitance. This bridge is illustrated in Figure 10–12. The balance depends upon the losses of the inductor. The Hay bridge is generally employed for measuring inductors with a low dissipation factor (D). Since $D = 1/Q$, a low D inductor is also a high Q inductor.

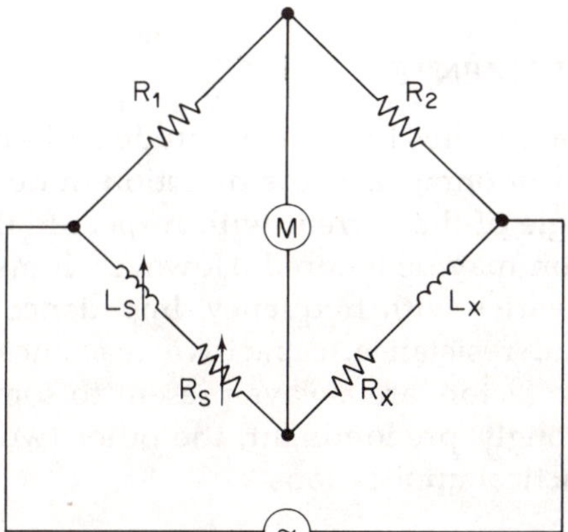

$$L_X = \frac{R_2 L_S}{R_1}$$

$$R_X = \frac{R_2 R_S}{R_1}$$

Figure 10–11 Resistance Ratio Bridge

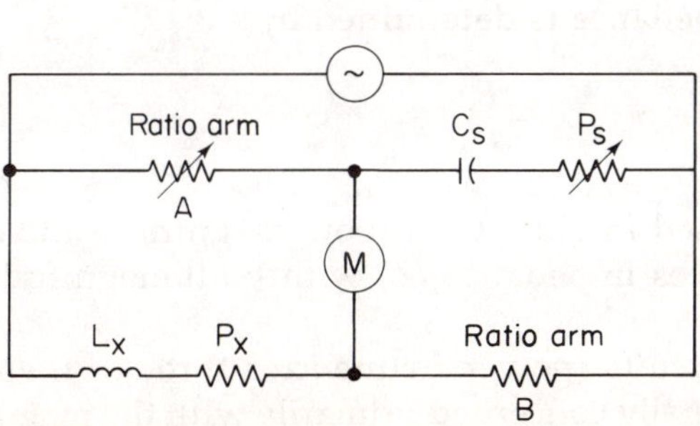

$$D = \omega C_S P_S \quad (\omega = 2\pi f)$$

$$L_X = \frac{ABC_S}{1 + D^2}$$

$$P_X = \frac{AB(\omega C_S)^2 P_S}{1 + D^2}$$

Figure 10–12 Hay Bridge for Inductance

When the dissipation is in excess of 0.05, the Hay bridge is usually exchanged for a Maxwell bridge as illustrated in Figure 10–13. In the Maxwell bridge, variable resistor P_s is in parallel with C_s. Otherwise there is little difference between the Hay and the Maxwell bridges. Balance now depends upon the storage factor (Q) of the coil. When the circuit is balanced, the formulas shown in the figure are true.

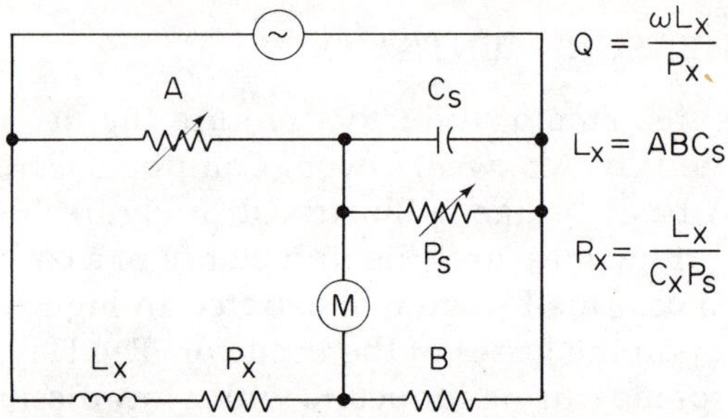

Figure 10–13 Maxwell Inductive Bridge

IMPEDANCE MEASUREMENTS

Impedance is a term that implies both a magnitude and a direction. The magnitude is expressed in ohms, and the direction in degrees. The direction is the phase angle of the current with respect to the voltage. In some cases, the direction may be ignored. However, it must be kept in mind that impedance varies with frequency. Impedance is composed of some combination of resistance, capacitive reactance, and inductive reactance. All three factors are always present to some extent, but when one factor is strongly predominant, the other two need not be considered in most practical applications.

Voltmeter-Ammeter Method

The magnitude of an impedance is determined by

$$Z = \frac{E}{I}$$

When E and I are expressed in polar form (both magnitude and direction), the solution produces impedance (Z) with both magnitude and direction.

In low-frequency circuits (power frequencies through the audio range) the interest is generally concerned primarily with the magnitude of Z. A test circuit similar to that illustrated in Figure 10–14 will pro-

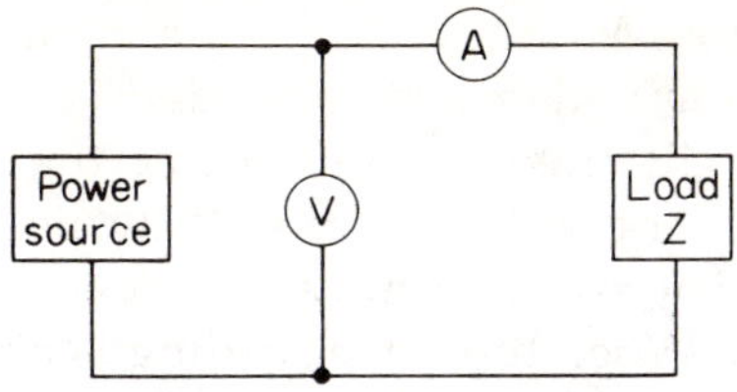

Figure 10–14 Measuring Impedance with Voltmeter-Ammeter

vide information for calculating the magnitude of Z for an unknown impedance.

The ammeter indicates the current through the load, while the voltmeter indicates the voltage across the load. The magnitude of impedance is then obtained by $Z = E/I$.

The impedance indicated by this measurement and calculation is true only at the test frequency. Another fallacy of this test is caused by the fact that the ammeter is in the path of total current. The reactance of the ammeter may have a significant effect on the impedance.

Voltmeter Method

A slight variation on the previous method of impedance measurement improves the accuracy by decreasing the meter interference. This method is illustrated in Figure 10–15. Here a small resistor is placed in series with the unknown impedance. This should be the smallest resistor that will provide a significant voltage reading for E_1. (For best results, the voltmeters used for both E_1 and E_2 should be electronic-type voltmeters.) E_1 is converted to current by dividing E_1/R.

E_2 is the voltage across the unknown impedance, Z. The magnitude of Z is obtained by dividing E_2/I.

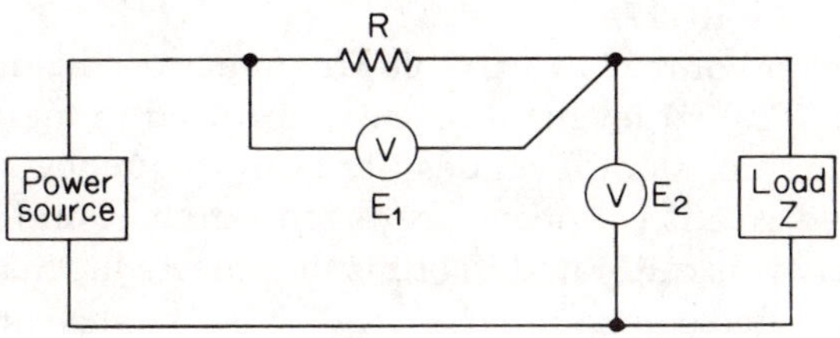

Figure 10–15 Measuring Impedance with Voltmeter

Impedance Meter

The basic impedance meter is designed to indicate only the magnitude of an impedance. The simplified circuit of such a meter is illustrated in

Figure 10–16. The meter, *M*, in this circuit is generally an electronic voltmeter. The unknown impedance is connected in series with a variable resistor, *R*, which is calibrated in ohms. When the switch is in position *B*, the meter indicates the voltage across *Z*. In position *A*, it indicates the voltage across *R*. The switch is moved back and forth between *A* and *B* while *R* is adjusted. When the same reading is obtained in both positions, $Z = R$. The ohmic indication on the dial of *R* now indicates the magnitude of *Z*.

The ac source for this circuit is generally an audio oscillator. If it has a variable frequency, which may be used to determine whether the impedance is inductive or capacitive, further manipulation is necessary. Place the switch in position *B* and increase the frequency of the oscillator. If the impedance is inductive, the meter reading will increase. If the impedance is capacitive, the meter reading will decrease. During this check, care must be exercised to keep the oscillator output voltage at a constant value.

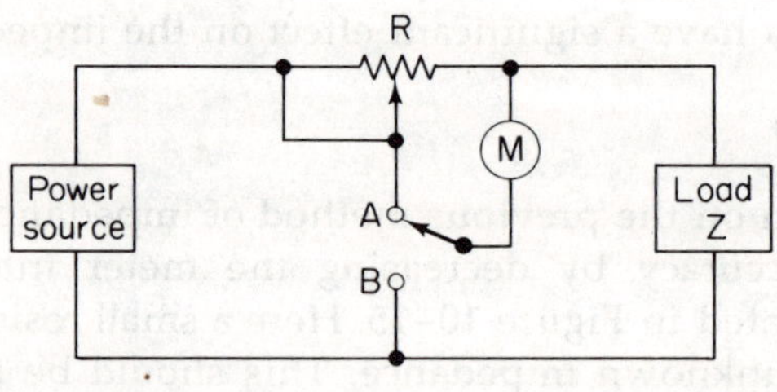

Figure 10–16 Bridge Impedance Meter

Vector Impedance Meter

The vector impedance meter is designed to provide a direct readout of impedance magnitude and direction. The test circuit of such a meter is illustrated in Figure 10–17.

The ac power source is a variable-frequency oscillator. Certain circuit refinements, called leveling circuits, are used to maintain the applied voltage at a constant level despite both frequency and load. The unknown impedance is connected between points *A* and *B*. R_1 and M_1 are parallel to *Z*. M_1 is calibrated in ohms to indicate the magnitude of *Z*.

The phase detector compares the phase angles of the current through *Z* and the voltage across *Z*. M_2 provides an angle reading in degrees. This is the impedance phase angle in respect to zero. The combined readings of M_1 and M_2 provide a complete picture of the impedance as follows:

$$Z\Omega \angle\theta$$

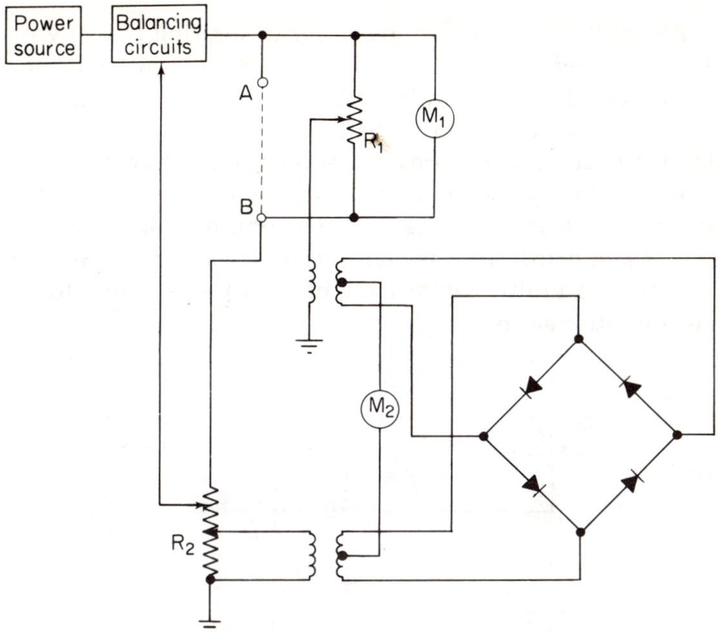

Figure 10–17 Measuring both Magnitude and Direction of *Z*

POWER MEASUREMENTS

Measuring power is relatively simple when the load is pure resistance. It is the product of the voltage across the load, times the current through the load. This is the average power under stable dc conditions.

When reactive components are introduced into an ac circuit, current and voltage may not be in phase. Now $E \times I$ is only apparent power, and it is expressed in VA. This represents the energy furnished to the circuit, but not the energy used by the circuit. The energy used is represented by the true power, and true power is

$$E \times I \times \cos \theta$$

where θ is the *phase angle* between the voltage and the current.

At power frequencies and audio frequencies, power is measured with a normal power meter that is calibrated directly in W. At frequencies from low RF to VHF, thermocouple ammeters may be used in conjunction with dummy loads to sample the output power. In the UHF range and higher, the most practical methods of power measurement employ a bolometer or a temperature-sensitive element.

At Line Frequencies

Figure 10–18 illustrates a test circuit for measuring power with a *dynamometer wattmeter*. The ac input is applied to terminals *A* and *B*; the load is connected between terminals *B* and *C*. This places the two current coils (*I*) in series with the load, and the voltage coil (*E*) in parallel with the load. Coil *E* is the movable coil, and the deflection is a function of the product of the instantaneous values of the current through the load and the voltage across the load. The pointer is attached to the movable coil (*E*), and the scale is calibrated directly in W. Most meters of this type have a multiplier resistor in series with coil *E* to extend the range of measurements.

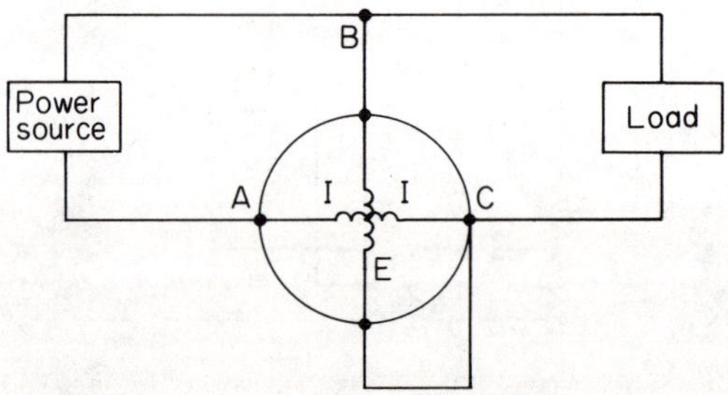

Figure 10–18 Measuring Power with a Wattmeter

Figure 10–19 illustrates a test circuit for measuring three-phase power. The three-phase ac input is connected to terminals 1, 2, and 3; the three leads to the load are connected to terminals *A*, *B*, and *C*. The pointer is coupled to both *E* coils, and movement is determined by the combined power of all three phases.

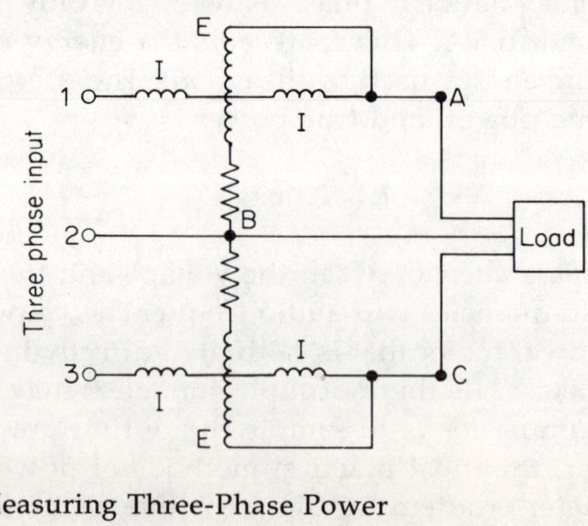

Figure 10–19 Measuring Three-Phase Power

At Audio and Low Radio Frequencies

This group of frequencies covers the spectrum from power frequencies to about 20 kHz. The most common reason for measuring power at these frequencies is to determine the output power of amplifiers and oscillators.

One easy method of measuring this power is to place a dummy load on the circuit and calculate the power of the dummy load. This is illustrated in Figure 10–20. In this case, the dummy load consists of a carbon resistor with a known resistance. This resistance must match the output impedance of the audio circuit in order to obtain maximum power transfer. The current through the dummy load is registered on the ammeter. Determining power is a simple application of $P = I^2R$.

A variation of this method is to use an electronic voltmeter instead of the ammeter. The meter is used to measure the voltage across the dummy load. Power is determined by $P = E^2/R$.

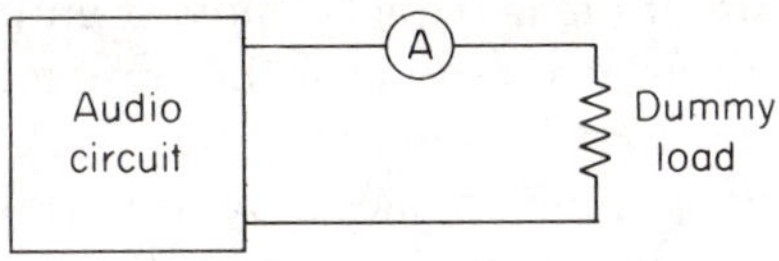

Figure 10–20 Measuring Audio Power

Output Power Meter

In many cases, exact measurement of RF power is not required. It may be more important to have an approximate measurement quickly than to take more time to obtain an exact measurement. These situations are well served by the output power meter. It is not highly accurate, but it is portable, fast, and easy to use. Figure 10–21 illustrates the use of such a meter.

This method uses a carbon pile to match the output impedance of the RF source. A portion of the voltage drop across the carbon pile is used to operate the meter, which may be calibrated in W.

The upper frequency limit is determined by the capacitance of the voltage divider. As frequency increases, the reactance decreases. The higher the frequency, the greater the mismatch of the impedance and the greater the error in the measurement.

Low power can be applied to this circuit continuously, but high power should be applied only for brief periods.

This is a general-purpose meter, and it may be used for a wide range of impedance, frequency, and power.

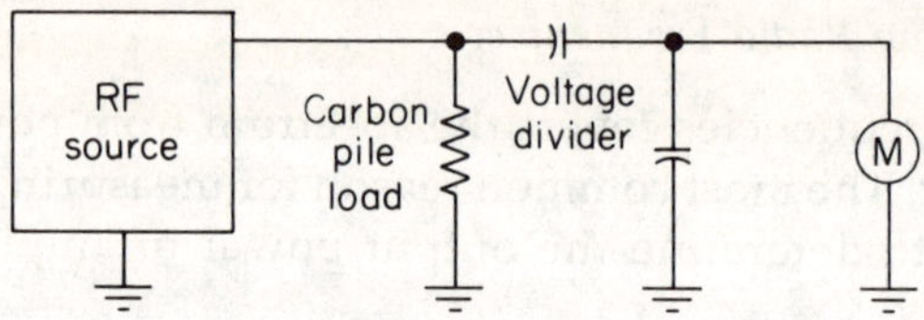

Figure 10–21 Use of Output Power Meter

Indirect Power Measurements

At ultrahigh frequencies, it is not practical to determine power by direct measurements. The RF energy is converted to some other form of energy, and the secondary energy is evaluated. The secondary energy must be directly related to the RF energy, and the associated time interval must be taken into account.

Lamp Load Method

The *lamp load* method of measuring RF power employs a bank of incandescent lamps as a dummy antenna for the transmitter. A photographic illumination meter is used to measure the resulting brilliance of the lamps. The RF source is then replaced by a dc power source, which is adjusted until the lamps glow at the same brilliance. A wattmeter in the dc source measures the dc power. This power is equal to the output power of the transmitter. The impedance of the lamps must match the transmitter output impedance. Such a test circuit is illustrated in Figure 10–22.

Instruments of this type have a variety of arrangements of lamps to match a range of impedances. The accuracy is within ±10% for frequencies between 200 MHz and 750 MHz.

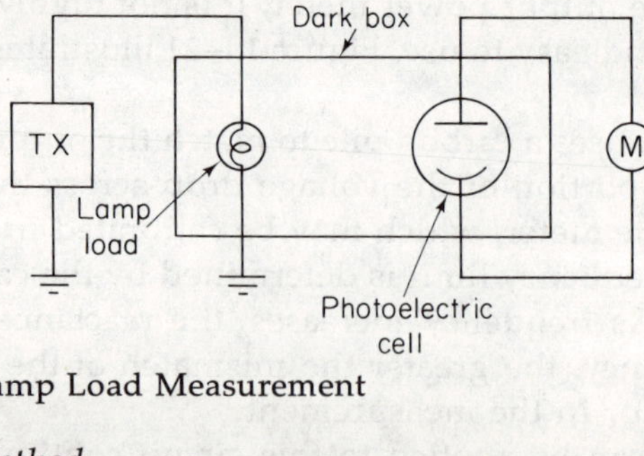

Figure 10–22 Lamp Load Measurement

Resistor Load Method

Power may be measured indirectly by measuring the temperature of a dummy load resistor. As always, the load must be noninductive, and it must match the impedance of the RF power source. One way to accomplish this is to pass a stream of air over the load resistor and measure the increase in the air temperature. This is illustrated in Figure 10–23.

Two thermocouples are connected to a meter that indicates the difference in thermocouple outputs. One thermocouple reacts to the ambient temperature; the other reacts to the same air after it is heated by the dummy load. The difference in temperature represents a temperature increase.

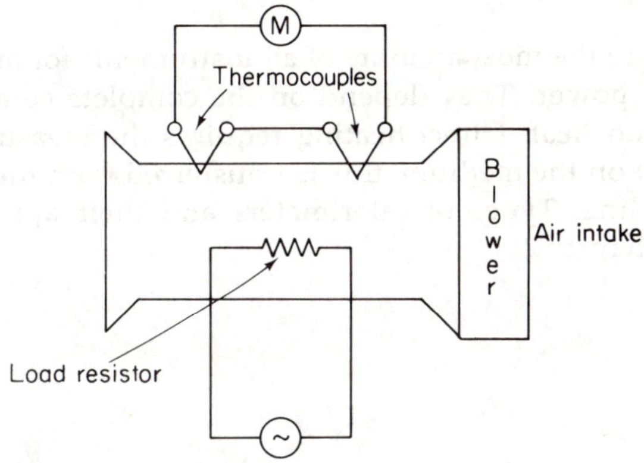

Figure 10–23 Resistor Load Measurement

Use of Bolometers

Bolometers are the most common active elements used to measure low power at high frequencies. The *barretter* increases its resistance with an increase in temperature. The *thermister* decreases its resistance when temperature increases. These elements are physically small for high sensitivity.

When RF power is applied to a bolometer element, the energy absorbed by the element causes a change in its resistance. An instrument that uses a bolometer as one leg of a bridge circuit can be used to measure power by the substitution method. A sample test circuit is illustrated in Figure 10–24.

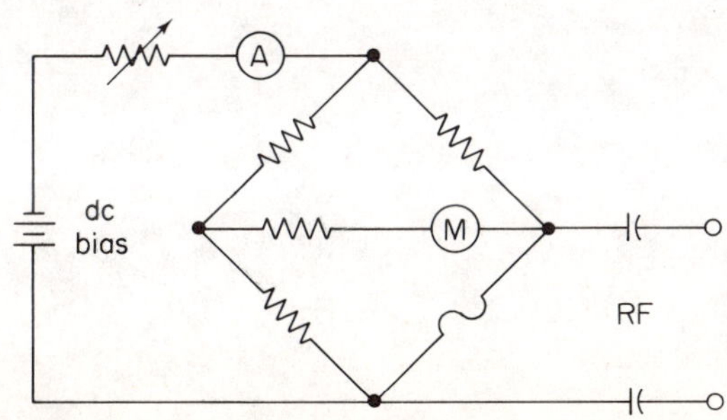

Figure 10–24 Bolometer Power Measurement

The balanced bridge technique balances the bridge with dc bias. Application of RF energy causes the bolometer to change resistance and unbalances the bridge. The dc bias is gradually removed until balance occurs again. The RF power is then equivalent to the dc power removed.

The unbalanced bridge technique converts the unbalanced reading directly into watts of RF power.

Use of Calorimeters

Calorimeters are the most accurate of all instruments for measuring high values of RF power. They depend on the complete conversion of the RF energy into heat. Direct heating requires the measurement of the heating effect on the medium; this is transferred from the load that terminates the line. Types of calorimeters and their applications were covered in Chapter 2.

chapter eleven

MEASURING FREQUENCY

Frequency measurement is a broad subject. Unless a particular range in the frequency spectrum is specified, the interest may range from near dc to beyond visible light. Electronic instruments are available for measuring frequencies in any desired segment of this broad spectrum. This chapter treats sample techniques from very low frequencies through the UHF band. Most of the instruments used here were discussed in Chapter 3.

MEASURING ROTATION AND VIBRATION FREQUENCIES

Many areas of electronics are vitally concerned with *rotational* and *vibratory* or reciprocating frequencies. This aspect can be of prime importance to any person who is concerned with the maintenance of motors or generators. Many other specialized areas require this type of measurement. A variety of instruments are available for these measurements: tuning forks, stroboscopes, vibrating reed meters, electromechanical counters, oscilloscopes, and frequency counters.

Using the Tuning Fork

The *tuning fork* is primarily intended for checking the rotational frequency of motors and generators, but it is not necessarily limited to these applications. One vital point to remember is that a particular fork can be used for only one frequency. This is the frequency stamped on the handle, the frequency for which it was manufactured.

Allow the machine being tested to reach operational speed under normal load conditions. Direct a light so that it illuminates the target area. The target area may be black and white segments on a motor governor, radial spokes in a flywheel, or other similar devices.

Hold the tuning fork by the handle, and strike one of the tines against the edge of the hand. This sets up the vibrations at the natural frequency of the fork. The slots in the plates on the tines of the fork are now opening periodically. View the target through these slots. If the segments of the target appear to be stationary, the machine is operating at the proper frequency. When the segments seem to move backward, against the known direction of rotation, the speed is too low. When the speed is too high, the segments appear to rotate forward, with the known direction of rotation.

There are conditions that cause the target to jump about in an erratic fashion. This indicates a variable rotational speed, and the most likely cause is a faulty governor. Any indication except a still target is indicative of improper speed and calls for either repair or adjustment or both. Adjustments are frequently incapable of producing a perfectly stationary target. An acceptable condition is reached when the target rotates slowly in the direction of known rotation.

Using the Stroboscope

Direct the lamp toward the target area of the mechanism for direct illumination. If an identification mark is not already clearly visible, make one with either chalk or grease pencil. Additional marks will cause no problem since only one will be visible at the proper speed or any submultiple thereof.

In most cases, the approximate speed of rotation is known. The *stroboscope* controls should be set to match this speed. Allow the mechanism to reach operating speed in normal load; then activate the scope. Vary the main frequency control that determines the flashing rate of the light until the reference mark on the moving part appears to stand still. The calibrated scale of the stroboscope will then indicate the speed in revolutions per minute.

Occasionally it becomes necessary to measure speeds that are completely unknown. In this case, set the stroboscope to its highest flasher frequency. While observing the target, reduce the flashing rate until a single mark appears and becomes stationary. The scale now shows the fundamental synchronism that corresponds to the rotational speed of the moving part. If the flashing rate is further reduced, other frequency points will produce a stationary mark. These are points where the rotational frequency is an exact harmonic of the flashing frequency.

A high-grade stroboscope represents a substantial investment in measuring instruments. The flasher tube life can be extended by careful use. This calls for keeping the flashing rate as low as possible during all measurements and keeping the activated interval as short as possible. When the instrument must be operated for a long period of time, this can be offset by taking measurements at lower flashing rates.

When measuring rotational speeds greater than the flashing rate, two points of synchronism must be obtained. As the flasher rate is reduced, record the frequency that produces the first stationary image; call this frequency *A*. Reduce the flasher rate to the next point where a stationary mark is visible; call this frequency *B*. The unknown rotational speed is then calculated by

$$RS = \frac{AB}{A - B}$$

For example, when frequency *A* is 3500 and frequency *B* is 3000 rpm (revolutions per minute), then

$$RS = \frac{3500 \times 3000}{3500 - 3000} \quad \frac{10{,}500{,}000}{500} = 21{,}000 \text{ rpm}$$

It is also feasible to measure rotational speeds below the lowest frequency of the stroboscope. In this case, multiple images will be observed. Now the true rotational frequency is the indicated frequency divided by the number of images. For example, with an indicated rpm of 1800 and three visible markers, the true rotational speed is

$$RS = \frac{1800}{3} = 600 \text{ rpm}$$

A word of *caution* is in order. When observing a stationary target with a stroboscope, the mind is readily convinced that the mechanism is actually stopped. Serious injury can result from touching the moving part; a conscious effort must be made to avoid this action.

Use of Frequency Counters

Various types of frequency counters have been designed or adapted for use as electronic *tachometers*. These instruments may be used to make accurate measurements of the rotational speed of machinery. A tachometer pickup may be used to produce trigger pulses that are coupled directly into the counter. The pickup is generally designed to generate one signal per revolution or 60 signals per revolution. With one signal per revolution, the counter indicates revolutions per second. With 60 signals per revolution, the count is registered in rpm.

MEASURING AUDIO FREQUENCIES

The most common electronic instruments for measuring audio frequencies are frequency counters and *oscilloscopes*. Although the oscilloscope was covered in some detail in Chapter 4, its application in frequency and phase measurements will be considered here because both are vital in audio work. Other applications appear at appropriate points through the remainder of this book. Bridge circuits are also used in measuring audio frequencies.

Using Bridge Meters

Many frequency meters make use of a common bridge circuit. This is a relatively uncomplicated arrangement, and it is very accurate, although accuracy tends to drop off at the high end of the audio band. A simplified frequency measuring bridge is illustrated in Figure 11–1.

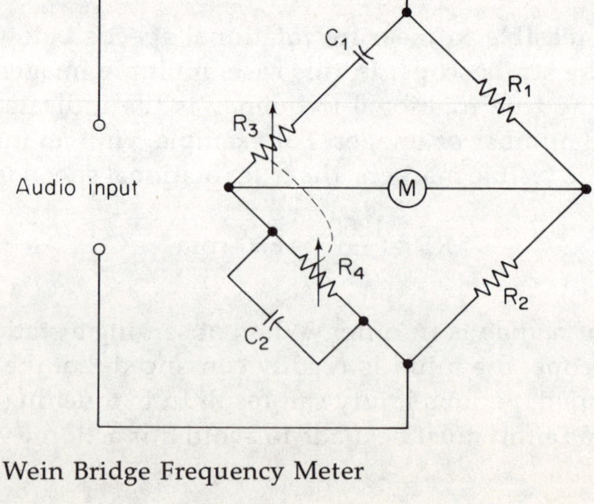

Figure 11–1 Wein Bridge Frequency Meter

This bridge is energized by application of an audio signal, and balance is achieved by turning a single dial that varies the resistance of both potentiometers. Balance is indicated by a null indication, and at that point, the frequency of the audio signal may be read from the calibrated dial. The frequency at balance is equivalent to $1/2\pi R_3 C_1$.

In practice, the fixed capacitance, represented by C_1 and C_2, may be several capacitors, usually two sets of three capacitors. A range switch selects the proper capacitors for the section of the audio band being measured.

In audio work, the null indicator is quite often a pair of headphones, and this provides suitable sensitivity in most cases. When greater sensitivity is of prime concern, the indicator should be a meter or other electronic device.

Using the Oscilloscope

At this point, oscilloscope use is analyzed specifically as it applies to audio frequencies. However, the techniques described here are equally effective with any frequency that will produce the illustrated patterns.

Oscilloscopes are available with *automatic readout.* These instruments make frequency measurements a very easy operation. Simply display the signal of unknown frequency, set the time base to display exactly one complete cycle, and read the frequency as it is displayed on the screen. This is illustrated in Figure 11–2. This readout shows two volts per division for a peak amplitude of five volts. The time is 200 ns per division, and the sine wave is a 500.1 kHz signal. Frequency can be measured to an accuracy of 0.00005%.

Oscilloscopes without the automatic readout require a bit more analysis on the part of the operator. There are several approaches, and all are reasonably simple. One method involves reading time and con-

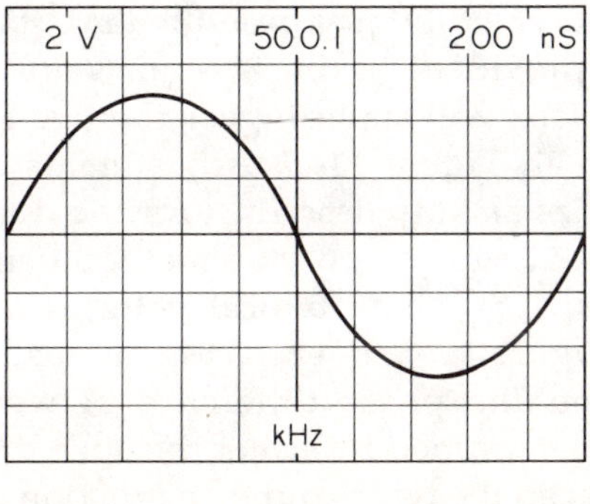

Figure 11–2 Automatic Readout

verting to frequency. Another method uses the principle of comparing two signals and obtaining frequency by a ratio.

In the time measurement method, one cycle of the unknown frequency signal is displayed on the screen, and the time is accurately measured. The time may be obtained from the setting of the time base control, but most scopes have range marks. These marks can be superimposed on the image and counted to obtain the exact time for one cycle. If the signal is not a perfect sine wave, care must be exercised to measure from the leading edge of one cycle to the leading edge of the next. Frequency is obtained by substituting time into the following equation and solving:

$$f = \frac{1}{t}$$

Figure 11–3 shows a display with markers and illustrates the solution for frequency. When the time is in seconds, the calculated frequency is in Hz (hertz); time in ms (milliseconds) gives frequency in kHz (kilohertz); and the frequency is in MHz (megahertz) when the time is in μs (microseconds). When measuring audio frequencies, it is convenient to read the time and calculate in (ms) milliseconds because the frequency is never higher than 20 kHz.

Multiple-trace oscilloscopes are easy to use for frequency measurements. Display a signal of known frequency on one trace and the signal to be measured on the other. Keep both time bases the same, and adjust until a definite ratio is obtained for complete cycles of both signals. The unknown frequency is obtained by

$$f_x = \text{Ratio} \times \text{Known Frequency}$$

A sample display and solution is illustrated in Figure 11–4.

The comparison method just described may be used with a single-trace oscilloscope provided the input signals are coupled through an electronic switch. This will superimpose the two signals on the single trace as illustrated in Figure 11–5. Adjustments and solving for frequency is handled as already described for the dual-trace oscilloscope.

Another well-known comparison method using a single trace involves *Lissajous patterns.* Some typical patterns and frequency ratios are illustrated in Figure 11–6. In each case, the pattern is a result of the frequency ratio and the phase difference between the two signals. Frequency can be determined by either the open or closed pattern since the difference between the two is purely a function of the phase.

Lissajous patterns are created by placing one signal on the horizontal deflection plates and another of a different frequency on the vertical deflection plates. In the following examples, the standard frequency is on the horizontal plates, and the unknown frequency is on the vertical plates. The standard frequency is referred to as *f*, and the

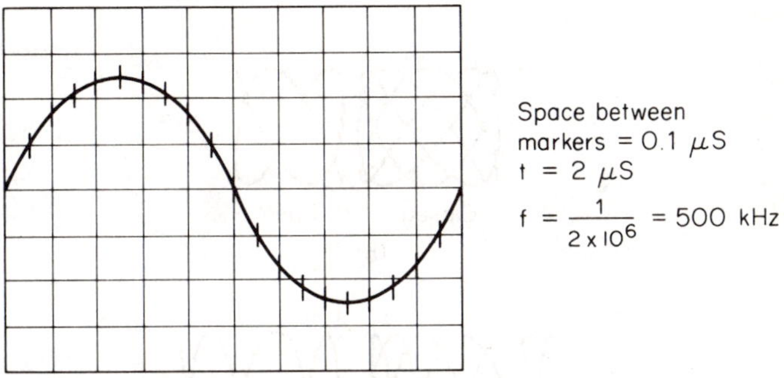

Figure 11–3 Converting Time to Frequency

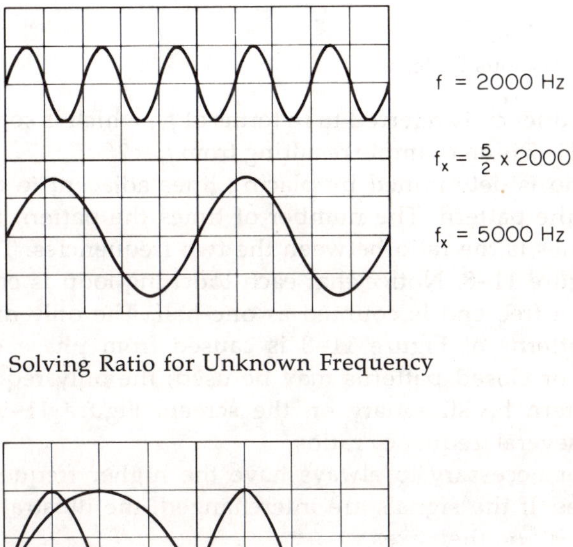

Figure 11–4 Solving Ratio for Unknown Frequency

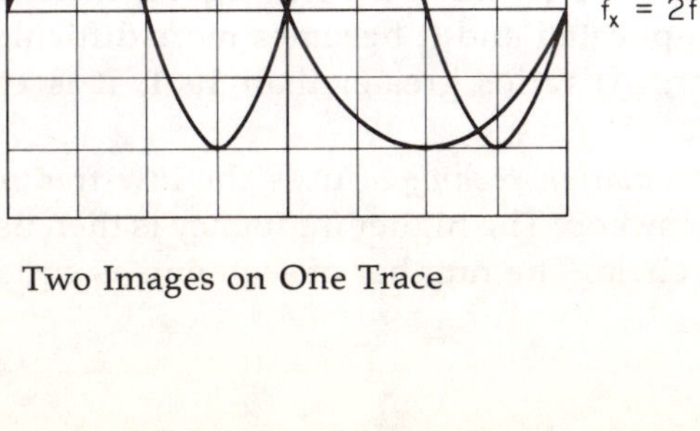

Figure 11–5 Two Images on One Trace

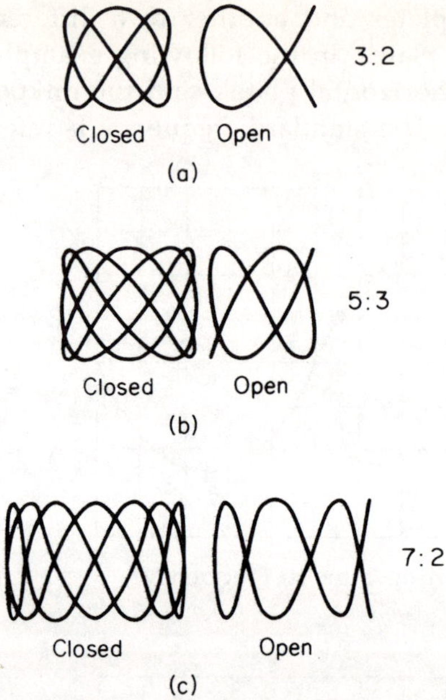

Figure 11–6 Lissajous Patterns

unknown frequency is referred to in terms of f_x, which is some multiple of f. Figure 11–7 is an example resulting from $f_x = 2f$.

The ratio is determined by placing lines adjacent to the top and left edge of the pattern. The number of times the pattern touches the respective lines is the ratio between the two frequencies. This is illustrated in Figure 11–8. Notice that each touching loop is counted as a one, and each free end is counted as one-half. The only difference in the three patterns of Figure 11–8 is caused from phase differences. Either open or closed patterns may be used; the only requirement is that the pattern be stationary on the screen. Figure 11–9 illustrates patterns for several frequency ratios.

It is not necessary to always have the higher frequency on the vertical plates. If the signals are interchanged, the illustrated patterns would rotate 90° on their axes.

As the frequency ratio between the two signals increases, the pattern grows more complicated and it becomes more difficult to stop the pattern on the screen. At ratios greater than 10:1, it is easier to use intensity modulation.

The *intensity-modulation technique* uses the low-frequency signal to establish a circular sweep. The higher frequency is then used to blank out segments of the circle. The number of segments is equal to the fre-

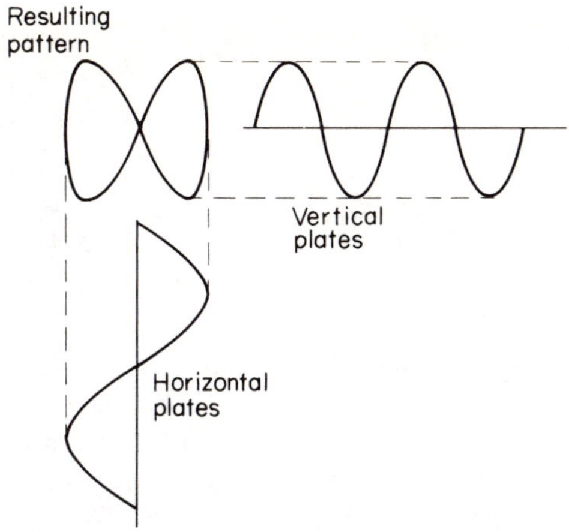

Figure 11–7 Lissajous Pattern, 2:1

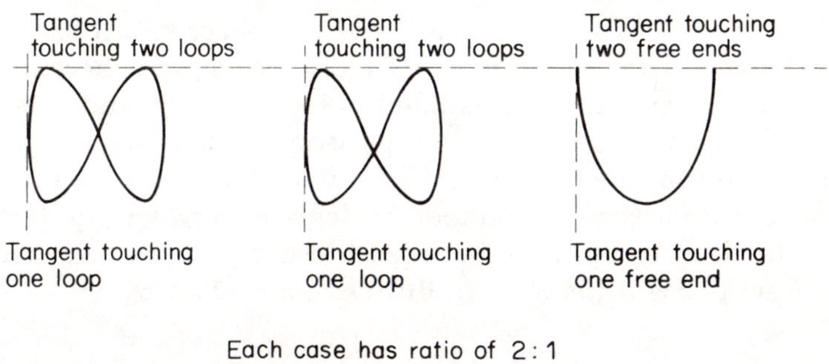

Figure 11–8 Determining Frequency Ratio

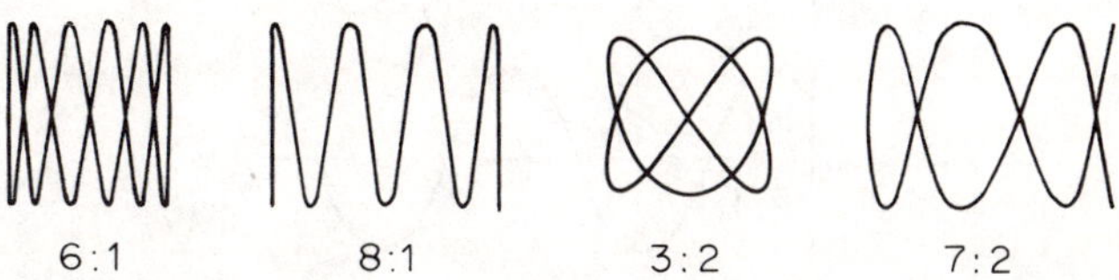

Figure 11–9 Patterns at Various Ratios

quency ratio. This technique is illustrated in Figure 11–10. The low frequency is applied through a phase-shifting network in a manner that causes the signal on adjacent plates to be 90° out of phase, and the signal on opposite plates to be 180° out of phase. This produces the circle. The high-frequency signal is then applied to the intensity-modulation terminal of the scope. Each negative alternation of the high-fre-

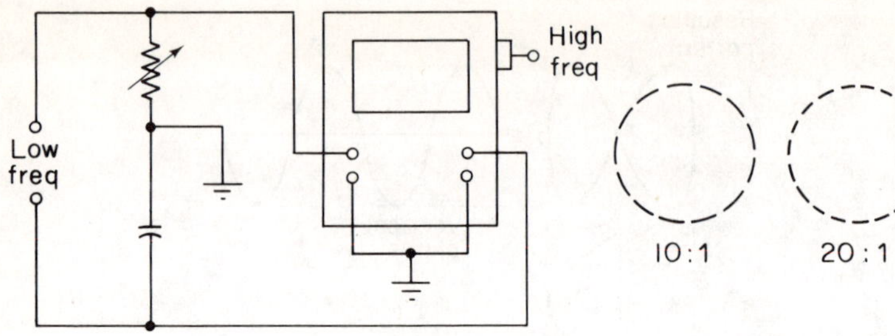

Figure 11–10 Ratio by Intensity Modulation

quency signal will reduce the intensity and cause a break in the circle. Each cycle of the low-frequency signal produces a circle, and each segment of the circle represents one cycle of the high-frequency signal.

MEASURING PHASE RELATIONSHIP

In many cases, it is necessary to know the phase relationship of two audio signals. With a dual-trace oscilloscope, this is checked by displaying the two signals, as illustrated in Figure 11–11, and determining the relative phases by direct comparison. Visual inspection of any identical point on the two waveshapes will show which signal is leading and the approximate number of degrees between the two waveshapes. Measuring the time between these points enables calculation of the exact phase difference. In this case, *b* leads *a* by 45°.

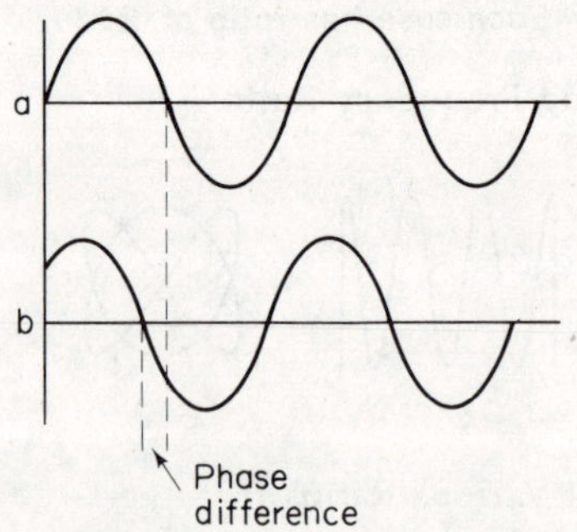

Figure 11–11 Phase Comparison

The same technique can be used with a single-trace oscilloscope by coupling both input signals through an electronic switch, as was illustrated in Figure 11–5.

Lissajous patterns provide another method of measuring phase difference on a single-trace oscilloscope. Figure 11–12 illustrates pat-

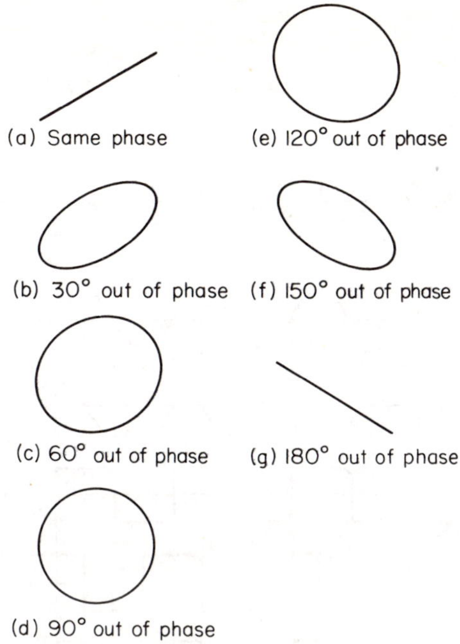

Figure 11–12 Patterns Produced by Phase Difference

terns produced by two signals of the same frequency and amplitude. The shape and attitude indicate the difference in phase. Patterns were selected for each 30° of phase difference. As the phase gradually changes through any given 30°, the pattern will fall between the pattern illustrated for each end of that phase segment.

Using Phase-Angle Meters

It is readily apparent that the accuracy of determining phase difference with an oscilloscope rests primarily with the operator. The judgment factor varies with the situation. While approximate phase differences are no problem, consistent precision measurements are impossible. The *phase-angle meter* is intended for use where a greater degree of precision is required.

The crossover phase-angle meter measures the time between zero and the crossover points on the two signals being compared. Figure 11–13 illustrates the method of comparison. Operation is simple. The two signals pass through wave-shaping circuits and then into a coincidence detector. *Maximum coincidence* occurs when the two signals are exactly *in phase.* At this time, there is no current through the meter, and it indicates a zero phase angle. *Minimum coincidence* occurs when

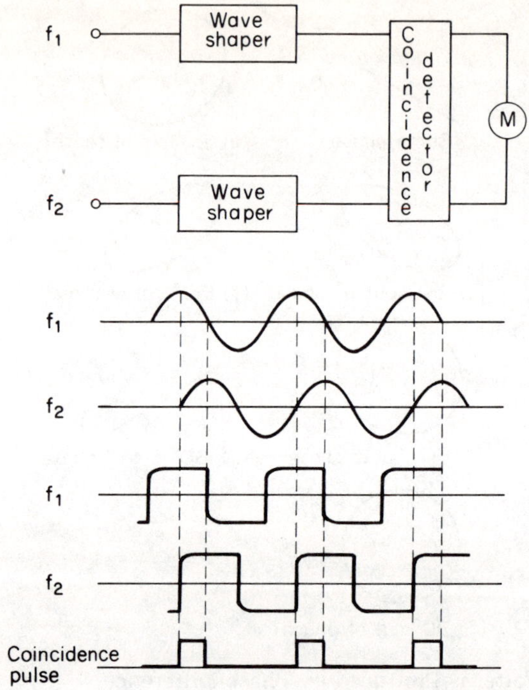

Figure 11–13 Measuring Phase Angle

the phase difference is *180°*. This situation supplies maximum current to the meter, and the indicator moves to indicate a phase angle of *180°*. This is a direct and relatively accurate measurement, and it is free from errors of interpretation.

MEASURING RADIO FREQUENCIES

Radio frequencies cover a broad spectrum ranging from the top end of the audio band to the superhigh microwave frequencies. There are many methods of measuring these frequencies. The technique and the equipment employed are governed by the range of frequencies being measured and the degree of accuracy required.

Using Absorption Wavemeters

This type of wavemeter absorbs energy from the circuit under test. Total test circuit power must be at least two watts for an accurate reading on this instrument.

Select a plug-in coil with a frequency range that covers the frequency to be measured. Occasionally it may be necessary to measure a frequency that falls in an unknown range. In this case, measurement is attempted with each coil until the right one is located.

In addition to the exposed coil on the wavemeter, there are usually exposed circuits in the equipment being tested. Dangerously high voltages are probably present in these circuits, so extreme care should be exercised to avoid touching these exposed circuits.

Hold the wavemeter pickup coil near the test circuit, and adjust the control until the light begins to glow. Loosen the coupling by moving the wavemeter away from the test circuit. Contrive to adjust the control and loosen the coupling until the light is a faint, barely visible glow. The calibration scale that matches the coil being used will now indicate the frequency of the test circuit.

Using Reaction Wavemeters

Circuits with total power of less than two watts cannot be tested with the absorption wavemeter. The absorbed power in this case is insufficient to light the lamp. The frequency of low-power circuits can be measured by observing the reaction on an oscillator. The easiest method of doing this is to use the *reaction wavemeter*. It is an oscillator with a meter as an indicator. It requires metering in the circuit of the oscillator, amplifier, or antenna.

Select the plug-in coil that has a range covering the test circuit frequency and attach it to the wavemeter. Move the wavemeter coil near the test circuit so the coil will be in the electromagnetic field. Maximum energy pickup is obtained near the ends of the coils in the tank circuit. Slowly turn the dial until a reaction is obtained on the meter indicator. Then slowly loosen the coupling by moving the wavemeter away from the test circuit. Continue to move the dial and loosen the coupling until maximum reaction is obtained at the point of minimum coupling. The frequency of the test circuit can now be read from the calibrated scale on the dial control. Make sure that the scale being read corresponds to the plug-in coil being used.

Another technique in using the reaction wavemeter involves the use of a radio receiver. This procedure is a little more complicated, and the human ear becomes the indicating device. Tune the radio receiver to zero beat with the frequency of the test circuit. Move the wavemeter near the test circuit, and tune the dial until the audio beat note is heard in the receiver. Loosen the coupling between the test circuit and the wavemeter. Continue to loosen the coupling and tune the dial until a distinct zero beat is obtained with minimum coupling. The test circuit frequency can now be read from the calibrated scale on the dial drive control. Be sure to read the scale that corresponds to the coil being used.

Using the Grid-Dip Meter

The *grid-dip meter* is one of the few test instruments for use with a de-energized circuit. It can be used to measure and adjust the frequency of a tank circuit without the necessity of applying power to the test circuit.

A switch applies plate potential to the grid-dip oscillator. A plug-in coil couples the grid-dip oscillator to the tank circuit being tested. The frequency of the meter is changed slowly until a dip occurs on the indicating meter. The meter is in the grid of the oscillator, and a dip indicates that the test circuit is absorbing energy from the meter oscillator. The meter oscillator is now tuned to the resonant frequency of the tank circuit under test. This common frequency is indicated by the calibrated dial.

The grid-dip meter may also be used as a detector for measuring frequencies of energized RF circuits. Headphones are connected into the grid circuit of the meter oscillator. The signal from the test circuit is coupled through the probe to the meter oscillator. As the dial is tuned, an audible note can be detected in the headphones each time the meter oscillator frequency approaches the fundamental or any harmonic frequency of the test circuit. The lowest frequency that produces a zero beat is the fundamental frequency of the circuit being tested. The frequency is indicated on the calibrated *dial.*

When the plate potential is removed from the oscillator, the grid-dip meter may be used as an absorption wavemeter. Similar to a conventional wavemeter, the meter reading will be at the maximum when the meter is close coupled and tuned to the frequency of an RF source.

Using Radio Receivers

A calibrated radio receiver is an excellent frequency meter under certain circumstances. Well-designed calibrated receivers can be used to measure frequencies to an accuracy of 0.04%. Factory-calibrated receivers may exceed this accuracy figure.

The carrier-actuated tuning indicator in the receiver (if present) should be turned off. Tune the beat-frequency oscillator (BFO) to produce a beat-frequency note while applying a known frequency to the receiver. Leave the BFO in this position, and tune the receiver to obtain a zero beat with the signal being measured. The calibrated dial of the receiver now indicates the frequency of the test signal.

When the transmitter being tested is too close to the receiver, the transmitter may block the receiver. This can be averted in most cases by disconnecting the receiver antenna. If the receiver is still blocked, turn off the power amplifier in the transmitter.

A slight deviation on this technique provides a quick and easy way to tune a transmitter to a desired frequency. Obtain a zero beat from a known frequency; then tune the transmitter until it produces a zero beat in the receiver.

Using Heterodyne Frequency Meters

Extreme accuracy may be obtained in measuring frequency with a *heterodyne frequency meter*. The tuning element is usually equipped with a vernier dial that permits very accurate settings. In most cases this meter has its own antenna, which eliminates the need for direct connections to the equipment being tested. Both local and remote signal frequencies can be measured. Figure 11–14 illustrates the coupling arrangement for making measurements.

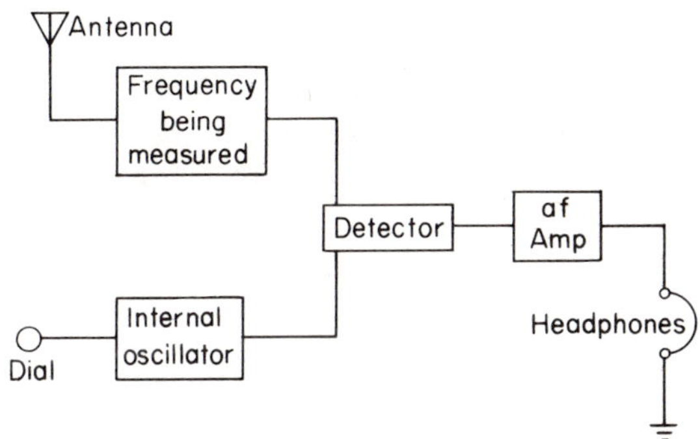

Figure 11–14 Measuring Frequency with Heterodyne Frequency Meter

Make sure the antenna is in a position to receive the signals to be measured. Set the dial to the approximate incoming frequency. Adjust the tuning dial until a beat note is heard in the headphones. This will occur when there is an audio difference between the incoming signal frequency and the frequency of the internal oscillator. Keep turning the dial until the note is reduced to zero. At this time, either the fundamental frequency or the harmonic frequency of the internal oscillator is the same as the frequency being measured. If the original dial setting was reasonably close, it is pretty definitely the fundamental frequency.

Some heterodyne frequency meters have indicators that register in frequency; most do not. Usually the setting of the vernier dial must be interpreted by reference to a calibration book. It is important that this book contain the same serial number as the frequency meter. Figure 11–15 illustrates the appearance of the vernier dial.

The dial in this instrument is calibrated in hundreds, tenths, and

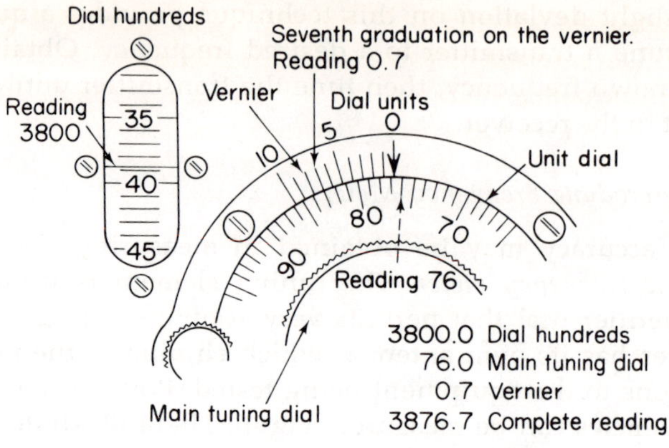

Figure 11–15 Vernier Dial of Heterodyne Frequency Meter

units. Consult the hundreds dial and record the highest full number below the actual indication. Record the highest full unit using the zero on the tenths dial as a marker. Record the tenths at the point where there is exact alignment between a tenths vernier and a unit marker. Total dial reading is the sum of these three numbers. In Figure 11–15, the readings are: 3800, 76, and 0.7 for a total of 3876.7. The calibration book lists a number very close to this with a *corresponding frequency.*

Calibrating a Heterodyne Frequency Meter

The National Bureau of Standards broadcast stations provide extremely accurate frequencies for calibrating a heterodyne frequency meter. Figure 11–16 illustrates the equipment arrangement. A radio receiver with an attenuator in the antenna circuit is used to intercept the standard frequency signal. The example shows a receiver tuned to the 10 MHz transmission frequency of WWV. The dial setting on the receiver should

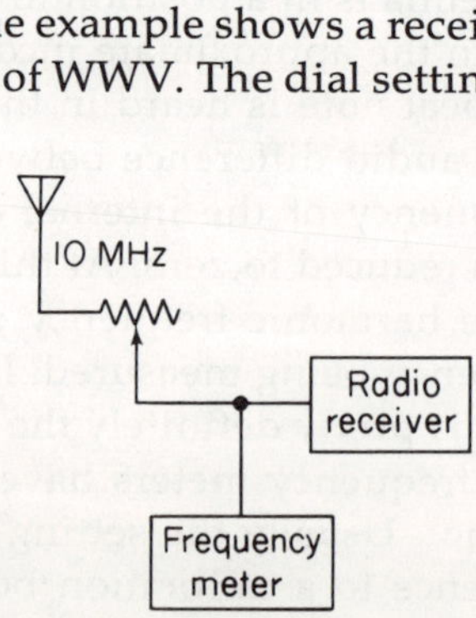

Figure 11–16 Calibrating a Heterodyne Frequency Meter

be very accurate. Accuracy of the setting is improved by using a receiver with an internal beat-frequency oscillator.

The frequency meter is connected as indicated, and the crystal calibrator switch is turned on. If the crystal is exactly on frequency, it will zero beat with the WWV signal. A low growl in the headphones indicates a difference of a few hundred hertz. The trimmer capacitors are tuned as necessary to obtain a good zero beat.

It may be necessary to use the attenuator in the receiver antenna line to adjust the amplitude of the received signal. A little practice will help you to recognize the best amplitude for heterodyning with the crystal frequency.

Using Frequency Counters

A *frequency counter* provides a direct measurement of input frequencies. A time base is selected to determine how long the sample count will continue. All input pulses (or cycles) during this period are counted. A functional diagram is illustrated in Figure 11–17.

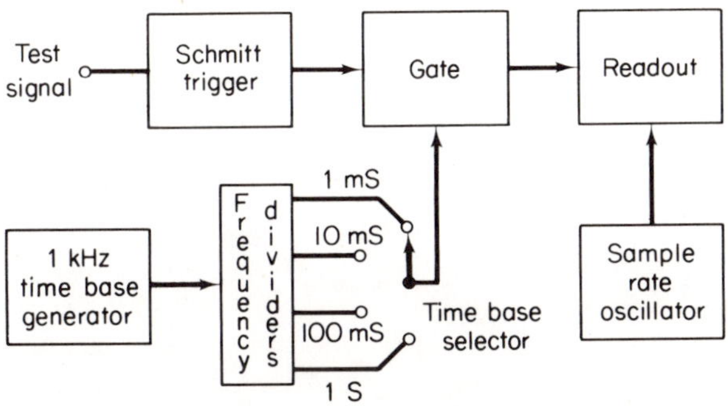

Figure 11–17 Measuring Frequency with a Counter

The input test signal (the frequency to be measured) is coupled into a Schmitt trigger. The output of the Schmitt trigger is uniform pulses at the same frequency as the test signal. These standard signals are gated to the counter for readout during specified timing periods.

The timing period that opens and closes the gate is determined by the position of the time-base selector switch. This switch generally has four positions, which may be labeled as either time or frequency.

The time-base generator is a 1 kHz oscillator; one cycle of this frequency is 1 ms in duration. Frequency dividers reduce the frequency and consequently increase the time base. 1 Hz corresponds to a time of 1 s, 100 Hz is equivalent to 10 ms, etc.

The selected time base determines the time the gate will be held open. During this timing period, each input cycle produces a pulse, which passes through the gate for counting. The number of pulses in a specified timing period is then converted to a frequency readout for display. The display remains until the start of the next timing period.

The recurrence rate of sample periods (gating periods) is determined by the sample-rate oscillator. This oscillator also erases the display at the beginning of each sample period.

chapter twelve

MEASURING WAVEFORMS

This chapter deals with the measurement and analysis of waveforms. Emphasis is placed on the oscilloscope because it is the most commonly used instrument. Since frequency, phase, and time measurements in relation to the oscilloscope were treated in Chapter 11, these subjects will not be repeated here.

Nearly all waveform measurements make use of a CRT display for viewing the waveform. The oscilloscope is the most popular and versatile among the CRT instruments. Stroboscopes, curve tracers, and spectrum analyzers are indispensable in certain areas, but they are all specialized instruments and therefore have more limited use than the oscilloscope.

Since a signal must be present before it can be viewed, the oscilloscope alone is not sufficient. When there is no natural signal it is necessary that an artificial signal be generated and used. This situation often arises when checking nonoperational equipment, or individual circuits on the test bench. It is then necessary to inject a signal in order to produce an observable waveform. This makes signal generators an indispensable part of instrument inventory.

INTERPRETING DISPLAYS

Once a visual waveform has been obtained, the next crucial step is proper interpretation of the patterns that are under observation. Distortion in a waveform is not always a bad thing. In fact, in a great many cases, distortion is essential to proper operation. Before it is possible to analyze a waveform, the characteristics of the signal on one axis must be known, and a sample of the expected signal (on the other axis) must be firmly fixed in your mind.

Horizontal Axis

It is generally common practice to use a sawtooth or sine wave of known frequency on the horizontal axis. This signal must then be synchronized with the fundamental or a submultiple of the signal under observation.

The sawtooth is most popular because it provides horizontal deflection that is linearly proportional to time. This generates a true plot of the test signal versus time. It must be kept in mind that all instruments, even advanced versions of the oscilloscope, have definite limitations. These limitations vary from one oscilloscope to another, and they affect the plot of the test signal. For this reason a perfectly good waveshape may appear different on two different oscilloscopes.

Oscilloscope Limitations

The oscilloscope is limited by phase distortion, input impedance, and maximum permissible input signal level. These limitations determine the frequency response and the sensitivity of both the horizontal and vertical amplifiers. The degree of limitation is determined by the design considerations, which force a compromise between production cost and instrument utility. The operator must be alert to the fact that these limitations may alter the shape of an observed waveform. This oscilloscope effect must be taken into account when analyzing the quality of a waveform.

Vertical Axis

Except in special applications, the signal to be observed is applied to the vertical axis of the oscilloscope. This will be referred to as the test signal, rather than the unknown signal. The operator must know many things about the test signal in order to make a meaningful analysis. He should know the general appearance of the ideal waveform at that test point, as well as the proper frequency, amplitude, time duration, and recurrence rate. In other words, a knowledge of the proper signal is necessary in order to evaluate the validity of the signal under observation.

The maintenance manual for the electronic equipment under test is the best source of information on proper waveforms. Of course, such manuals are not available for circuits under design. However, the designer knows what he expects from his circuits and should make this information available to the person performing the tests. The basic rule still applies: a knowledge of the proper signal is necessary in order to evaluate the test signal.

Proper Waveforms versus Distortion

Distortion is generally a loose term. Sometimes it is so loose that it loses all meaning. Under some circumstances, a perfectly formed square wave can represent intolerable distortion. *Distortion,* for this discussion, means a *deviation* from the normal, proper signal. The proper output of an electronic circuit has no distortion regardless of the waveshape. A distorted waveform from one circuit may well be a perfectly valid waveshape from another circuit.

The Class A amplifier conducts all the time. This means that a properly operating Class A amplifier should faithfully reproduce its input waveform. The only change between input and output should be signal amplitude and, in some cases, phase inversion. All other differences between input and output are distortions caused by faults in the amplifier circuit.

The Class B amplifier conducts only 50% of the time. Therefore, it is expected that the output will be considerably different from the input. If there is ever a case where a Class B output is the same as the input, then that stage is producing an intolerable degree of distortion. This amplifier is biased at cutoff; any input with both positive and negative components will have part of the input waveform clipped off and eliminated. Figure 12–1 illustrates the Class B effect on a sine wave input. The positive portion of the input overcomes the bias and causes the amplifier to conduct. This portion of the input is amplified and inverted through the amplifier. The negative portion of the input increases the amplifier bias, which is already at the cutoff level. The amplifier does not conduct during this time and produces a zero output.

Class AB amplifiers conduct more than 50 but less than 100% of the time. This is a common circuit for waveshaping by both cutoff and saturation limiting. And of course it can be used for either type of limiting. A sample input and resultant output are shown in Figure 12–2. As the input swings positive, the amplifier current increases and accurately reproduces the inversion of the input. A point is reached where the increasing amplitude causes no further change in amplifier current. The steady current squares off the peak of the output because the amplifier is saturated. When the input passes the peak and starts to drop, it will control the amplifier current again as soon as it drops below the saturation level. Input is again faithfully reproduced until the amplifier is driven into cutoff. The cessation of current, while in cutoff, squares off the positive peaks of the output.

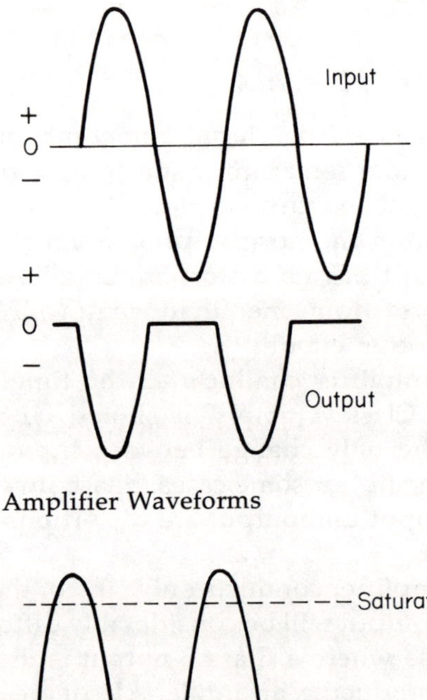

Figure 12–1 Class B Amplifier Waveforms

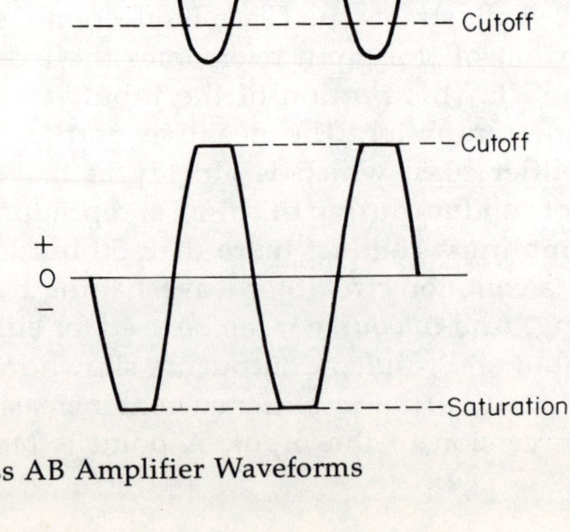

Figure 12–2 Class AB Amplifier Waveforms

Causes of Distortion

Any waveform except a pure sine wave is composed of signals of more than one frequency. The fundamental sine wave is a simple waveform; all others are complex waveforms. If a circuit can pass the fundamental frequency of a simple wave, the only distortion it can cause is in the amplitude. Since the complex waveform has a fundamental and one or more harmonics, it can be distorted in both amplitude and shape. If the frequency response of the circuit is insufficient to handle all harmonics, the harmonic or phase relationship will be altered with a resultant change in waveshape.

A complex waveform may suffer phase distortion when passing through an RC coupling. The capacitive reactance varies inversely with the frequency and tends to affect the phase of the higher harmonics.

Stray coupling between circuits is a frequent cause of distortion. For this reason, leads and cables should be kept as short as possible and well isolated from one another. When using the oscilloscope, these safety measures should be observed: short leads, proper probes, good grounds, and shielded cables.

Impedance mismatch between two circuits will cause distortion. This is obvious in an audio circuit where the impedance output of the final audio amplifier fails to match the impedance of the speaker. The resultant squeaks and squeals are evidence of the distortion of the audio signal.

Incorrectly tuned circuits, especially resonant circuits, will cause signal distortion. This will result in excessive attenuation of some harmonics and overemphasis of others. Evidence of distortion appears in a radio receiver when the tuning controls are not properly set. In this case, the distortion overattenuates one of the sidebands.

Feedback circuits are frequently used to broaden the frequency response of amplifiers and to increase the stability. These same circuits cause a great deal of distortion when they are not functioning properly.

Some other causes of distortion in marginally operational circuits are:

1. Over- or underdamping.
2. Slightly incorrect time constants.
3. Aging circuit components.
4. Incorrect circuit voltages.

When observing a distorted signal, the operator must determine the source of the distortion. It could be in the signal source, the test circuit, adjacent equipment, or in the oscilloscope that is displaying the signal. Careful analysis is essential in order to avoid snap decisions.

Sine Wave Distortion

The waveshapes in Figure 12–3 were taken from the output of Class A amplifiers. The input in each case was a pure sine wave.

The flattening effects on the negative peaks in Figure 12–3(a) can be caused by a number of things, such as insufficient bias, excessive load resistance, and leaky coupling capacitors.

The opposite effect is shown in Figure 12–3(b). Here the distortion has altered the shape of the positive peaks. The most probable causes of distortion are excessive bias and insufficient load resistance.

Both peaks in Figure 12–3(c) are distorted. This indicates an overloaded amplifier. The input signal is driving the amplifier to both cutoff and saturation.

In Figure 12–3(d) and (e) the positive peaks have been distorted. These are effects from output transformers. This type of distortion indicates that the signal frequency is too near the low-frequency end of the transformer frequency response curve.

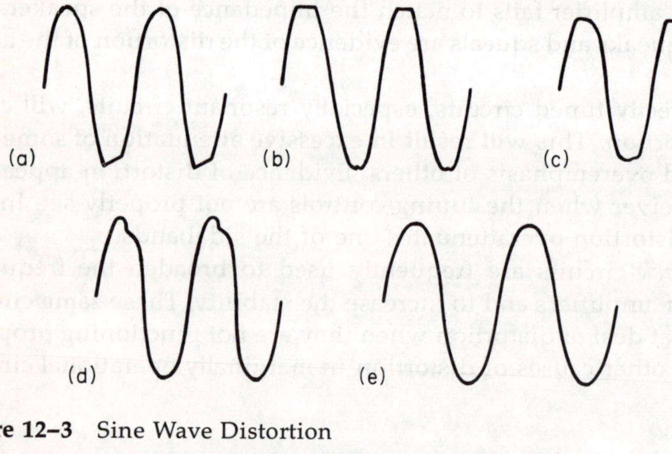

Figure 12–3 Sine Wave Distortion

Figure 12–4 illustrates a distorted sine wave after rectification by a bridge rectifier. This rectified output is taken before filtering, and the distortion indicates a faulty rectifier. This could be caused by faulty location of the center tap on the high transformer secondary. A more likely cause is dissimilar characteristics of the rectifying elements or a defect in one of the rectifiers.

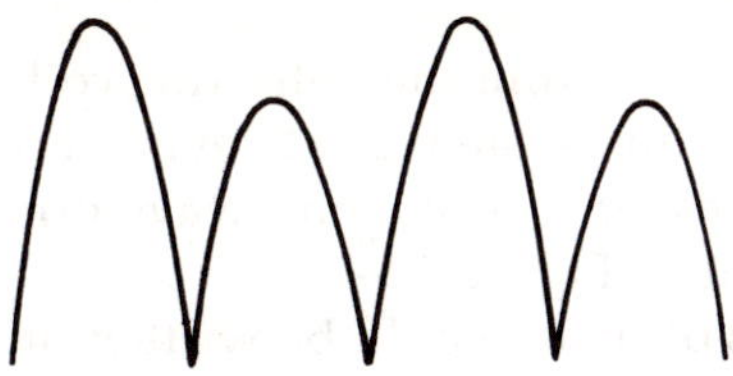

Figure 12–4 Rectifier Distortion

Square-Wave Distortion

Square waves are composed of a fundamental frequency and an infinite number of odd harmonics. Any serious distortion of the 3rd, 5th, or 7th harmonic will change the waveshape in a very definite manner. Serious impediment to the higher harmonics can also be detected. These characteristics make the square wave an excellent test signal for checking frequency response in many types of amplifier circuits.

Phase distortion is especially injurious to impulse sounds, that is, sounds of a transient nature. This type of distortion is readily revealed when the amplifier is tested with square-wave signals. Square-wave test analysis also provides information regarding damping and general amplifier characteristics. Figure 12–5 shows a square-wave input and outputs with several types of distortion. The cause of the distortion is indicated under each output.

Phase response is extremely important in television amplifiers. The use of square-wave inputs with the oscilloscope is a widely recognized method of checking these circuits.

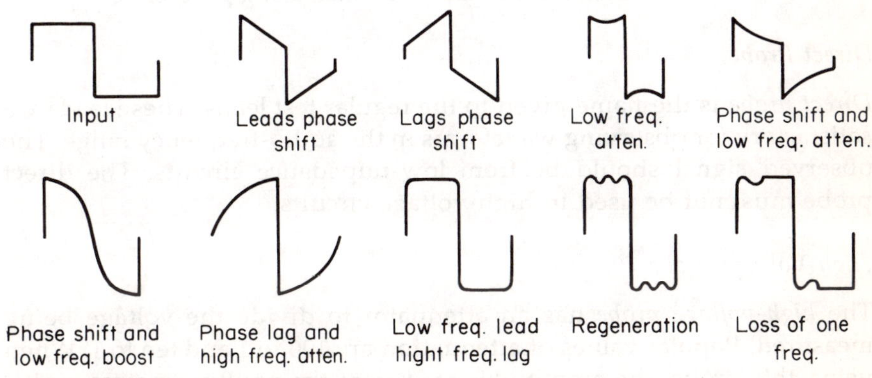

Figure 12–5 Amplifier Distortion of Square Waves

Rectangular Waveform Distortion

Figure 12–6 is a waveform from the deflection coils and damper circuits in the horizontal deflection system of a television receiver. This is nearly an ideal voltage waveshape. The same waveform with various types of distortion is shown in Figure 12–7.

Several volumes of books could be written on the analysis of sample waveforms, and they would still be only samples. Such works have their place, but a busy shop is not the place. The final analysis rests with the technician or engineer who is making the tests. He must know all he can about the appearance of an acceptable signal in each case and be alert for any deviation between the signal observed and the correct signal. The foregoing examples were intended to emphasize the importance of this procedure.

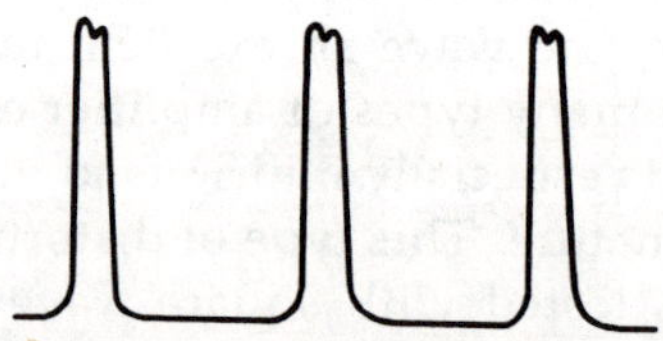

Figure 12–6 Horizontal Deflection Waveform

OSCILLOSCOPE PROBES

The oscilloscope is intended to handle a wide range of frequencies, amplitudes, and waveforms. There are a variety of probes for making inputs compatible with oscilloscope circuits. Four general types of probes, alone or in combination, can handle most measuring problems.

Direct Probe

Direct probe is the name given to the regular test leads. These leads are satisfactory for observing waveforms in the audio-frequency range. The observed signal should be from low-impedance circuits. The direct probe must not be used in high-voltage circuits.

High-Voltage Probe

The *high-voltage probe* has an attenuator to divide the voltage being measured. Popular values of attenuation are 100 to 1 and ten to 1. When using this probe, the amplitudes read from the oscilloscope are multiplied by the attenuation factor. The high-voltage probe should be used in all measurements involving amplitudes in excess of 600 V.

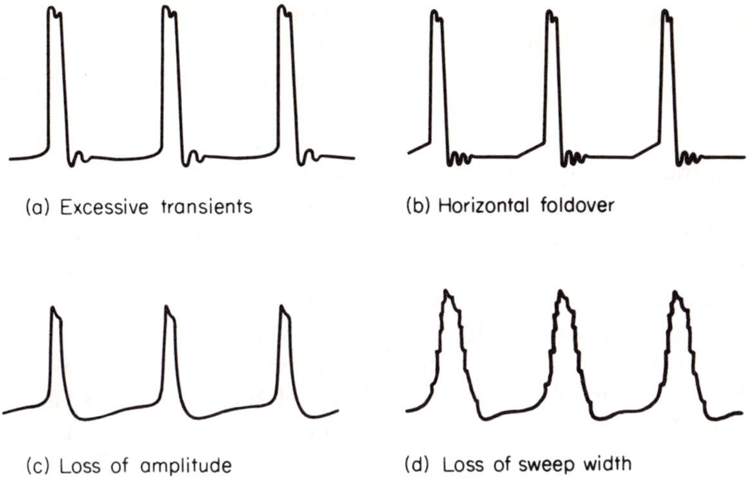

Figure 12–7 Distorted Deflection Waveforms

Detector Probe

The *detector probe* is used in radio-frequency circuits containing modulated waveforms. This probe demodulates the waveform before it is applied to the oscilloscope.

Low-Capacitance Probe

The *low-capacitance probe* enables the observation of high-frequency complex waveforms. This probe has a high impedance as well as a low capacitance.

MEASURING VOLTAGE

In many cases, such as those encountered in high-speed transients and signals with harmonics, the oscilloscope is the only accurate instrument for measuring voltages. The accuracy of a measurement depends upon the same two factors that affect accuracy when measuring voltage with other instruments: (1) calibration of the instrument and (2) the degree of precision in taking the reading.

Calibrating the Oscilloscope

Some oscilloscopes have internal adjustments for calibrating; others have variable attenuator controls. In either case the oscilloscope must be calibrated for a deflection factor of a specific number of volts per centimeter. The lighted graticules conveniently divide the viewing area into centimeters. When the ratio of volts per centimeter is known, it is a simple matter to convert centimeters of vertical deflection directly to volts.

This calibration must be accomplished before using the oscilloscope to measure voltage. Without a knowledge of the deflection factor, the voltage amplitude cannot be determined.

After calibration, the type of probe being used must be taken into consideration. Attenuation probes decrease the amplitude of the applied signal by a definite ratio. The same figure must be applied to the oscilloscope reading as a multiplication factor. Suppose that an oscilloscope has a deflection factor of 10 V/cm and the vertical deflection is 5 cm. If a direct probe is being used, the applied voltage is 10 ×5 or 50 V. With the times 10 probe, this same deflection would be read as 10 × 5 × 10 or 500 V. With the times 100 probe, it is 10 ×5 × 100 or 5000 V.

Direct Voltage

Measuring direct voltage is simply a matter of noting where the trace appears with no input, and then observing the number of centimeters that are displaced when the voltage being measured is applied. At times it may be more convenient to reduce the horizontal sweep to a single dot, and observe the vertical displacement of this dot as the direct voltage is applied to the input. In any case the number of centimeters vertically between the no-input position and the input position converts directly to volts of input direct voltage.

Now there is the problem of reading precision. Measurements seldom fall on a precise number of centimeters. Nearly always there is a deflection of x cm plus a fractional portion of another centimeter. This is a resolution problem that requires reading skills, and like all skills, reading skill improves with practice.

In addition to the graticule grid, most oscilloscopes have finer division marks along the center graticule both horizontally and vertically. This is illustrated in Figure 12–8. The oscilloscope screen represented here has eight divisions vertically and ten divisions horizontally. Each division is 1 cm. Along the center graticules, small markers divide each centimeter into five equal parts. The sine wave displayed on this screen has an amplitude of 2.2 cm. With a little practice, reading accuracy should be within 0.1 cm. This possible error can represent several volts when measuring a large potential, but it is a small percentage of the total amplitude. With a small voltage, the percentage of error is just as great even though the number of volts is smaller.

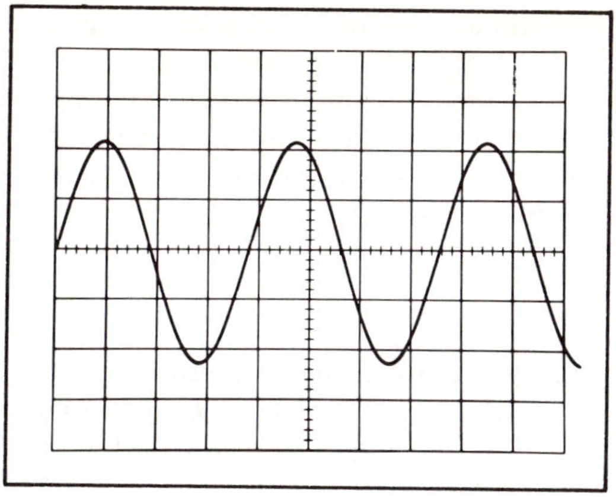

Figure 12–8 Oscilloscope Graticules

Sine Waves

As shown in Figure 12–8, the oscilloscope presents the picture of the complete waveshape. The operator has a choice of reading the voltage amplitude in terms of peak value, peak-to-peak value, or instantaneous value. Since most alternating voltage values are given as effective values, it is frequently necessary to calculate the effective value from a peak-value reading. This is a simple conversion since the effective value is equal to 0.707 × peak value. For example, suppose that the oscilloscope in Figure 12–8 is calibrated for a deflection factor of 5 V/cm. Vertical deflection is 2.2 cm. for a peak value of 5 ×2.2 or 11 V. The effective value is 0.707 × 11 or 7.777 V.

Composite Signals

Unless the sine wave being measured is taken directly from a power supply, chances are that it is composed of both direct and alternating voltages. For instance, a waveshape from an audio amplifier is in reality the direct operating voltage that is changing amplitude at audio rate. However, this change in amplitude has no effect on the accuracy of measurement of the sine wave.

On the oscilloscope, the direct voltage causes a vertical offset of the sweep. If it is negative, the sweep appears below the center line; if positive, the sweep appears above the center line. Using the vertical centering control to bring the sweep back to center in no way affects the sine wave amplitude. However, any time this recentering becomes necessary, the test signal does not have a zero reference; it is riding on a reference of some direct-voltage value.

Not only sine waves, but most waveforms observed are likely to be composite signals. If the direct-voltage component is of interest, it can be measured by counting the number of divisions that the sweep is displaced from the center line.

Complex Waveforms

With most complex waveforms, the value of interest is the total waveform—the peak-to-peak value. In most service manuals, values of waveforms such as square waves, pulses, and sawtooth waveshapes are listed as peak-to-peak values.

Radio-Frequency Signals

The frequency of RF signals is often too high for the oscilloscope to present an accurate display. When this occurs, the voltage amplitude can still be measured by using an appropriate RF probe on the oscilloscope input. A schematic of such a probe is shown in Figure 12–9. Such a probe rectifies the RF and produces a direct voltage only slightly less than the peak value of the RF signal. This direct voltage is applied to the oscilloscope and read in the same manner as any other direct voltage.

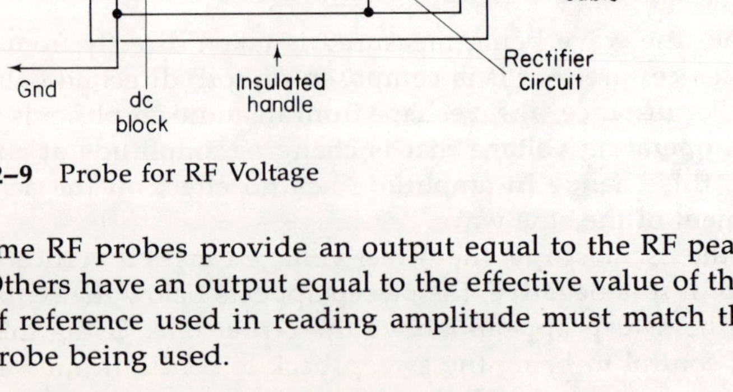

Figure 12–9 Probe for RF Voltage

Some RF probes provide an output equal to the RF peak-to-peak value. Others have an output equal to the effective value of the RF. The frame of reference used in reading amplitude must match the output of the probe being used.

Pulse Shape

The shape of a pulse is of primary concern when dealing with pulse testing. The pulse width, rise time, and fall time are as important as the amplitude of the pulse. The *pulse width* is a measure of the time duration of a pulse at the half-voltage point. *Rise time* is the time required for the leading edge of a pulse to rise from its 10% voltage point to its 90% voltage point. *Fall time* is the time required for the trailing edge of a pulse to decrease from the 90% voltage point to the 10% voltage point. These definitions are illustrated in Figure 12–10.

These pulse characteristics are especially critical in timing, gating, and triggering circuits. The expanded sweep feature of the oscilloscope is essential to this type of close observation.

MEASURING CURRENT

The oscilloscope uses voltage for deflection, and it has no provision for measuring current directly. However, this is a small problem when the relationships of Ohm's law for voltage, current, and resistance are considered.

The procedure requires that a resistor of known value be connected in series with the test circuit. A voltage measurement is then made across the resistor. From this point, it is a simple matter to convert the voltage reading to current. This procedure is illustrated in Figure 12–11.

With the resistor shown in Figure 12–11, each volt on the oscilloscope converts directly to 1 mA of current. If a one-ohm resistor is used, each volt is equivalent to one ampere. Assuming that the resistor wattage rating is high enough to withstand the current, the smaller the resistor value, the more accurate the measurement.

MEASURING SPECTRA

All signals, except pure sine waves, contain multiple-frequency elements of differing amplitudes. Pulse modulation does not consist of a single frequency that is turned on and off. It is a fundamental carrier wave modulated by the waveform of the pulse trigger signal. This creates a spectrum of frequencies that requires a special instrument for adequately viewing and measuring. The *spectrum analyzer* is a CRT instrument especially designed for this purpose.

Although less versatile than the oscilloscope, the spectrum analyzer is a useful instrument with many practical applications. It is an

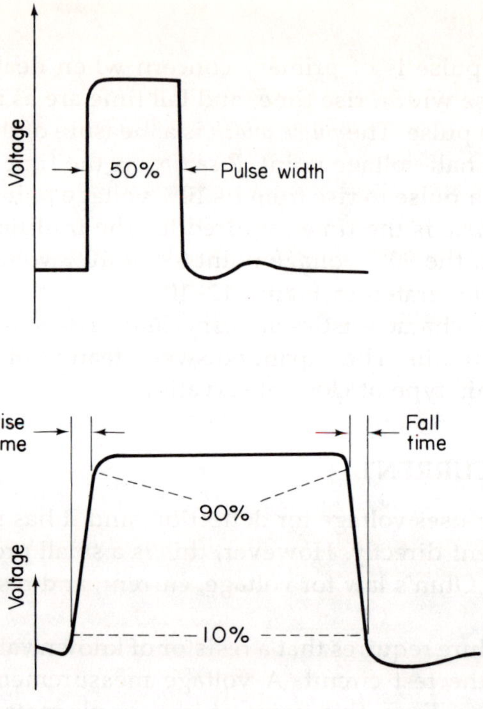

Figure 12–10 Pulse Characteristics

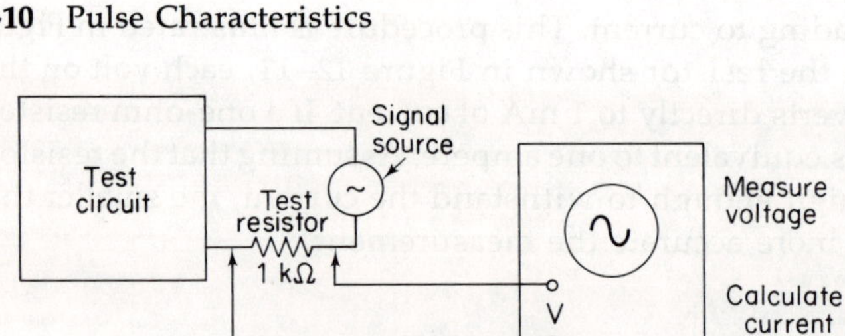

Figure 12–11 Measuring Current

indicator for frequency standards; it can be used to measure pulse width, frequency, noise spectra, PRF, standing-wave ratios, and modulation. It is used to check RF leakage, frequency difference between signals, frequency drift of oscillators, parasitic oscillations, and proper operation of transmitters and signal generators. It provides a panoramic CRT presentation for observing such items as frequency relationships, frequency pulling, pulse shape, and sidebands. In short, the spectrum analyzer can be a very valuable instrument in many lines of electronic endeavor.

Measuring Frequency

When a signal is modulated by another signal of a different frequency, it produces a fundamental frequency and several sideband frequencies. Collectively these frequencies are called a *spectrum*. The RF input to the spectrum analyzer is mixed with the frequency of a local oscillator to produce an intermediate frequency. The local oscillator sweeps through a band of frequencies. During each sweep, each frequency of the incoming signal momentarily produces the proper intermediate frequency. The result is a panoramic display of the entire spectrum with each signal in a position corresponding to its frequency.

Controls on the analyzer enable precise adjustments of the spectrum width and center frequencies. The horizontal display on the CRT is centered about the selected fundamental frequency, low frequencies on the left and high frequencies on the right. Most spectrum analyzers use a built-in wavemeter that places a marker on the CRT. The frequency of any signal on the screen can be determined by moving the marker to that signal and reading the frequency of the wavemeter. Figure 12–12 is a sample of a spectrum display of the output from a pulse-modulated oscillator.

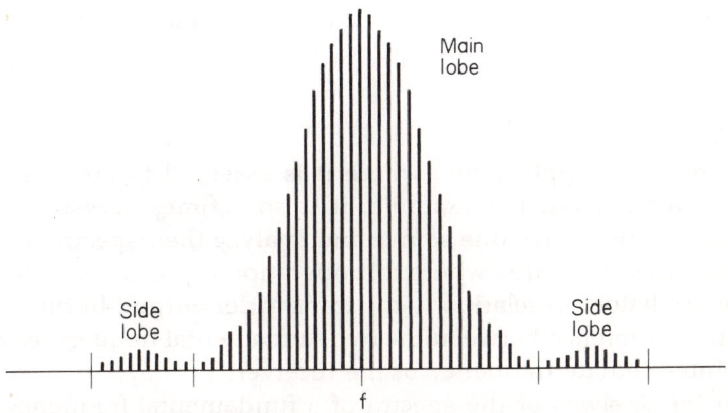

Figure 12–12 Pulse-Modulated Spectrum

The *f* on the diagram marks the center frequency of the main lobe as well as the fundamental frequency of the oscillator. Remember that frequency is plotted on the horizontal axis. The difference in frequency between any two adjacent lines is equal to the fundamental frequency of the modulator.

This drawing shows only two side lobes, but there may be two or more lobes on each side of the main lobe. These side lobes occur because the modulating pulse can never be a perfectly rectangular pulse.

The harmonic frequencies in the imperfect modulating pulse cause both amplitude and frequency modulation of the oscillator frequency. Excessive amplitude of a side lobe or an excessive number of side lobes indicate trouble with the modulating signal source.

Measuring Amplitude

The pattern observed on the spectrum analyzer, and represented in Figure 12–12, represents energy distributed among several different frequencies. The frequencies are plotted on the horizontal axis, and the amplitude of each frequency is plotted on the vertical axis. The amplitude of any given line represents the proportion of energy present at that frequency.

Amplitude measurement procedures and results are largely dependent upon the design of the instrument in use. There are three common methods of detection used in spectrum analyzers, but normally only one of these methods is used in any given instrument. A knowledge of the method of detection is vital to an interpretation of the relative amplitudes displayed on the CRT. These three methods of detection are: (1) linear, (2) logarithmic, and (3) square-law.

With square-law detection, the amplitude represents the power of the signal at that frequency. With either linear or logarithmic detection, the amplitude represents voltage.

Selecting Frequency Bands

There are many applications where it is essential to observe a broad band of frequencies. For example, it is sometimes necessary to view signals from more than one source and analyze their spectra in relation to one another. One area where this operation is prevalent is the output of local oscillators in relation to the transmitter output. In this case, the frequency difference between the two fundamental frequencies is equal to the intermediate frequency of the receiver.

In the analysis of the spectra of a fundamental frequency signal, the best results are obtained by selecting a narrow frequency band. This type of observation provides detailed information not available with the broad-band picture. For this reason, accurate display of a high-resolution narrow-band spectrum should be a prerequisite in the selection of a spectrum analyzer.

Interpreting the Display

The width of the center lobe should be only slightly greater than the bandwidth of the receiving equipment. If the bandwidth is too wide, transmitting power is being wasted. Energy used by signals outside the receiver band does nothing for the usable received signal.

The minimum amplitude points adjacent to the main lobe should be deep, approaching zero in amplitude. If not, this indicates frequency modulation of the oscillator.

The main lobe should have only one distinct peak. Two or more peaks indicate that the oscillator is operating in more than one mode or that the frequency is being pulled by mismatched elements in the transmission line. Other faults could cause these indications, but those mentioned here are the most likely.

chapter thirteen

MEASURING ELECTROMAGNETIC WAVES

The subjects encompassed by this chapter include transmission lines, transmitters, antennas, waves in space, and, to some degree, reception. This is a broad area, and therefore we shall concentrate on the measurements of primary importance. These are: modulation, standing waves, and field strength. That is not to say that other quantities are of no interest, but rather that many other quantities can be obtained as byproducts of these primary measurements. Wavelength, impedance, frequency, and power loss can be obtained from standing-wave measurements. Antenna gain, directivity, radiation pattern, and relative power can be obtained from field-strength measurements.

The instruments used in these measurements have been treated in previous chapters. See Chapter 5 for signal generators; most of the other instruments appear in Chapter 6.

MEASURING AMPLITUDE MODULATION

In *amplitude modulation*, the ideal situation is 100% modulation of the carrier. When a carrier wave is modulated by a sine wave, 100% modulation is obtained when the modulating signal average power is half the average power of the carrier wave. This causes the average power of the modulated carrier to be 1.5 times the average power of the carrier before modulation. This increase in power is divided equally between the upper and lower sidebands. During peaks of a 100% modulated wave, the amplitude of the carrier is twice that of the unmodulated carrier. The modulated-wave peak power is four times as great as the peak power of the unmodulated wave.

In voice modulation, only the highest peaks can be allowed to produce 100% modulation. Otherwise intelligence is lost, and intolerable distortion results. Voice peaks will cause 100% modulation when the modulation signal causes a 25% increase in the average power of the carrier. Therefore, the average power required for voice modulation is less than that required for modulation with a sine wave.

Antenna Current Method

The power increase due to modulation is indicated by the increase in *antenna current*. This increase in current can be measured, and the percentage of modulation can be calculated from the results. Figure 13–1 illustrates one method of measuring the antenna current.

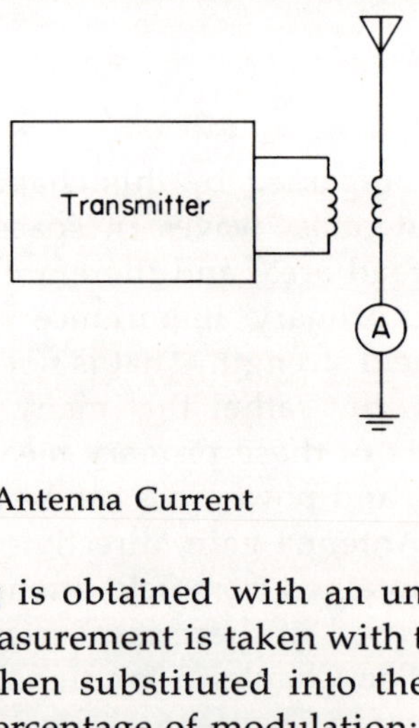

Figure 13–1 Measuring Antenna Current

A current reading is obtained with an unmodulated carrier (I_1), and another current measurement is taken with the modulation present (I_2). These values are then substituted into the following formula in order to calculate the percentage of modulation:

$$\text{Modulation Percentage} = 100\sqrt{2\left(\frac{I_2^2}{I_1^2} - 1\right)}$$

Take an example where unmodulated current is 5 A and the modulated current is 6 A, as shown here:

$$\text{Modulation Percentage} = 100\sqrt{2\left(\frac{6^2}{5^2} - 1\right)}$$
$$= 100\sqrt{2(0.44)}$$
$$= 100\sqrt{0.88}$$
$$= 100 \times 0.938$$
$$= 93.8\%$$

According to previous statements, 93.8% modulation should represent almost a 50% increase in power. That can be checked by assuming a value of antenna impedance and calculating the power before and after modulation. Assuming 50 ohms of impedance provides the following:

Before Modulation:

$$P = I^2 Z$$
$$= 5 \times 5 \times 50$$
$$= 1250 \text{ W}$$

After Modulation:

$$P = I^2 Z$$
$$= 6 \times 6 \times 50$$
$$= 1800 \text{ W}$$

Antenna Voltage Method

Modulation percentage may also be calculated from the *antenna voltage* measurements. The voltage may be measured using an RF probe and a peak reading electronic voltmeter as illustrated in Figure 13–2.

Two voltage readings must be taken: one without modulation (E_1) and one with modulation (E_2). Remember that these are peak values. The values are then substituted into the following equation:

$$\text{Modulation Percentage} = 100\,\frac{(E_2 - E_1)}{E_1}$$

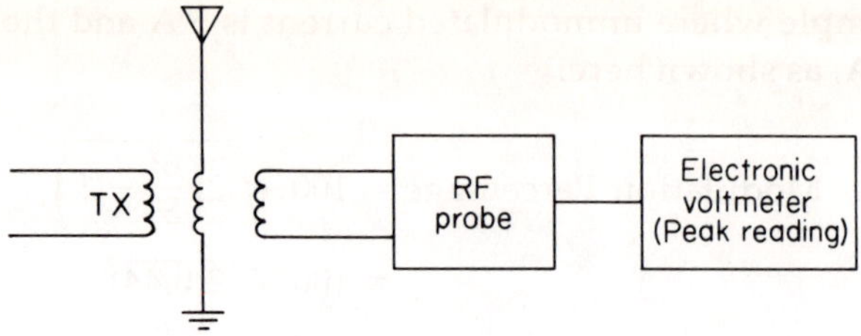

Figure 13–2 Measuring Antenna Voltage

For example, a 100% increase in peak voltage amplitude indicates 100% modulation as follows:

$$\text{Unmodulated Peak Voltage} = 50\text{V}$$

$$\text{Modulated Peak Voltage} = 100\text{ V}$$

$$\begin{aligned}\text{Modulation Percentage} &= 100\,\frac{(E_2 - E_1)}{E_1}\\ &= 100\,\frac{(100-50)}{50}\\ &= 100 \times 1\\ &= 100\%\end{aligned}$$

Measuring with Slide-Back Voltmeter

The voltage and current methods of modulation measurement have several inherent problems which contribute to inaccurate results. For instance, a complex modulating wave causes a continuous variation in the modulation percentage. Distortion can also cause a false reading and result in inaccurate calculations. Unequal positive and negative peaks in the modulated wave are also contributing factors.

Much more reliable results can be obtained by using the *slide-back electronic voltmeter* to measure the voltages. The meter circuit is illustrated in Figure 13–3.

The tuned circuit is coupled to the transmitter, and the resulting measurements of voltage are used to calculate the percentage of modulation. Two voltage readings are needed, and they are taken as follows:

1. Adjust R_1 for 0 on the voltmeter.

2. Adjust R_2 for 0 on the ammeter.

3. Adjust C_1 for slightly more than half-scale deflection on the ammeter.

4. Readjust R_1 for 0 on the ammeter.

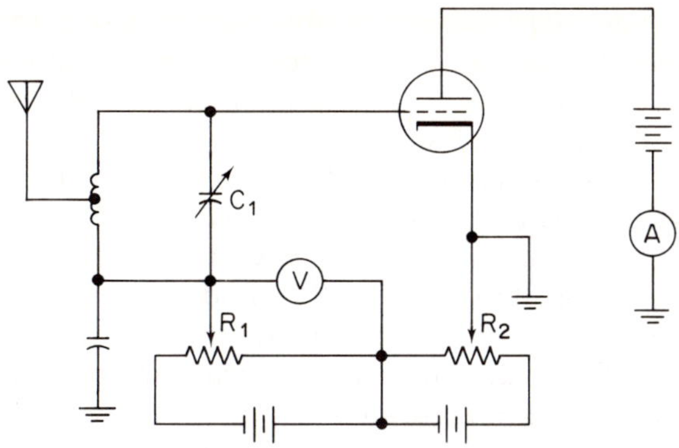

Figure 13–3 Slide-Back Voltmeter

The reading (E_1) on the voltmeter at this time is the peak voltage in the tuned circuit, and it is proportional to the peak value of the unmodulated carrier. Record this value of voltage, and without disturbing the tuned circuit, proceed as follows:

1. Modulate the carrier wave.
2. Readjust R_1 for 0 current.

The new reading (E_2) on the voltmeter is proportional to the peak value of the modulated carrier. Record this value of voltage, and calculate the modulation percentage with this equation:

$$\text{Modulation Percentage} = 100\,\frac{(E_2 - E_1)}{E_1}$$

Again, a 100% change in voltage ($E_2 = 2E_1$) represents 100% modulation. This is a well-established method of measuring amplitude modulation, and the results are reliable; they closely represent the actual modulation conditions.

Measuring with the Oscilloscope

The *oscilloscope* is widely used as an amplitude-modulation measuring instrument. It presents the waveforms for visual monitoring and is fairly accurate in measurements of modulation percentage. It also shows over- or undermodulation. On a five-in. scope, an error of about 10% can be expected, but this is close enough for most purposes.

The type of oscilloscope pattern observed is dependent on how the oscilloscope is connected. It may be connected for a wave-envelope pattern, a trapezoidal pattern, or a double-ellipse pattern.

The *wave-envelope pattern* is obtained by using the oscilloscope sawtooth sweep voltage on the horizontal plates and applying the RF signal to the vertical plates. This arrangement with the possible resulting pattern is illustrated in Figure 13–4.

In order to accomplish actual measurements, it is necessary to obtain a stationary pattern. This occurs only when the sweep frequency is the same as the modulation frequency or some submultiple thereof. At 100% modulation, the peak of the modulated carrier is twice the amplitude of the unmodulated carrier, and minimum amplitude reaches zero but does not remain there for a significant period of time. This waveform is illustrated in Figure 13–5.

The modulating signal, in this case, is a symmetrical sine wave, and with 100% modulation, amplitudes *a*, *b*, and *c* are all equal. With any symmetrical wave, the percentage of modulation is equal to either *a* or *b* divided by *c*. If the modulating wave is not symmetrical, *a* and *b* will be of different amplitudes. In this case, the larger of the two amplitudes is divided by *c* to obtain the percentage of modulation.

Trapezoidal patterns may be obtained using the modulating signal

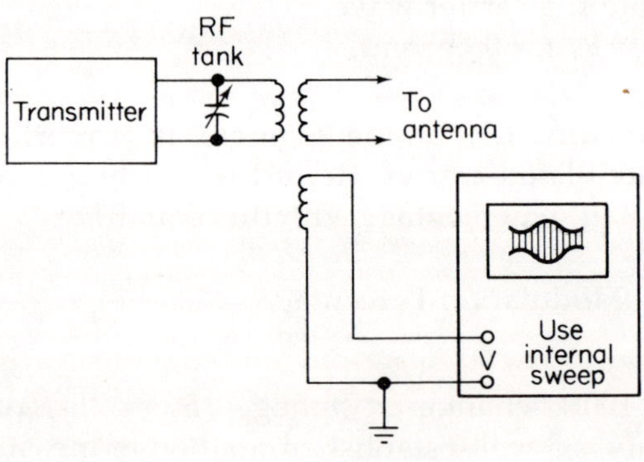

Figure 13–4 Obtaining Wave-Envelope Pattern

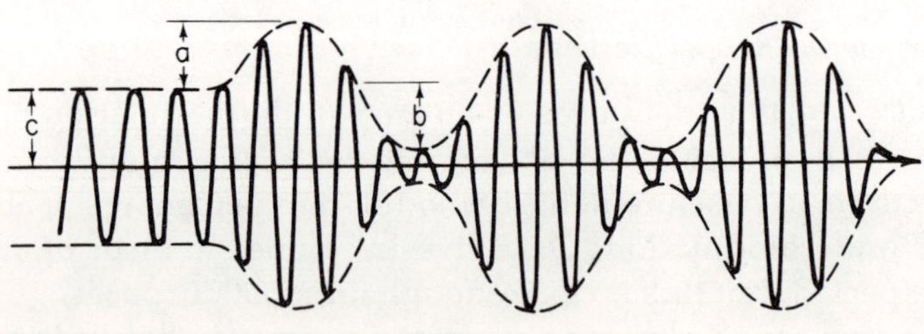

Figure 13–5 Carrier with 100% Modulation

on the horizontal deflection plates and the modulated carrier wave on the vertical plates. Scope connections and possible patterns are illustrated in Figure 13–6.

The oscilloscope pictures in Figure 13–6 show the unmodulated carrier and the carrier with different percentages of modulation. The amplitude of the unmodulated carrier (*a*) is a reference to show amplitude change due to the modulating signal. The narrow end (*b*) of the trapezoidal pattern represents the modulation troughs, and the wide end (*c*) represents modulation peaks. These amplitudes need not be reduced to voltage measurements; the number of divisions of deflection is sufficient for calculations. The percentage of modulation may be calculated as follows:

$$\text{Modulation Percentage} = 100\left(\frac{c-b}{c+b}\right)$$

With 50% modulation, the ratio between minimum and maximum amplitude is 1 to 3 regardless of the divisions used. When these values are substituted into the equation, the results are as follows:

$$\begin{aligned}\text{Modulation Percentage} &= 100\left(\frac{c-b}{c+b}\right)\\ &= 100\left(\frac{3-1}{3+1}\right)\\ &= 100 \times 0.5\\ &= 50\%\end{aligned}$$

In the case of over 100% modulation, the narrow end of the trapezoidal pattern flattens out, becomes elongated, and grows brighter than the remainder of the pattern. This indicates that the minimum amplitude is going all the way down to zero and is remaining there for a finite period of time. If maximum amplitude is checked against the unmodulated carrier amplitude under these conditions, it will be more than twice the amplitude of the unmodulated carrier.

A *double-ellipse pattern* for measuring amplitude modulation can be obtained by connecting the oscilloscope into the receiver circuit. Oscilloscope connections and possible patterns are illustrated in Figure 13–7.

An ellipse consisting of a single sharply defined line is produced by the unmodulated carrier. The shape is due to the phase difference between the signals applied to the horizontal and the vertical deflection plates. The phase shift is accomplished by the 50 kΩ resistor along with the input capacitor in the horizontal amplifier of the oscilloscope.

When the carrier is modulated, the pattern becomes a double ellipse in a ribbon pattern. Any degree of modulation less than 100% will produce a blank spot inside the center ellipse. As the percentage of modulation increases, the center blank space decreases, and it completely disappears at 100% modulation. Over 100% modulation causes a bright spot to appear in the center ellipse.

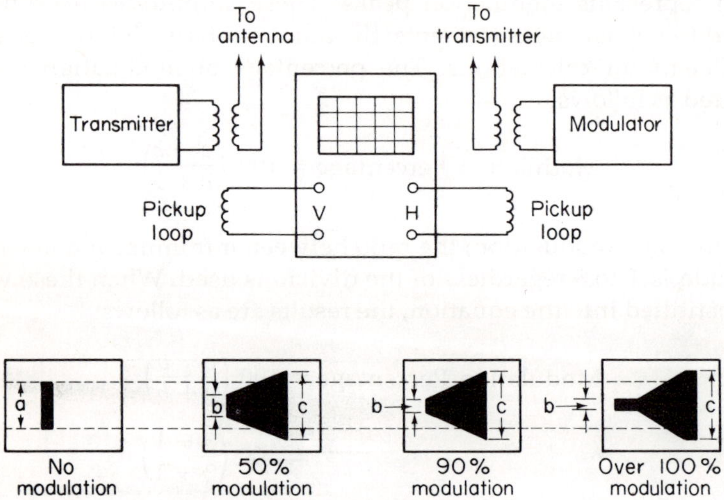

Figure 13–6 Obtaining Trapezoidal Patterns

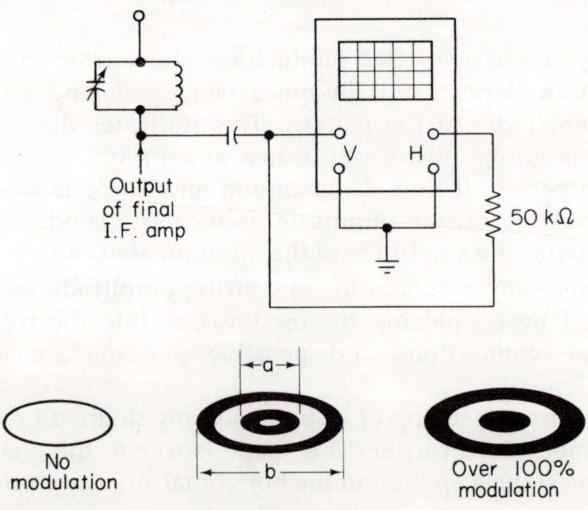

Figure 13–7 Obtaining Double-Ellipse Patterns

Distance a in the figure represents the minimum amplitude, and distance b represents the maximum amplitude. Modulation percentage may be calculated as follows:

$$\text{Modulation Percentage} = 100\left(\frac{b-a}{b+a}\right)$$

The ratio of a to b in the drawing is about 1 to 2, and this represents about 33% modulation.

MEASURING FREQUENCY MODULATION

In *frequency modulation*, the carrier amplitude remains constant, and the modulating signal causes the carrier frequency to change. Using the unmodulated carrier frequency as the rest frequency, a point of reference, the modulated signal frequency varies about the rest frequency. The *rate of deviation* is the number of total changes per second, and the rate is the same as the frequency of the modulating signal. The *degree of deviation* is the amount of frequency shift to either side of the rest frequency.

The degree of deviation determines the bandwidth of the transmitted signal and the number of sidebands it contains. In FM all the intelligence is in the sidebands, and it is desirable to retain as many sidebands as possible. The limit is a legal, rather than a physical, limitation. The Federal Communications Commission allows 75 kHz maximum deviation on the entertainment broadcast bands. This is narrowband FM transmission. Broad-band FM transmission is allowed a much wider frequency deviation, going up to several hundred MHz. At any rate, the degree of maximum deviation is predetermined. For lack of a better term, maximum allowable deviation is often referred to as 100% modulation.

Since it is desirable to transmit as many sidebands as possible, the amplitude of the modulating signal should be adjusted to provide a degree of deviation as near as possible to the maximum allowable deviation.

Measuring Frequency Deviation with a Radio Receiver

Probably the easiest method of measuring frequency deviation utilizes an FM receiver with a beat-frequency oscillator as the measuring instrument. The only coupling necessary is through the air waves between transmitting and receiving antennas as shown in Figure 13–8.

In this test, the modulating signal is replaced by a variable direct voltage. Since the amplitude of the modulating signal determines the

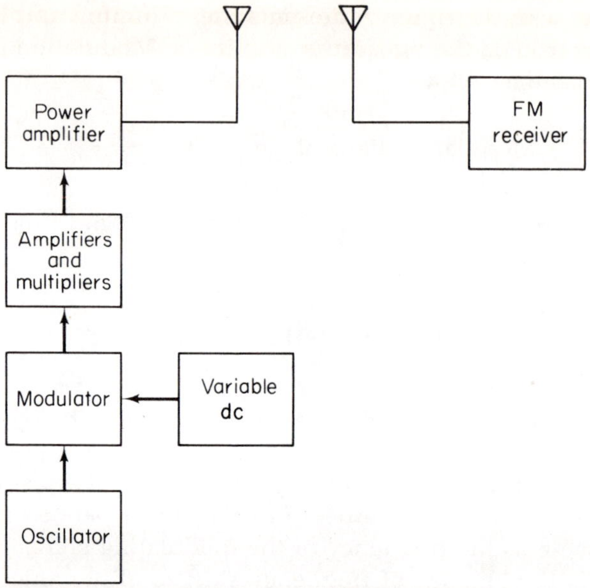

Figure 13–8 Measuring Frequency Deviation

degree of deviation, the frequency is of no importance. The FM receiver is tuned to a frequency corresponding to either the minimum or maximum allowable excursion. The direct voltage is then applied and slowly increased until a zero beat is obtained in the receiver. The amplitude of the direct voltage is now equal to the maximum allowable peak amplitude of the modulating signal.

Since transmission of a carrier during tuning and testing operations is an undesirable practice, a dummy antenna should replace the regular antenna. If this reduces the output to a value too low for proper reception, a pickup loop may be used to replace the receiving antenna. A suitable signal can then be obtained by loosely coupling the output of the power amplifier to the pickup loop.

Once the degree of deviation has been determined, additional information can be obtained by calculating the modulation index and referring to the modulation index table. (See Appendix E for a sample index table.)

The modulation index is calculated by dividing frequency deviation by the frequency of the modulating signal in the following manner:

$$M_i = \frac{f_d}{f_m}$$

For example, with a frequency deviation of ±75 kHz and a modulation frequency of 15 kHz, the modulation index is as follows:

$$M_i = \frac{f_d}{f_m}$$

$$= \frac{75}{15}$$

$$= 15$$

Locating an index of 15 on the modulation index table in Appendix E reveals the following information:

1. There are 20 effective sidebands above and below the center frequency.

2. The bandwidth is 40 times the modulation frequency. In this case, the bandwidth is as shown below:

$$\begin{aligned} \text{Bandwidth} &= 40\text{ f} \\ &= 40 \times 15\text{ kHz} \\ &= 600\text{ kHz} \end{aligned}$$

Under these conditions, many of the effective sidebands are not being transmitted. This is always the case in narrow-band FM transmission. The sidebands become ineffective when their amplitude drops to 1% of the amplitude of the unmodulated carrier.

When calculating the modulation index, it must be considered that most modulation signals are not simple sine waves. They are complex waves that vary in both frequency and amplitude. The modulation frequency used in calculations must be the highest frequency of the modulating signal.

Measuring FM with Spectrum Analyzer

The spectrum analyzer may be used to measure the center frequency of the carrier, the frequency of the modulating signal, bandwidth, degree of deviation, modulation index, and the number of significant sidebands. This can all be done without any physical contact with the transmitter and during normal transmission.

The transmitted signal can be picked up on the spectrum analyzer's antenna. The spectrum analyzer, of course, is tuned to the transmitter frequency. The center vertical-line position shows the center frequency, the frequency of the unmodulated carrier. The distance between any two adjacent vertical lines is the frequency of the modulating signal.

Sidebands are effective as long as they increase and decrease in a cyclic fashion; each vertical line is a sideband. When the sidebands reach the point of ineffectiveness, the amplitude gradually drops and continues to drop until the sidebands disappear in the noise. The beginning of the final decrease is considered the edge of the band. A similar band edge is seen on either side of the center frequency. The number of vertical lines between these extremes is the number of effective sidebands. The bandwidth is the frequency of the upper edge of the band minus the frequency of the lower edge of the band.

MEASURING STANDING WAVES

A *standing wave* is a condition resulting from a mismatch of impedance on a transmission line. (Transmission lines include antennas.) If a line is terminated in its characteristic impedance, it theoretically has an infinite electrical length. All the energy leaving the source moves in traveling waves from source to load. The load absorbs all the energy, and none is reflected back toward the source. This is an ideal condition, and such a line would have no standing waves.

In practice this ideal condition is seldom, if ever, achieved. The source sends out waves of energy which travel to the load. The load absorbs part of this energy and reflects the remainder. The reflected energy travels back down the line toward the source. The incident traveling wave moves from the source toward the load; the reflected traveling wave moves from the load toward the source. The reflected wave is lost energy caused from improper matching of impedance.

The incident waves and the reflected waves travel in opposite directions along the same line. At times they are in phase and additive; at other times they are out of phase and cancel. The combined result is a stationary wave with peaks and nodes that can be measured. The waves are formed in the manner illustrated in Figure 13–9.

The stationary wave is a current wave, and the transmission line is terminated in an open. The reflected wave is 90° out of phase with the incident wave, and all the energy is being reflected. Voltage waves exist in the same manner, only they are 90° out of phase with the current waves. Either voltage or current waves can be measured. Measuring these standing waves is an easy way to obtain valuable information about the condition and efficiency of the equipment.

Standing-Wave Ratio

The instruments used in measuring standing waves detect and measure the values of voltage or current. No distinction is made between positive peaks and negative peaks. Thus, there are two nodes and two peaks in each wavelength. The *standing-wave ratio* (SWR) is a ratio between the peak value and the node value. The lower this ratio the better the line; a 1 to 1 ratio denotes a perfect line with no standing waves.

The standing-wave ratio may be obtained from either voltage or current waves, but most commercial meters work with voltage. The SWR is also the ratio between the characteristic impedance of the line and the actual impedance of the load.

Detecting Standing Waves

Standing-wave detectors may be as simple as a neon bulb with a coupling arrangement. In fact, a fluorescent light tube is a good detector and needs no coupling arrangement. The construction and use of some of these devices have been described in Chapter 6. Such devices can yield valuable general information about the conditions of mismatch. For instance, they can be used to measure wavelength in the line. Wavelength in the line divided by the line propagation factor is wavelength in space. Wavelength (γ) is readily converted to frequency since

$$\gamma = \frac{v}{f}$$

where v = velocity of light, 300×10^6 m/sec
f = frequency in Hz

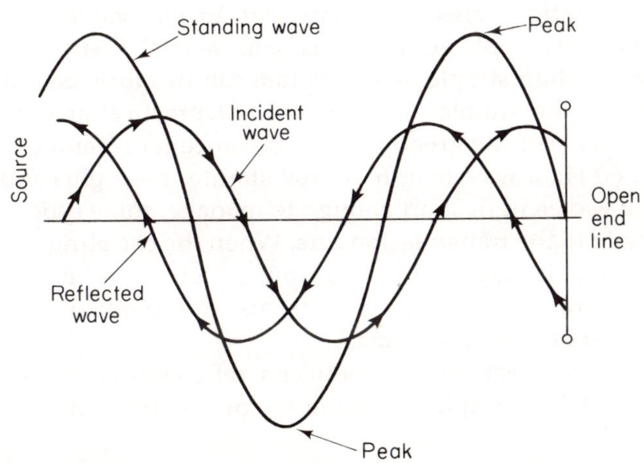

Figure 13–9 Formation of Standing Waves

Measuring with Lecher Lines

A simple *Lecher line* as illustrated in Figure 13–10 can be used to measure peak and node values. The coupling loop is moved along the transmission line, and the node and peak values of current are recorded. The values can be converted directly to SWR by

$$\text{SWR} = \frac{\text{peak}}{\text{node}}$$

Since SWR can be either voltage or current, the current SWR is maximum current divided by minimum current. The voltage SWR is maximum voltage divided by minimum voltage. For example, with a current peak of 5 A and a current node of 1 A, the SWR = 5/1 or 5.

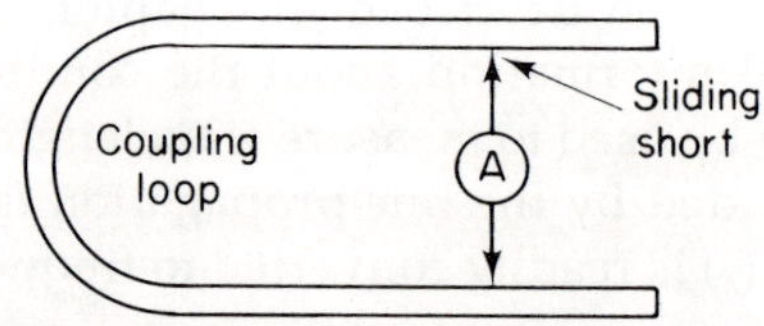

Figure 13–10 Lecher Line

Measuring SWR with Commercial Instruments

Commercial instruments for measuring SWR may carry the name of SWR detector, SWR meter, or reflectometer. They may be calibrated to indicate SWR ratios, reflection coefficient, or just values of voltage or current. For most applications, these commercial instruments are far more desirable than simple detectors that can be fabricated in the shop. For one thing, the simple detectors are not practical at the lower frequencies because of the great physical distance represented by a wavelength. (At 60 Hz a wavelength is 3109 statute miles plus about 546 ft.) Even with a wavelength short enough to manage, some type of coupling must be made to the transmission line. When the coupling is made with a simple device, it may load the equipment and cause erroneous indications. The commercial instruments are designed and shielded to overcome most of these problems.

There is a direct relation between reflection coefficient (Γ) and SWR, so that it is a simple operation to convert from one to the other, as follows:

$$\Gamma = \frac{\text{Reflected Voltage}}{\text{Incident Voltage}}$$

$$\text{SWR} = \frac{1+\Gamma}{1-\Gamma}$$

$$\Gamma = \frac{\text{SWR}-1}{\text{SWR}+1}$$

For example, an incident voltage of 10 V and a reflected voltage of 2 V will yield the following results:

$$\Gamma = \frac{2\text{ V}}{10\text{ V}} = 0.2$$

$$\text{SWR} = \frac{1+0.2}{1-0.2} = \frac{1.2}{0.8} = 1.5 \qquad \text{or 1.5 to 1}$$

Reversing the process produces these results:

$$\Gamma = \frac{\text{SWR}-1}{\text{SWR}+1}$$

$$= \frac{1.5-1}{1.5+1} = \frac{0.5}{2.5} = 0.2$$

A method of using a commercial instrument for measuring both SWR and reflection coefficient is illustrated in Figure 13–11. Some suitable means of shielded coupling is provided to couple the pickup probes or loops to the transmission line. Each pickup loop is in series with an impedance (Z) that matches the impedance of the transmission line.

The signal on either pickup loop is rectified and coupled to a two-position switch. In one position, the switch samples the energy of the incident wave; in the other position, it samples the energy of the reflected wave. The voltmeter may be used in the obvious manner to indicate the values of incident and reflected voltages. If it is so used, the operator reads these values and calculates either the reflection coefficient or the SWR.

On some instruments, the meter is calibrated in either the reflection coefficient or the SWR. In this case, the potentiometer is adjusted for some prescribed value with the switch set to incident. When the switch is moved to the reflected position, the meter reading drops to the proper value of either the SWR or the reflection coefficient.

Reflectometers are multipurpose instruments designed to indicate

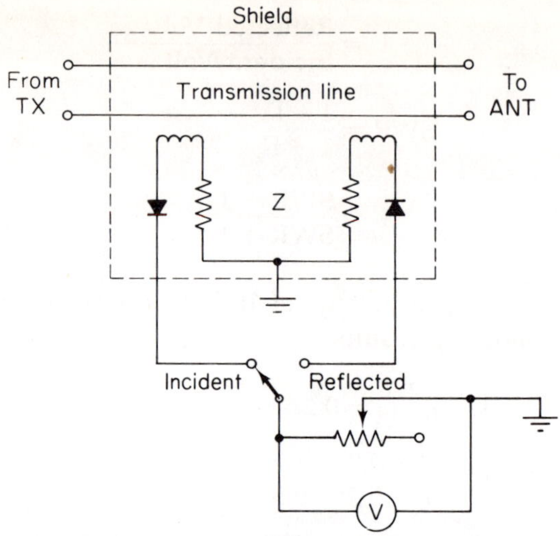

Figure 13–11 Measuring SWR and Reflection Coefficient

ratios directly. One method of coupling is illustrated in Figure 13–12. The ratio of incident to reflected energy is detected and coupled to a ratio meter. The detectors are usually either barretters or silicon crystals. They operate on low power and may have square-law characteristics. The ratio meter resolves the ratio and indicates this directly. The readout may be in impedance, reflection coefficient, or standing-wave ratio.

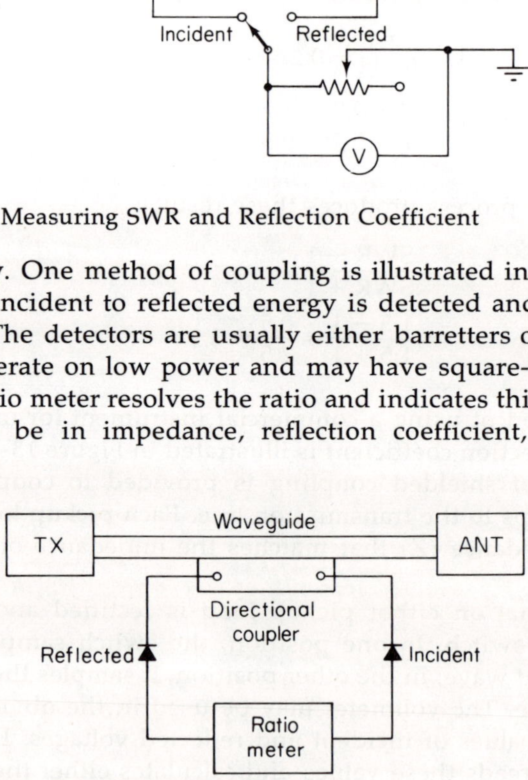

Figure 13–12 Reflectometer Coupling

MEASURING FIELD INTENSITY

Field intensity (or strength) measurements are necessary to determine the radiated power from an antenna, the coverage of the radiation field, the intensity of the field at various points, and the gain of the antenna. For some applications, it is essential to know the exact intensity in terms of microvolts per meter. In this case, the absolute values of field intensity must be measured and plotted. At times relative values of intensity will serve the purpose.

Measuring Absolute Field Intensity

A field intensity of 50 μV/m means that the electromagnetic field will induce 50 microvolts in a certain type of standard antenna that is one meter in length and of a specified diameter, when it is placed at that point in the electromagnetic field. Absolute field strength requires the actual measurement of the induced voltage in terms of microvolts per meter. The equipment needed for such measurements include the following:

1. A calibrated antenna.
2. A commercial field intensity meter.

Or

1. A calibrated antenna.
2. A well-shielded communications receiver.
3. A calibrated RF signal generator.
4. An indicating meter.
5. A calibrated attenuator.

The commercial field-intensity meter is already calibrated to provide indications in microvolts per meter. If this is not available, the receiver must be calibrated in the following manner:

1. Obtain a meter reading using the calibrated antenna.
2. Remove the antenna and connect the calibrated RF signal generator to the antenna jack.
3. Adjust the signal generator output until a reading is obtained which is identical to the first reading.

The absolute field strength is then computed by dividing the voltage for the reading by the antenna correction factor. For instance, the RF signal generator reveals that 100 μV produces the proper reading, and the calibrated antenna has a correction factor of 2. The absolute field strength is 100/2 or 50 μV/m.

The calibrated attenuator introduces another factor that must be included in the calculations. This is a standard attenuation factor, and it is common to all attenuators. The μV/m is multiplied by the attenuation factor. If the calculated indication at a certain point is 50 μV/m and the attenuator has reduced the signal by a factor of 3, the absolute field strength is 3×50, or 150 μV/m.

After the meter is calibrated, the signal generator is replaced with the calibrated antenna. From this point on, a measurement consists of reading the meter and converting the reading to microvolts per meter.

Measuring Relative Field Strength

Commercial instruments are available for measuring relative field strength. These too are called field-strength meters or field-intensity meters. They consist of an antenna, a detector, and a meter for indications. The antenna is called a pickup antenna because it is not calibrated. The whole device, except the antenna, should be well shielded.

The indicator is normally a logarithmetic-type meter. The object is to take readings proportional to a decibel reference, and the reading of changes in field strength is generally directly in decibels.

Procedure for Measuring Field Strength

The procedures for measuring field strength are essentially the same for both absolute and relative field strength. The main difference is in the equipment being used. However, the terminology used here is oriented to absolute measurements.

Field-strength measurements on skywaves are not reliable. Skywave field intensities vary considerably from one day to the next. They even vary from hour to hour as the characteristics of the ionosphere change with the passing of time.

Ground waves and direct waves produce a pattern of field intensity that remains fairly constant. The procedures discussed here are for plotting these patterns.

The first step is to plot radial lines on a terrain map of the area. These lines extend out from the transmitter site in all directions and should be no more than 45° apart as illustrated in Figure 13–13.

The next step is to travel outward along a radial, and plot the field strength. The first group of readings is intended to determine what areas are included in specified groups of field intensities. Suppose that these intensities are 100 μV/m, 50 μV/m, 25 μV/m, and 5 μV/m.

Readings should be taken at *half-mile* intervals until they begin to approach the 100 μV/m level. Then at quarter-mile intervals and at reduced intervals until the point is located where the field intensity is exactly 100 μV/m. This point is plotted on the map.

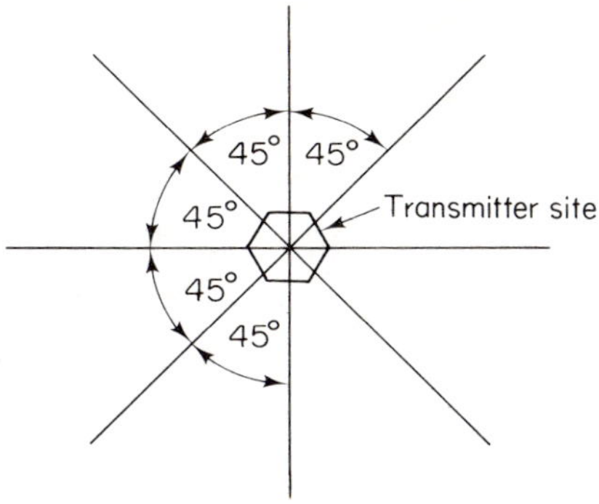

Figure 13–13 Plotting Radials

Continue to move out in the same direction. Take repeated readings in like fashion until points of 50, 25, and 5 μV/m are located and plotted.

Return to the transmitter site, and set up a course along another radial and plot the same points of intensity. Each radial is taken in turn until the plots are obtained on all radials. The points of equal intensity are now joined by a continuous line. These contour lines enclose the areas that are within specific field-intensity regions. A sample contour plot is illustrated in Figure 13–14.

This figure shows an imaginary plot of field intensity about a radio installation in Jackson, Mississippi. Contour line 1 encloses the territory within the 100 μV/m field intensity zone; line 2 marks the boundary of the 50 μV/m zone; line 3 is the boundary for the 25 μV/m zone; and line 4 marks the limits of the 5 μV/m area. Notice that the pattern is directive and concentrated in a southeast direction.

Any person who has taken part in such a plot will readily recognize these lines as completely arbitrary. An actual plot would be far less uniform because of obstructions which prevent traveling in a straight line. There are rivers, lakes, forests, swamps, marshes, and buildings that continuously force detours. Secondly this plot is much too large to be real. Since direct and ground waves only reach out about 75 miles, an actual plot would shrink to about the size of Zone 1.

About the only way to get a perfectly uniform plot is to use a helicopter for transportation while taking the measurements. In this case, great care would have to be exercised as to the altitude. Ground and direct waves are roughly line of sight, and a few feet of altitude can make a big difference in the field strength.

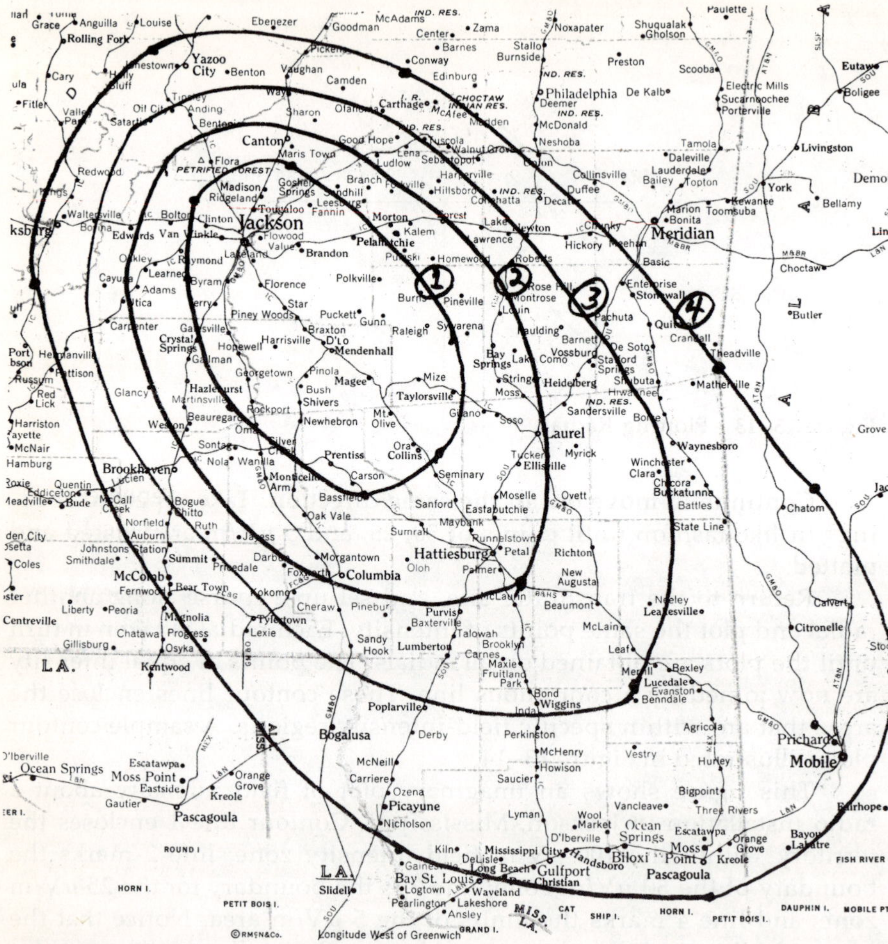

Figure 13–14 Sample Contour Plot

chapter fourteen

MEASURING ACTIVE COMPONENTS

All electronic circuits contain an active component in the form of an electron tube or a solid-state device. There are a variety of measurements that can be made on these components, both in circuit and out of circuit. Preliminary tests of a general nature determine whether or not a component is functioning. Special tests and measurements reveal the ability of the component to perform exacting tasks under specified conditions. Many tests can be performed through circuits that are simple to fabricate, and dynamic characteristics of amplifiers can be revealed by a few simple measurements and load line calculations. There are also professional instruments available for measuring active components.

MEASURING ELECTRON TUBES

Most new electronic devices are being constructed to use *solid-state components.* Someday, electronic tubes will be relics of the past, but this won't happen for many years. Of all the electronic equipment in existence today, a large majority of it is of electron-tube design. Building, testing, and supplying replacement tubes for this equipment are big business, and will remain so for some years to come. Thus, a large portion of the technician's job revolves around the *electron tube* and how it performs.

Most tube-testing instruments and procedures encourage the performance of four preliminary checks before proceeding to more detailed, special tests. These are: (1) shorted elements, (2) excessive gas, (3) leakage between cathode and heater, and (4) weak emission. These four checks have been adequately covered in Chapter 7, along with the tube measuring instruments. Several additional practical measurements are described here.

Measuring Mutual Conductance

Mutual conductance (or transconductance) is a measure of the effectiveness of the control grid in controlling the plate current. Specifically it is the ratio of the change in plate current (ΔI_p) to a change in the grid voltage (ΔE_g). Other values are, of course, held constant during this measurement. Mutual conductance is normally measured in micromhos (μ mho), and the symbol is g_m. The equation is:

$$g_m = \frac{\Delta I_p}{\Delta E_g}$$

The *static test* for g_m places the tube in a circuit similar to that illustrated in Figure 14–1. The switch, in the position shown, applies a negative potential to the grid but not enough to stop the plate current. The I_p under these conditions is recorded. The switch is then changed to apply a less negative potential to the grid. This change in grid voltage is a precise one, and its value must be known.

The change in bias causes a change in I_p which is noted on the meter. Mutual conductance is now calculated by the equation previously discussed.

The dynamic mutual conductance test places the tube in a circuit similar to that shown in Figure 14–2. In this dynamic test, a fixed value of dc bias is applied to the grid, and the plate furnishes current to a full-wave rectifier. The current through the rectifier goes to alternate plates on alternate half-cycles of the ac input. The meter indicates the difference in current from the two rectifier plates. As long as E_g remains fixed, the meter reading is zero.

An ac signal is now applied to the grid in addition to the dc bias. This signal is the same frequency as the ac on the transformer. As the grid swings positive, the tube conducts hard; as it swings negative, it conducts less. The difference between minimum and maximum plate current is registered on the meter. The meter is calibrated for direct readings in micromhos.

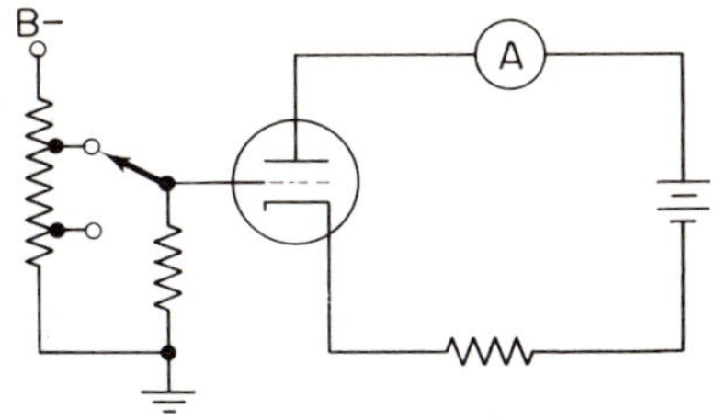

Figure 14–1 Measuring Static g_m

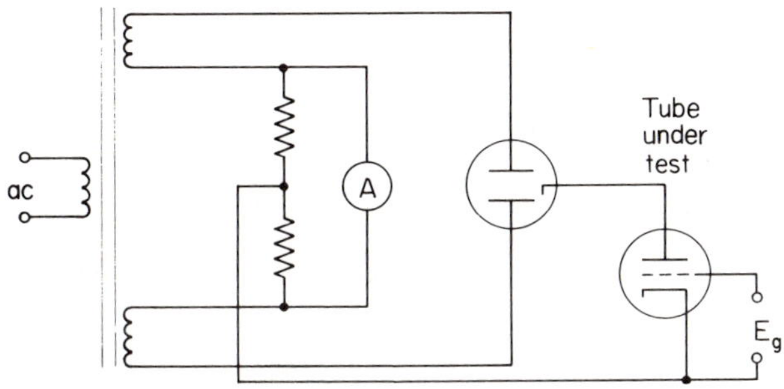

Figure 14–2 Measuring Dynamic g_m

dc Plate Resistance

The *dc plate resistance* (R_p) is not generally measured but is easily calculated from other measurements. R_p varies with the value of the negative bias on the grid: the higher the bias, the higher the resistance. R_p for any specified value of E_g is obtained by dividing the plate voltage (E_p) by the plate current (I_p), as follows:

$$R_p = \frac{E_p}{I_p}$$

For example, a given bias produces an I_p of 6 mA with an E_p of 150 V. The dc plate resistance is:

$$\begin{aligned} R_p &= \frac{E_p}{I_p} \\ &= \frac{150\ \text{V}}{6\ \text{mA}} \\ &= 15\ \text{k}\Omega \end{aligned}$$

ac Plate Resistance

The *ac plate resistance* is a measure of the effect that the changing plate voltage has on the plate current. It is obtained by setting a fixed bias on the tube, changing the E_p by a specific amount, and observing the effect on the I_p. The relationship is expressed as

$$r_p = \frac{\Delta E_p}{\Delta I_p}$$

For example, at a given value of bias, E_p changes from 225 V to 255 V. This 30 V change causes I_p to increase by 1.5 mA. The ac plate resistance is then as shown below:

$$\begin{aligned} r_p &= \frac{\Delta E_p}{\Delta I_p} \\ &= \frac{30 \text{ V}}{1.5 \text{ mA}} \\ &= 20 \text{ k}\Omega \end{aligned}$$

Amplification Factor

The *amplification factor* of a tube is a ratio of voltage changes with I_p held constant. It indicates the ability of a tube to amplify a signal, and it is symbolized by the Greek letter mu (μ), as follows:

$$\mu = \frac{\Delta E_p}{\Delta E_g}$$

For instance, an increase of 36 V in E_p will cause an increase in I_p. Increasing (negative) E_g by 2 V brings I_p back to its original value. Mu is then

$$\begin{aligned} \mu &= \frac{\Delta E_p}{\Delta E_g} \\ &= \frac{36 \text{ V}}{2 \text{ V}} \\ &= 18 \end{aligned}$$

Therefore; a 1 V change at the grid is equivalent to an 18 V change at the plate. The following relationship exists among the tube parameters:

$$\mu = g_m \times r_p$$

Other Tests and Measurements

The ordinary electron tube has several other parameters that may require measurements. To name a few, there are the interelectrode capacitance, loading effect, power output, cathode interface impedance, bulb temperature, electrode dissipation, and nonlinear characteristics. Some of these are extensions of tests already described; others require special equipment and procedures.

Tube-testing instruments are generally designed to handle many different types of tubes and provide a number of tests for each. Detailed procedures on roll and flip charts are included as part of the testing equipment. These procedures must be followed carefully to avoid damage to the tube and to provide reliable test results.

Some test instruments are programmed by punched cards. When a tube is inserted into the proper socket, and a card is inserted, the equipment decodes the card and automatically performs the indicated test or tests.

Special Tubes

Microwave tubes, gas tubes, phototubes, camera tubes, storage tubes, and cathode-ray tubes—all fall into separate specialized categories. Most of these tubes require highly specialized instruments for adequately measuring their performance. Directions for the types of tests to be performed and the correct testing procedures are included with the test equipment.

Hot-cathode *gas tubes* need to be tested for heater and emission characteristics, repeatability, breakdown, anode voltage drop, control cutoff, grid current, and thermal characteristics.

Cold-cathode *thyratrons* must be checked for ionization and deionization times and potentials.

Phototubes must be checked for normal current and voltage characteristics, sensitivity, dark current, collector efficiency, current amplification, and noise.

Some of the tests required for *camera tubes*, *storage tubes*, and *CRTs* are basic emission, beam focusing, leakage, gas currents, luminance, contrast, persistence, and deflection factors. The *color picture tubes* have a series of color tests that must be performed.

MEASURING SOLID STATE DIODES

Measurements of semiconductor diodes center around the characteristics of rectification and switching. Current leakage and breakdown voltage checks are of a vital nature. In appearance, semiconductor diodes are much simpler than electron diodes, but they require relatively complex measurements. All this leads to specialized testing instruments or test circuits that are especially designed for this purpose.

Measuring Front-to-Back Ratio

This type of measurement may be an out-of-circuit test performed with a simple ohmmeter. A simple resistance measurement can be used to test a diode's ability to pass current in only one direction. Check the current rating of the diode before subjecting it to an ohmmeter check. Some sensitive crystal diodes can be destroyed by ohmmeter current. Figure 14–3 illustrates the method of measuring the resistance.

The ohmmeter in Figure 14–3(a) is connected for measuring the front (or forward) resistance. Current from the ohmmeter is against the arrow, and a good diode should have a low forward resistance.

In Figure 14–3(b) the current is with the arrow, so that this meter is measuring the back (or reverse) resistance. This resistance should be much higher than the forward resistance. The exact ratio is determined by the type of diode, but in most cases the reverse resistance is several hundred times as great as the forward resistance. The power rectifier can operate satisfactorily with a ratio of 10 to 1.

Diode testers may accomplish this check by applying front and back voltage and measuring the corresponding currents. These testers normally read in terms such as good or bad.

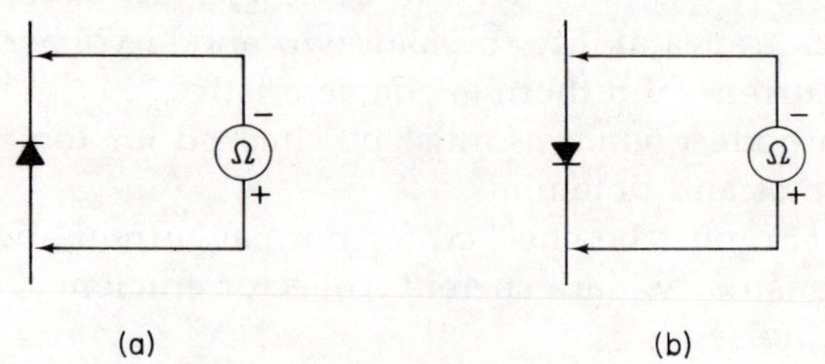

Figure 14–3 Measuring Diode Resistance

Measuring Forward Voltage Drop

The *forward voltage drop* is the voltage across the diode for a specified value of forward current. This can be performed on a tester or on a fabricated circuit. In either case, the circuit is similar to that illustrated in Figure 14–4. The applied dc potential is varied until the ammeter indicates the desired value of forward current. The forward voltage is then indicated by the voltmeter. This should normally be a small voltage,

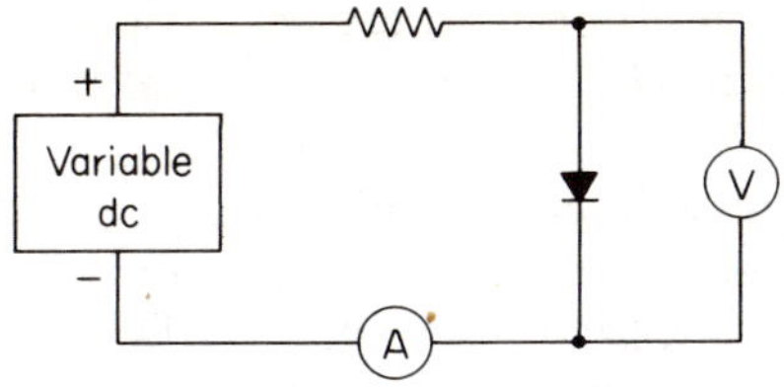

Figure 14–4 Measuring Diode Forward Voltage Drop

and its maximum value should be obtained from the manufacturer's data sheet.

Measuring Diode Reverse Leakage

The *reverse leakage* through a diode is the current resulting from the application of a specified level of reverse voltage. Figure 14–5 illustrates the test circuit. This is the same circuit used for testing forward voltage drop, but the diode is reversed. The variable dc is increased until the voltmeter indicates the specified level of reverse voltage. The current indicated by the ammeter at this time is the reverse leakage current of the diode. It should be a very small current; the maximum level is indicated on the data sheet.

When fabricating circuits for most diode tests, electronic voltmeters should be used. This is necessary to avoid circuit loading and to provide the high sensitivity necessary.

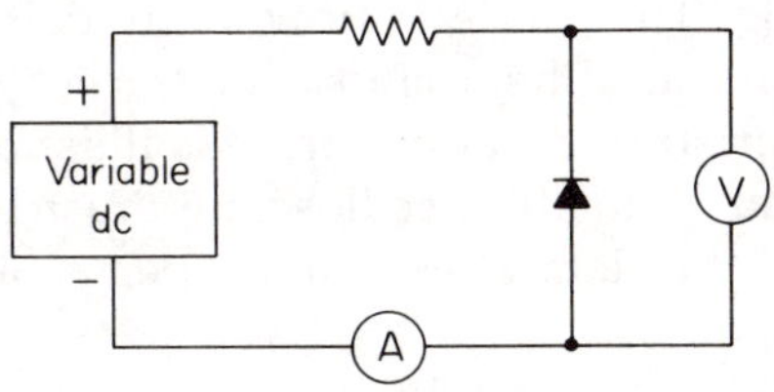

Figure 14–5 Measuring Diode Reverse Leakage

Measuring Dynamic Characteristics of Power Rectifiers

Measurements using only dc potentials are called *static tests* because the diode is not in an actual operating situation. A diode normally operates with ac. This causes more heat on the junction and causes the characteristics to change. When ac is applied, the diode approaches actual operating conditions, and the measurements are dynamic in nature. Dynamic measurements are best performed with a characteristic curve tracer or an oscilloscope. If an average oscilloscope is used, both horizontal and vertical channels must be voltage-calibrated. A test circuit is illustrated in Figure 14–6.

The total applied voltage is connected to the horizontal channel, and the horizontal deflection is proportional to the amplitude of this voltage. The voltage across the resistor is developed by the current through the diode. This voltage (proportional to diode current) is connected to the vertical channel.

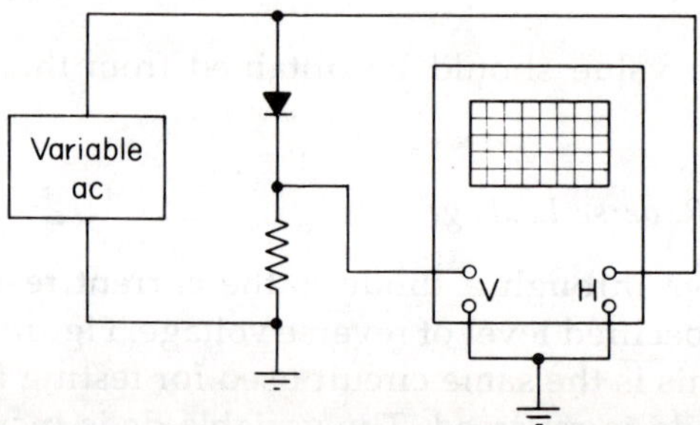

Figure 14–6 Dynamic Measurement of Power Rectifier

The resistor should be a noninductive precision resistor of either one ohm or 1000 ohms. This will enable direct conversion of the oscilloscope voltage reading into diode current. For a one-ohm resistor, one volt is equivalent to one ampere. For a 1000-ohm resistor, one volt is equivalent to one milliampere.

The variable ac potential is set to the maximum rated peak inverse voltage for the diode. The diode is now alternately biased in forward and reverse directions, and the oscilloscope traces the complete current and voltage characteristics. A sample trace is illustrated in Figure 14–7.

The values read from the oscilloscope must be compared to the values indicated on the data sheet for proper analysis of the results.

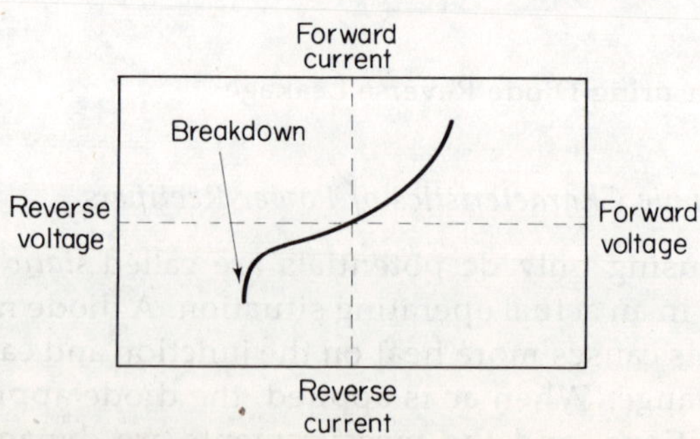

Figure 14–7 Current-Voltage Curve

Measuring Dynamic Characteristics of Small Signal Diodes

A small signal diode has rectifying properties but operates with reverse voltages much larger than the forward voltages. The test circuit must provide this voltage differential in a manner similar to that illustrated in Figure 14–8.

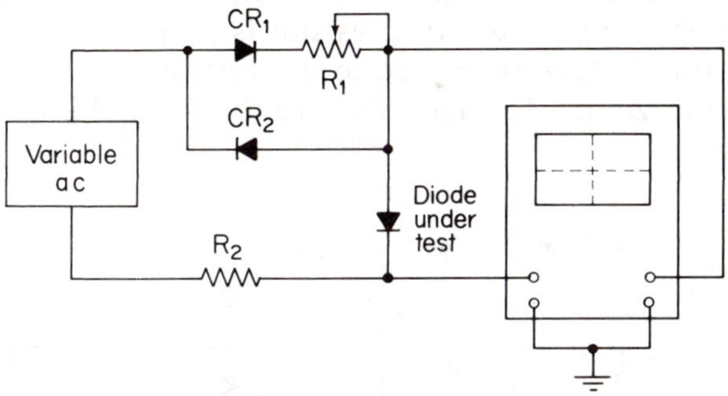

Figure 14–8 Measuring Small Signal Diode

The variable ac is adjusted for maximum rated peak reverse voltage of the diode under test. This potential is applied to the test diode through CR_2. CR_2 and the test diode are in series but are connected in opposite directions. When the negative attenuation of the ac is applied to the cathode of CR_2, this is forward bias for CR_2 and reverse bias for the test diode. A very small voltage drop occurs across CR_2, leaving the full reverse voltage across the test diode.

R_2 is a noninductive resistor with a precision value of either one ohm or 1000 ohms. This will enable direct conversion from oscilloscope indication to current as described under the rectifier test.

On the alternate half cycle, a positive potential is applied to the anode of CR_1. CR_1, R_1, and the test diode are in series, and R_1 is adjusted to limit the voltage to the maximum rated peak value of forward voltage recommended for the diode.

The voltage applied to the diode is coupled to the horizontal channel of the oscilloscope. The voltage across R_2, which is proportional to the diode current, is applied to the vertical channel. The oscilloscope traces the complete current-voltage characteristics curve. These characteristics can be measured and compared to the values indicated on the data sheet.

Measuring Diode Recovery Time

There is a measurable time lapse between the application of reverse bias to a diode and the time the reverse current reaches a steady state.

This is the *recovery time* for that diode, and it is the most critical characteristic of a diode to be used in a switching circuit. The duration of recovery time limits the minimum duration of the switching pulse that can be used. A sample test circuit for measuring the recovery time is illustrated in Figure 14–9.

A variable dc is applied as a forward bias to the test diode. The voltage is adjusted until the ammeter indicates the proper forward current. The voltage across the series resistor is proportional to the diode current, and this voltage is applied to the vertical channel of the oscilloscope. The horizontal channel is time-calibrated and uses the internal sweep voltage.

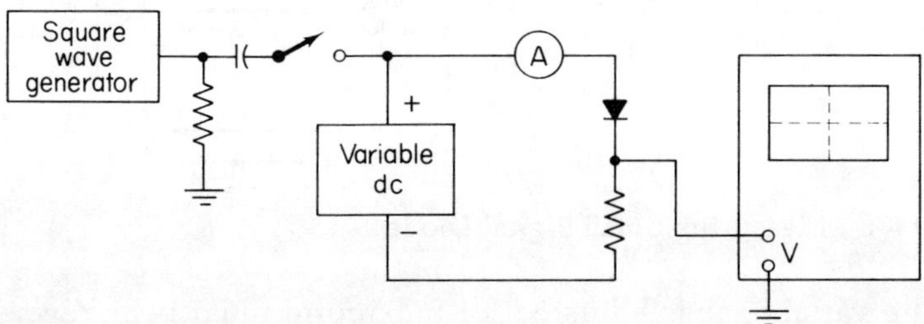

Figure 14–9 Measuring Recovery Time

The output of the square-wave generator is switched into the circuit without disturbing the previous settings. The negative alternation of the square-wave input places a strong reverse voltage across the diode. The current switches from normal forward to strong reverse, then gradually decreases to a stable reverse value at the end of the recovery time. The positive swing of the square wave switches the diode back to forward current. The resultant picture on the oscilloscope is illustrated in Figure 14–10.

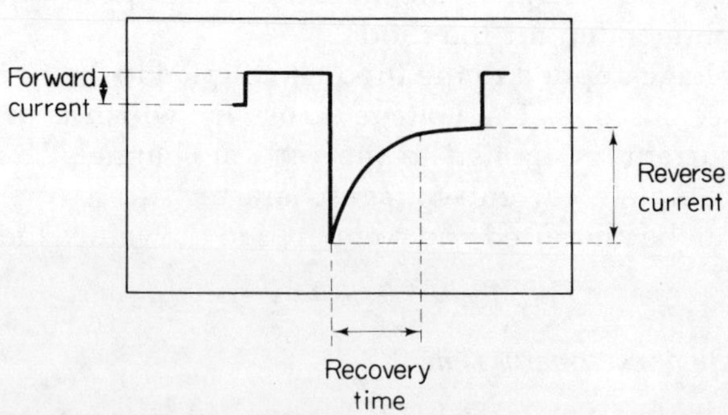

Figure 14–10 Switching Characteristics

Each positive transition of the square wave switches to forward current. Each negative transition switches to reverse current. A measurable transit time is required after each negative transition for the diode reverse current to stabilize. Trying to switch during the recovery period will result in distortion.

Measuring Characteristics on a Curve Tracer

A more complete picture of the dynamic characteristics of a diode may be obtained by using a *curve tracer.* This instrument can display the forward and reverse quantities with a high degree of accuracy. These quantities may then be compared to the recommended values given on the data sheet. The test circuit and several sample displays are illustrated in Figure 14–11.

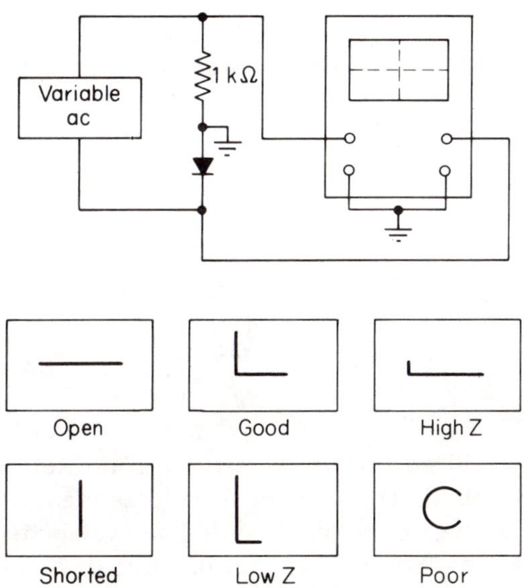

Figure 14–11 Measuring Diodes with a Curve Tracer

Measuring Zener Impedance

The impedance of a Zener diode varies with its physical size and with the junction current. The regulating ability of a Zener is directly related to its impedance. For reliable results, Zener impedance measurements should be made under precise conditions. One test circuit for this purpose is illustrated in Figure 14–12. The variable dc is adjusted until the dc ammeter registers about 20% of the maximum Zener current. The ac

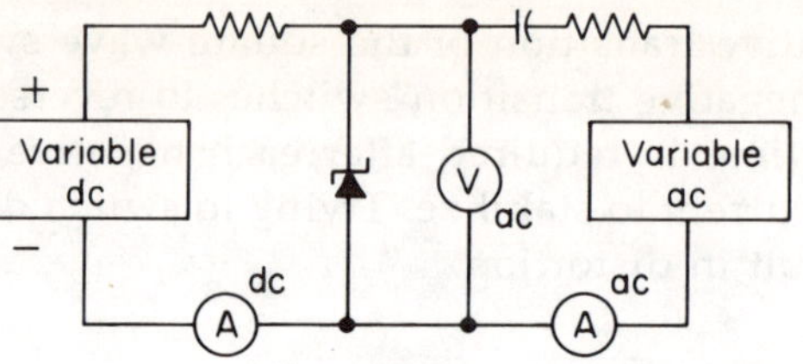

Figure 14–12 Measuring Zener Impedance

is then adjusted until the ac ammeter indicates about 10% of the maximum rated value.

The voltmeter now indicates the ac voltage across the Zener diode. Impedance is calculated from the ac voltage and current values, as follows:

$$Z = \frac{v}{i}$$

For example, when ac current is 3 mA and ac voltage is 12 V, the Zener impedance is

$$\begin{aligned} Z &= \frac{v}{i} \\ &= \frac{12\ \text{V}}{3\ \text{mA}} \\ &= 4\ \text{k}\Omega \end{aligned}$$

Measuring Zener Avalanche Voltage

The Zener diode will go into the *avalanche* condition when sufficient reverse bias is applied. The avalanche voltage measurement replaces the reverse leakage test of other diodes. The avalanche voltage can be measured with a circuit similar to that illustrated in Figure 14–13.

The variable dc potential is connected as reverse bias across the Zener. The voltage is slowly increased until avalanche occurs. This condition will be indicated by a large increase in current through the ammeter. The voltage indicated by the voltmeter at this time is the avalanche voltage.

Measuring Zener Dynamic Characteristics

The dynamic characteristics of a Zener diode can be measured by applying ac and using an oscilloscope. A sample circuit for this test is illustrated in Figure 14–14.

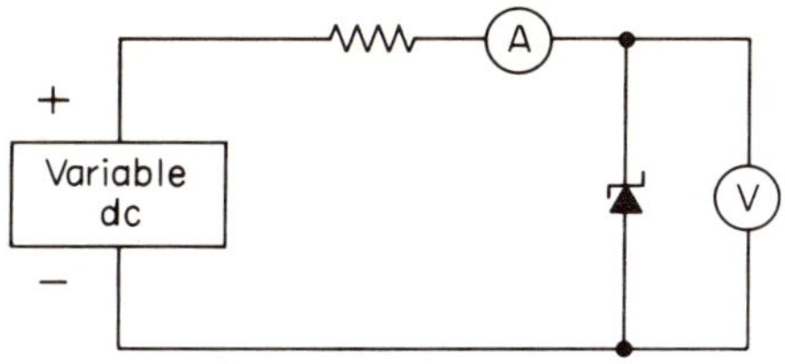

Figure 14–13 Measuring Avalanche Voltage

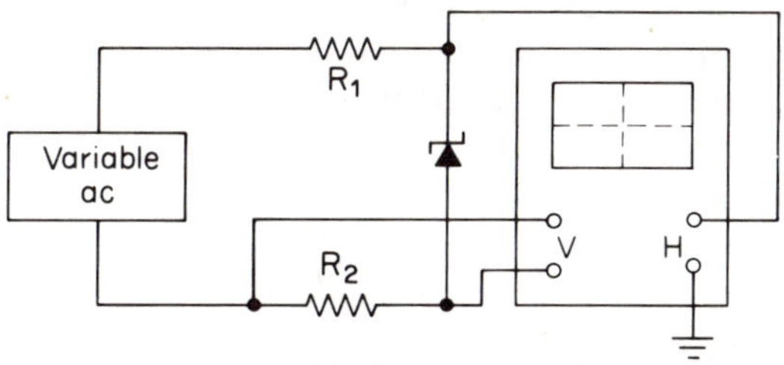

Figure 14–14 Measuring Zener Dynamic Characteristics

The voltage drop across R_2 is proportional to the Zener current. Its value should be chosen to enable a direct mental conversion from voltage to current. This voltage is coupled to the vertical channel of the oscilloscope. Total applied voltage is coupled to the horizontal channel. Both channels must be voltage calibrated for this test.

The variable ac is set to a potential large enough to cause the current to enter the Zener region. The values can now be read from the screen and compared against those indicated by the data sheet. The picture will be similar to that illustrated in Figure 14–15.

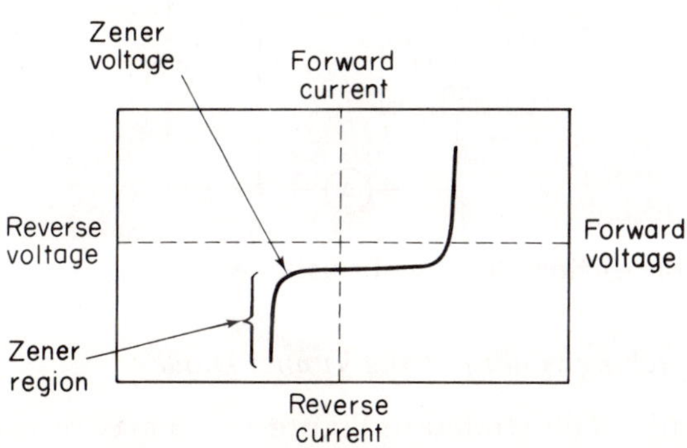

Figure 14–15 Zener Current-Voltage Characteristics

Measuring Negative Resistance of Tunnel Diodes

The most important characteristic of a *tunnel diode* is the negative resistance. Negative resistance can be measured with acceptable accuracy by using a simple test circuit similar to that illustrated in Figure 14–16.

With the voltage polarity as shown, the tunnel diode is reverse biased, and any current indication is reverse current. The variable dc is gradually increased until the voltage across the diode starts to drop. If a further increase in applied voltage causes an increase in current but a decrease of voltage across the diode, this is the negative-resistance region. The negative resistance can now be calculated by dividing the decrease in the diode voltage by the increase in the diode current as follows:

$$-R = \frac{\Delta v}{\Delta i}$$

For instance, a voltage drop of 5 V is accompanied by a current increase of 2.5 mA. The negative resistance is then

$$\begin{aligned} -R &= \frac{\Delta v}{\Delta i} \\ &= \frac{5\ \text{V}}{2.5\ \text{mA}} \\ &= 2\ \text{k}\Omega \end{aligned}$$

The series dropping resistor in this circuit should be on the order of 30 kΩ to provide adequate protection for the diode.

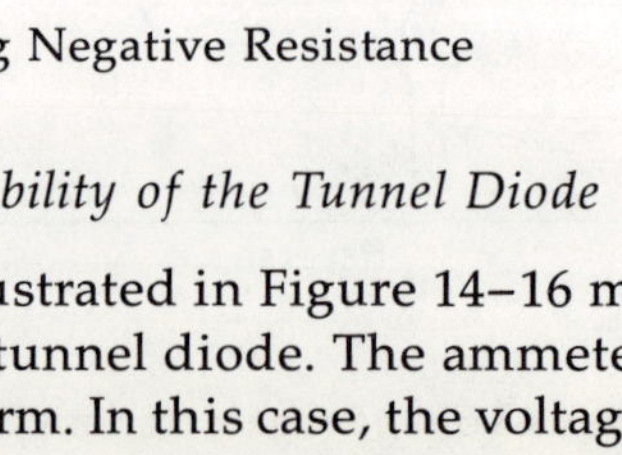

Figure 14–16 Measuring Negative Resistance

Measuring Switching Ability of the Tunnel Diode

The same circuit as illustrated in Figure 14–16 may be used to check the switching ability of a tunnel diode. The ammeter is not needed, but its presence will do no harm. In this case, the voltage is gradually increased until the voltmeter indicates a sudden rise in voltage. This shows that the diode has switched and is operating normally. Now the voltage is gradually reduced. When the critical voltage is reached, the diode switches again and causes a sudden decrease on the voltmeter.

The voltage reading that causes the sudden change, in each case, is the critical switching potential.

MEASURING BIPOLAR TRANSISTORS

The practical measurement of *bipolar transistors* centers around their amplification properties. These measurements and calculations hinge upon the use of transfer circuit constants, which are commonly referred to as parameters. Chapter 7 discussed the hybrid parameters and the test circuits in some detail under both open and shorted conditions. Additional information on parameters, abbreviations, and conversion formulas may be found in Appendix F, Transistor Data.

Measuring Collector Leakage Current

Collector leakage is an indication of deterioration, and even though the leakage is within tolerance, it may be a sign of pending failure. Starting with new transistors, those with the highest leakage current are likely to be the first to fail. A simple circuit for performing this measurement is illustrated in Figure 14–17.

The variable dc is adjusted to near the maximum rated value of collector voltage as registered on M_1. Since the leakage current is too low for accurate measurement with common meters, a voltage reading is taken and converted to current. The leakage current through the series resistance is proportional to the voltage across it. This should be a resistor on the order of 20 to 30 kΩ. The voltage across the resistor is measured by M_2, and the leakage current is calculated by Ohm's law. Of course this voltmeter could be calibrated in microamperes to enable a direct reading of leakage current. This is an advantage when meas-

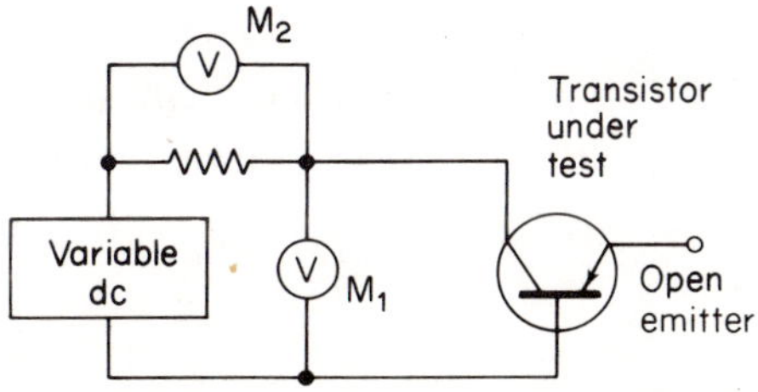

Figure 14–17 Measuring Collector Leakage Current

uring a large number of transistors at one time. This method is applied in some transistor testers.

Some testers use a sensitive ammeter to measure the leakage current. This may also be done with ordinary shop meters if reasonable care is exercised. The ammeter would be connected in series in place of the resistor. This leakage current must not exceed a given maximum for a specified value of bias.

Indications of leakage can also be determined by using an ohmmeter. The transistor is essentially two diodes connected back to back. Each diode should have low forward resistance and high reverse resistance. The method of measurement is indicated in Figure 14–18.

In each case with polarities as shown, the ohmmeter is indicating the reverse resistance. The readings should be high, in the range of 10 to 60 kΩ. Ohmmeter polarities must be reversed to measure the forward resistances. These will be low, on the order of 300 to 700 ohms. The front-to-back ratio is the important factor, not the precise resistance. This should be at least 30 to 1, and in some cases, it will be greater than 100 to 1.

There are two precautions to be observed during the ohmmeter test. First, an ohmmeter with a large internal voltage must not be used. Second, the ohmmeter must not be used on the $R \times 1$ range. For an adequate determination of front-to-back ratio, the same ohmmeter range should be used for both measurements.

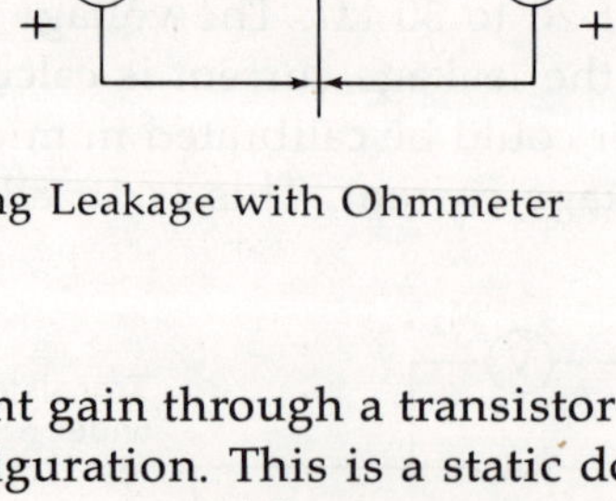

Figure 14–18 Measuring Leakage with Ohmmeter

Measuring Alpha

Alpha (α) is the current gain through a transistor which is connected in a common-base configuration. This is a static dc measurement, and

$$\alpha = \frac{\Delta I_C}{\Delta I_E}$$

A sample circuit for this measurement is illustrated in Figure 14–19.

Both collector and emitter currents are measured and recorded.

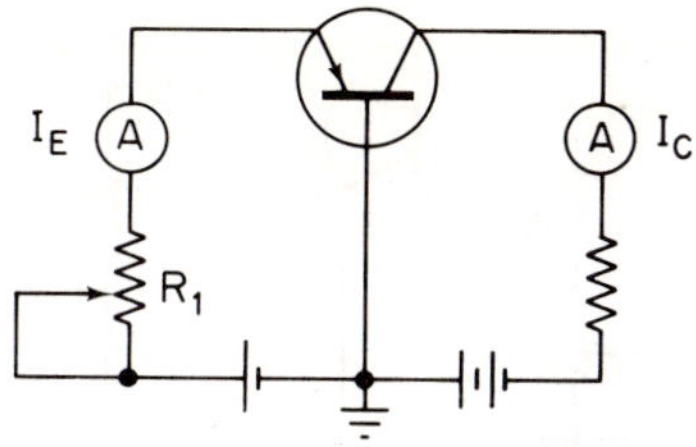

Figure 14–19 Measuring Alpha

Then emitter current (I_E) is changed by adjusting R_1. This causes a change in collector current (I_C). The current gain is calculated by using the change in the two currents in the alpha formula. For instance, a change in I_E of 4 mA causes a change in I_C of 3.5 mA. In this case, alpha is

$$\alpha = \frac{\Delta I_C}{\Delta I_E} = \frac{3.5 \text{ mA}}{4 \text{ mA}} = 0.875$$

Alpha is always less than unity, but it may be as high as 0.95 in some cases.

Alternating current gain is sometimes referred to as alpha, but there is a difference. Alternating current gain is a dynamic characteristic tested under actual operating conditions. It is sometimes called the small-signal characteristics of the transistor.

Measuring Beta

Beta (β) is the current gain of a transistor connected in a common-emitter configuration. There is a dc beta and an ac beta. The dc beta is sometimes designated as H_{FE}, and ac beta as h_{fe}. At any rate, the dc beta is a static gain, a relationship between a change in (dc) base current and a change in (dc) collector current. The ac beta is a dynamic relationship between a change in (ac) base current and a change in (ac) collector current. A test circuit for measuring dc beta is illustrated in Figure 14–20.

Both collector current and base current are measured and recorded; then the base current is changed a specific amount by adjusting R_1. The resulting change in collector current is noted, and beta is calculated as follows:

$$\beta = \frac{\Delta I_C}{\Delta I_B}$$

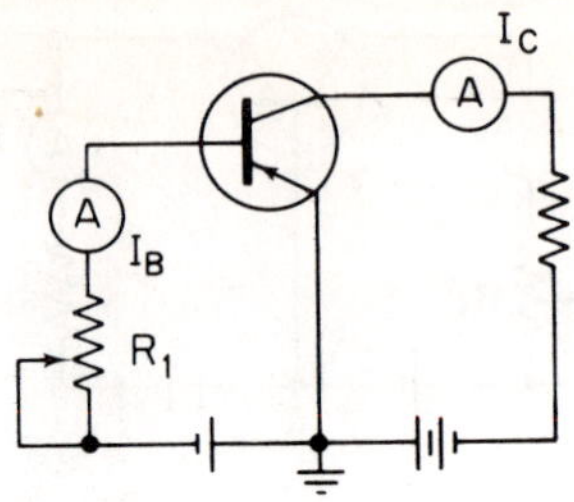

Figure 14–20 Measuring dc Beta

For instance, a change of 50 μA in I_B causes a change of 2.5 mA in I_C. In this case, beta is

$$\begin{aligned}\beta &= \frac{\Delta I_C}{\Delta I_B} \\ &= \frac{2.5 \text{ mA}}{50 \ \mu\text{A}} \\ &= 50\end{aligned}$$

This figure is a bit high, but the values of dc beta may range from 25 to 50.

There is a direct relationship between alpha and beta, and it is sometimes necessary to make the conversion from one to the other. Most manufacturers specify either alpha or beta but not both. Either of these values may be calculated when the other is known because

$$\alpha = \frac{\beta}{1+\beta} \quad \text{and} \quad \beta = \frac{\alpha}{1-\alpha}$$

The small signal gain of a transistor in a common-emitter configuration is ac beta. Measuring this quality is accomplished by applying an ac signal to the base input and measuring the gain at the collector.

Measuring Breakdown Voltage

Breakdown voltage is specified as some minimum value of reverse voltage on most transistor data sheets. The data sheet further specifies the quantity of leakage current to be drawn at this breakdown voltage. For example, the data sheet may specify collector-to-base breakdown voltage (BV_{CBO}) as 35 V with I_C of 0.1 mA and emitter open. A test circuit for measurement is illustrated in Figure 14–21.

The variable dc potential is adjusted until the specified I_C is indicated on the ammeter. The voltage is then read, and if it is less than

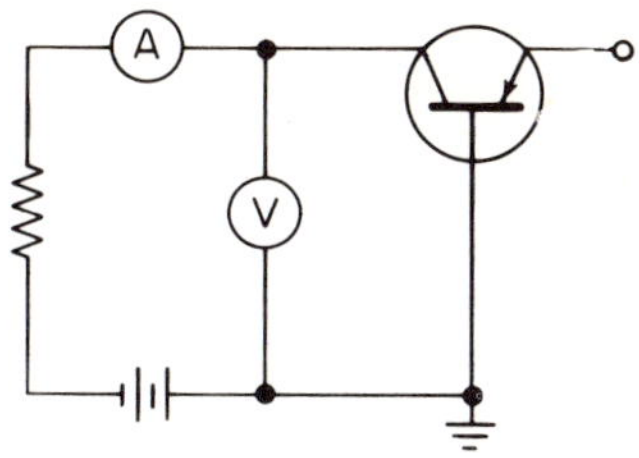

Figure 14–21 Measuring BV_{CBO}

the specified value, the transistor is faulty. Using the specifications previously mentioned, the collector-to-base voltage must be 35 V or greater when the I_C is 0.1 mA. This measurement determines the condition of both junctions at the same time.

There are several other types of breakdown voltages that may appear on a data sheet. Each of them is clearly identified in a manner that shows the type of test circuit that is to be used to take the measurement. Collector-to-base with emitter open has just been described. Its designation is BV_{CBO}. Collector-to-emitter with base shorted is BV_{CES}. BV_{CEO} means breakdown voltage from collector to emitter with the base open. BV_{CEX} indicates that the collector to emitter breakdown voltage is to be measured with the emitter-base junction reverse-biased.

Measuring Cutoff Frequency

The *cutoff frequency* of a transistor is the input frequency that will reduce the output power by 50%. A circuit similar to the one shown in Figure 14–22 may be used for this measurement.

The variable frequency signal generator is set to about 400 Hz and a voltage level sufficient to produce a convenient level of undistorted output. Output voltage (ac) is measured and recorded. Then the generator voltage is held constant, and the frequency is gradually increased.

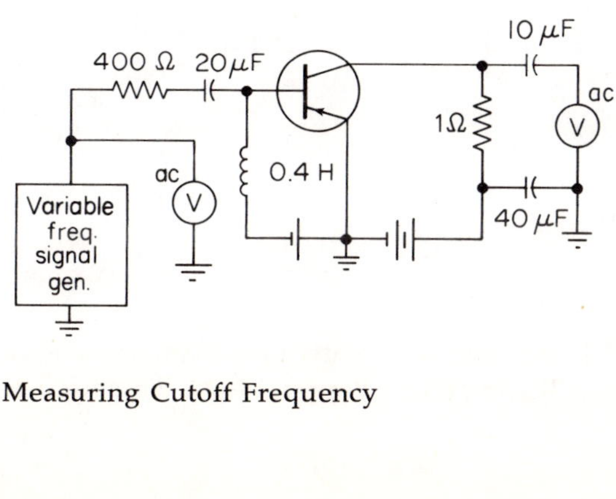

Figure 14–22 Measuring Cutoff Frequency

As the cutoff frequency is approached, the output voltage will start to decrease. Continue to increase the frequency until the output voltage drops to 0.707 of its original value. The input frequency at this time is the cutoff frequency of the transistor.

MEASURING FIELD-EFFECT TRANSISTORS

The *field-effect* transistor has become very popular because it incorporates the good features of both electron tubes and transistors without the predominant weaknesses of either. These transistors are now widely used in electronic equipment, test instruments, and integrated circuits.

The field-effect transistor (FET) is a solid-state device with high-input and high-output impedance. Like an electron tube, it is a voltage-controlled device, and may be operated at high-voltage levels. Like the transistor, it has no excess heat to dissipate, and it may be operated with the low voltages normally used with other transistors.

The metal oxide semiconductor field-effect transistor (MOS FET) is an insulated gate FET that can be manufactured in microscopic size. These transistors are representative of a major breakthrough in the area of large-scale integration.

Types of FET

The common FET is constructed in two types: N channel and P channel. The N channel corresponds to a PNP type transistor. The P channel corresponds to the NPN type transistor. Symbols for both types are illustrated in Figure 14–23.

The N channel FET is symbolized in Figure 14–23(a); the P channel symbol is shown in Figure 14–23(b). The gate corresponds to the base of bipolar transistors or to the control grid of an electron tube. The source replaces the emitter, and the drain takes the place of the collector.

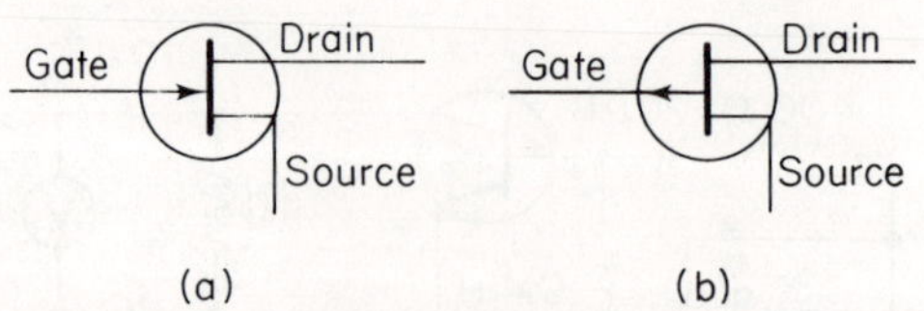

Figure 14–23 FET Symbols

Circuit Configurations

Either type of FET may be used in any one of three basic circuit configurations. These are illustrated in Figure 14–24. The circuits illustrate the

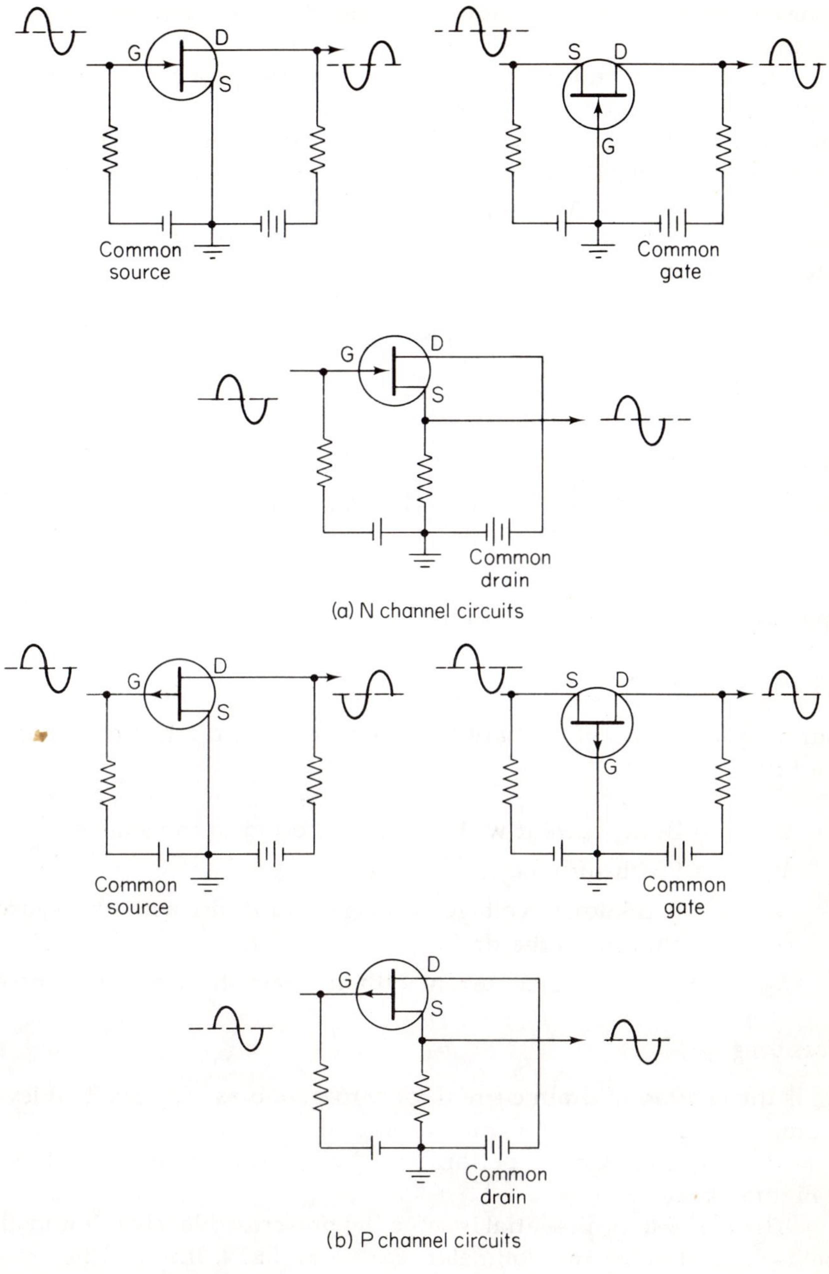

Figure 14–24 FET Circuit Configurations

proper bias for each configuration. The waveshapes are voltages and represent the relative amplitude and phase relation between the input and the output.

The common-source circuit may be compared to a bipolar common emitter or to an electron-tube grounded cathode. The common-gate circuit is similar to a bipolar common-base or the grounded-grid circuit of an electron tube. The common-drain circuit in an FET functions approximately the same as a common collector in a bipolar transistor or a grounded plate in an electron-tube. The voltages and currents are illustrated in Figure 14–25.

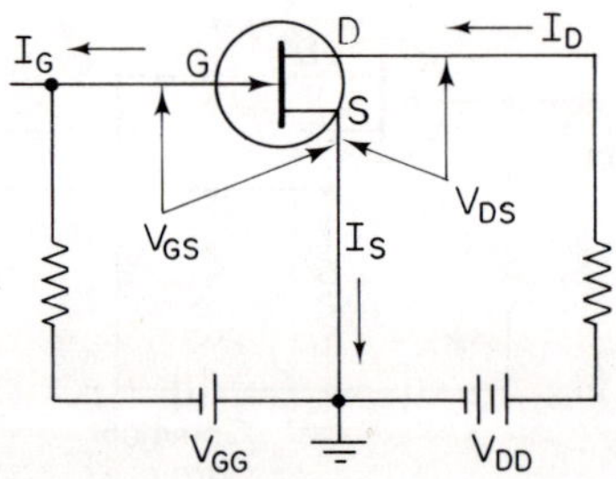

Figure 14–25 FET Voltages and Currents

FET Parameters

Four parameters are of primary concern in evaluating the performance of a FET:

I_{DSS} = Drain current with the gate shorted to the source.

V_P = Pinch-off voltage.

BV_{GDS} = Breakdown voltage from gate to drain with the source shorted to the drain.

I_{GSS} = Gate to source current with the drain shorted to the source.

Measuring I_{DSS}

I_{DSS} is the saturation drain current for zero-gate bias at a specified level of drain-source voltage. It is measured at the start of the saturation region. A sample circuit for performing this measurement is illustrated in Figure 14–26.

The variable dc potential is set to the prescribed level of drain voltage as indicated on the voltmeter. For some FETs, this will be −5 V. Allow the drain current to stabilize. It will settle down after a few seconds warm-up. The ammeter is now indicating the I_{DSS}.

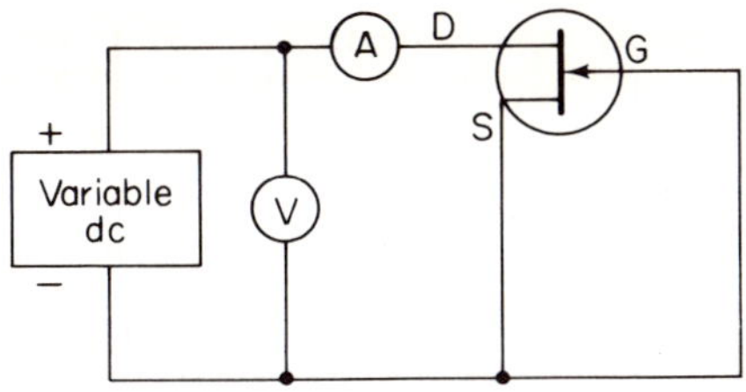

Figure 14–26 Measuring I_{DSS}

Measuring V_P

The gate-source pinch-off voltage is the level of gate bias at which the drain current is reduced to zero. A test circuit for making this measurement is illustrated in Figure 14–27.

Some minute value of drain current is generally specified as a practical cutoff value. This is usually about 1 μA. The variable dc potential is reduced until this prescribed current is indicated on the ammeter. The pinch-off voltage is now indicated on the voltmeter.

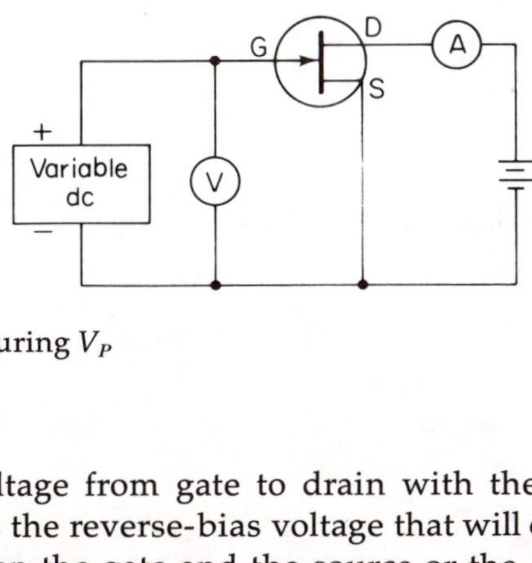

Figure 14–27 Measuring V_P

Measuring BV_{GDS}

The *breakdown* voltage from gate to drain with the source and drain shorted together is the reverse-bias voltage that will cause an avalanche breakdown between the gate and the source or the drain. A circuit for this measurement is illustrated in Figure 14–28.

The variable dc potential is adjusted until the specified value of breakdown current is indicated on the ammeter. The voltmeter is now indicating the BV_{GDS}.

Measuring I_{GSS}

The gate-to-channel (either source or drain) junction has a small quantity of leakage current when the junction is reverse-biased. The quantity I_{GSS} is the current that exists at a specified level of reverse bias when the

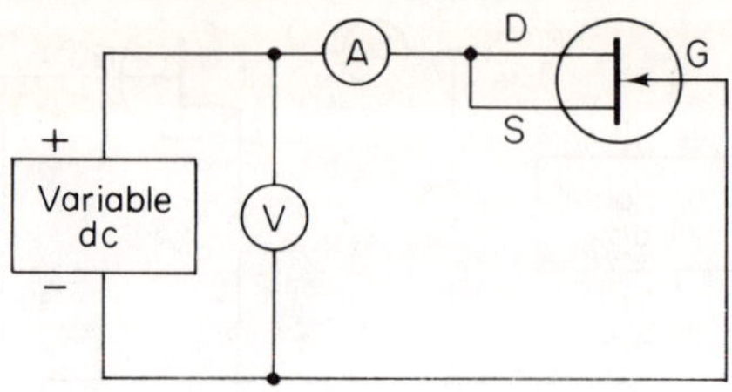

Figure 14–28 Measuring BV_{GDS}

drain and source are shorted together. A measuring circuit for this quantity is illustrated in Figure 14–29. The variable dc potential is set to the specified level of reverse bias as indicated on the voltmeter. This is probably a rather high voltage, just short of the breakdown value. The indication on the ammeter is now I_{GSS}.

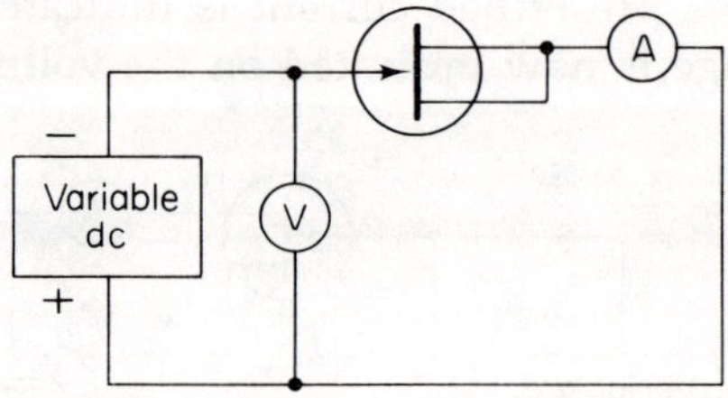

Figure 14–29 Measuring I_{GSS}

chapter fifteen

APPLICATIONS OF SIGNAL GENERATORS

The signal generator is not specifically a measuring instrument, but its use is essential to a great number of measurements. Signal generators are indispensable for taking dynamic measurements on portions of a circuit. They provide the signal for testing complete equipment during malfunctions and tuning operations. Many uses of signal generators have been illustrated in previous chapters. This chapter concentrates on specific applications of the signal generator in additional common testing situations.

MEASURING CIRCUIT CHARACTERISTICS

This section includes test circuits for measuring such characteristics as inductance, capacitance, impedance, and resonance. Circuits have been selected randomly with no attempt to concentrate on a specific type. However, the tests and measurements have a general application that makes them adaptable to many types of circuits.

Measuring Parallel Resonance

This test is a dynamic measurement of the resonant frequency of a parallel tank circuit. It is more realistic if the tank is measured in its own circuit. With the signal generator furnishing the signal, the test may be very simple, as indicated in Figure 15–1. The circuit illustrated is a low-frequency circuit as evidenced by the coupling capacitor in the voltmeter lead. The circuit is easily adapted to intermediate and radio frequencies by using an appropriate probe for the meter input.

The procedure here is very basic. Isolate the oscillator or amplifier to be tested from its neighboring circuits, but leave the normal operating potentials present. Connect the signal-generator output into the input circuit of the test stage. Connect the electronic voltmeter to the tank circuit being tested.

Start the signal generator on a low frequency, well below the suspected resonant frequency of the tank circuit. Gradually increase the frequency while watching the output voltage on the voltmeter. As the resonant frequency is approached the voltage will start to rise. This is caused from oscillations in the tank circuit.

There should be a fairly sharp peak when the resonant frequency of the tank is reached. As frequency increases above resonance, the voltage will drop. Tune the input frequency until the voltage is at its maximum point. The frequency indicated on the dial of the signal generator is the resonant frequency of the tank circuit.

Sometimes it is possible to measure frequencies above the upper limits of the signal-generator frequency. This is accomplished by using harmonics of the basic generator frequency to activate the test circuit. This method is not always reliable because of the low power in the harmonic frequencies. The power of the resonant harmonic may be too low to produce distinctive changes in the voltage reading.

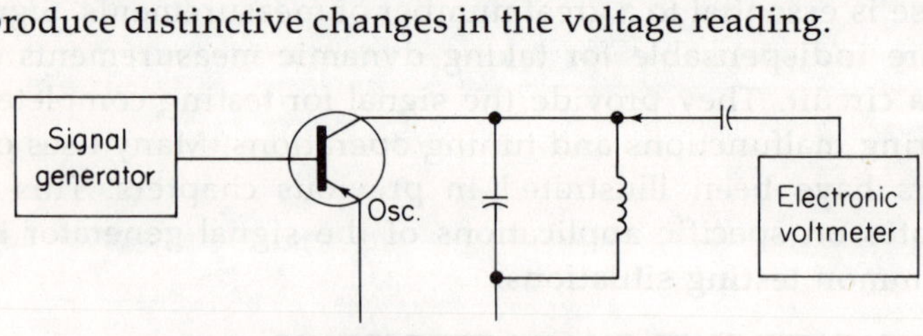

Figure 15–1 Measuring Parallel Resonance

Measuring Inductance and Capacitance

If either inductance or capacitance of a tank circuit is known, the other may be determined by carrying the resonance test one step farther. Using the resonant frequency and the value of the known component, the other can be calculated by

$$f = \frac{1}{2\pi\sqrt{LC}}$$

For example, when a 2 pF capacitor produces a resonant frequency of 5 MHz, the inductance of the coil is

$$L = \frac{\left(\frac{0.159}{f}\right)^2}{C}$$

$$= \frac{\left(\frac{0.159}{5 \times 10^6}\right)^2}{2 \times 10^{-9}}$$

$$= \frac{(0.032 \times 10^{-6})^2}{2 \times 10^{-9}}$$

$$= \frac{1024 \times 10^{-6}}{2 \times 10^{-9}}$$

$$= 0.0005 \times 10^{-3}$$

$$= 0.5\ \mu H$$

When the inductance is the known quantity, the same formula can be used to calculate the capacitance. It is only necessary to exchange places with the L and C.

$$C = \frac{\left(\frac{0.159}{f}\right)^2}{L}$$

Since the quantities are generally frequency in MHz, inductance in μH, and capacitance in pF, considerable work may be saved by keeping these units during calculation. The formulas would then become:

$$L = \frac{25{,}281}{C \times f^2}$$

and

$$C = \frac{25{,}281}{L \times f^2}$$

where L = inductance in μH
C = capacitance in pF, and
f = frequency in MHz

Measuring Distributive Capacitance

Every circuit has a certain amount of distributive capacitance. This becomes especially bothersome in transformer and inductive coupling circuits. In fact, every coil has enough distributive capacitance to become a parallel-resonant tank circuit at some frequency. This fact enables measurement of the distributive capacitance.

The test circuit is similar to that in Figure 15–1, but there is no physical capacitor across the coil. Resonant frequency is obtained in the manner previously described. The resonant frequency and the inductance of the coil are then used to calculate the distributive capacitance.

Measuring Series Resonance

The same equipment used for parallel-resonant circuits can be used for measuring the resonant frequency of a series circuit. The test circuit is similar to that shown in Figure 15–2. The signal generator and electronic voltmeter are both connected parallel to the resonant circuit. The resistor (R_1) in series with the signal generator is to limit the current when resonance is reached. The value of this resistor is not critical, but it is necessary to prevent generator overload as the impedance of the LC circuit approaches zero.

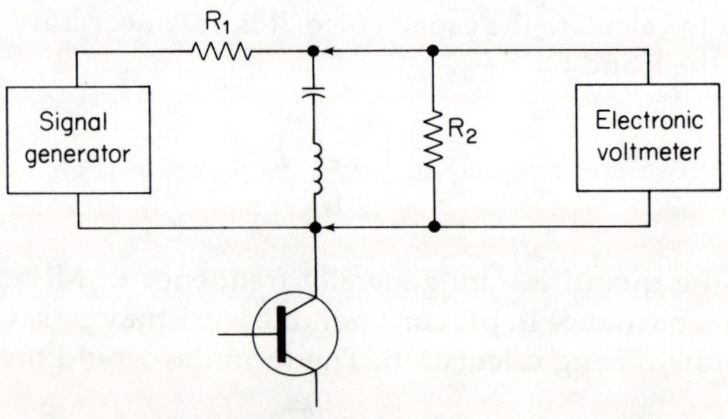

Figure 15–2 Measuring Series Resonance

The parallel resistor (R_2) is necessary to load the circuit and broaden the resonant response. Without this resistor the peak may be too sharp to detect on the meter. The value is not critical, but the smaller the resistor, the sharper the response. A 50 kΩ resistor will probably be about right.

Start the signal generator at a low frequency and gradually increase the frequency while watching for a decrease in the voltage indication. The voltage will start to drop as the resonant frequency is approached. Tune the generator for a minimum voltage indication. The output frequency of the signal generator is now the same as the resonant frequency of the LC circuit.

Avoiding False Indications

In any resonance measurement, it is possible to obtain resonance indications on harmonics of the resonant frequency. This is one reason for starting the generator at a frequency below resonance. It is also possible to obtain resonance indications as the generator is tuned through a submultiple of the resonant frequency. These indications are less likely and less pronounced than the harmonics, but they can be deceptive.

Both harmonic and submultiple indications can be avoided when the approximate resonant frequency is known. The signal generator is then set just below the suspected point of resonance. In many cases, the approximate frequency is not known. Then the only procedure is to start the frequency low and beware of small indications. After locating what is apparently the correct resonant frequency, it is a good idea to continue raising the frequency until another indication is obtained. If this is a harmonic indication, it will be much less pronounced than the first one. If all readings are questioned in this fashion, there is little chance of being taken in by a false indication.

Measuring Bandwidth

Once resonant frequency has been established, it is little additional effort to determine the bandwidth. However, when this is the intent, the input and output connections become more critical. Since loading broadens the resonance response curve, the loading must be minimized to enable accurate determination of the bandwidth. One method of accomplishing this is indicated in Figure 15–3.

The output of the signal generator is coupled through an isolation resistor (R_1) into the test circuit. This resistor minimizes the loading effect of the signal generator, but the resistor value is not critical. The electronic voltmeter has a minimum loading effect anyway, and in many cases the meter input will be coupled through an RF probe.

The signal generator is now tuned to the resonant frequency in the manner previously described. The bandwidth extends between the low half-power point and the high half-power point. These points

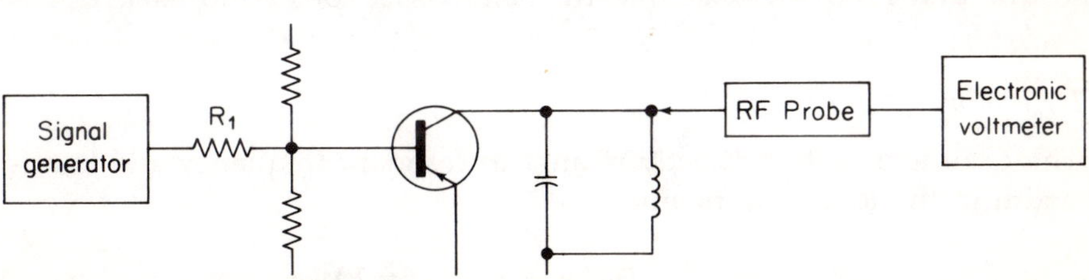

Figure 15–3 Measuring Bandwidth

are determined by locating a frequency which provides a voltage reading that is equal to 0.707 × the maximum. The voltage is maximum at the resonant frequency (f_r). Note this reading and calculate 0.707 of its value.

Raise the frequency of the signal generator until the voltage indication drops to the calculated 0.707 point. This is the high half-power point. Note this frequency and designate it as f_1.

Reduce the frequency of the signal generator back to resonance, and beyond, until the 0.707 value is encountered again. This is the low half-power point. Note this frequency and designate it as f_2. The relationship of these three frequencies is illustrated in Figure 15–4.

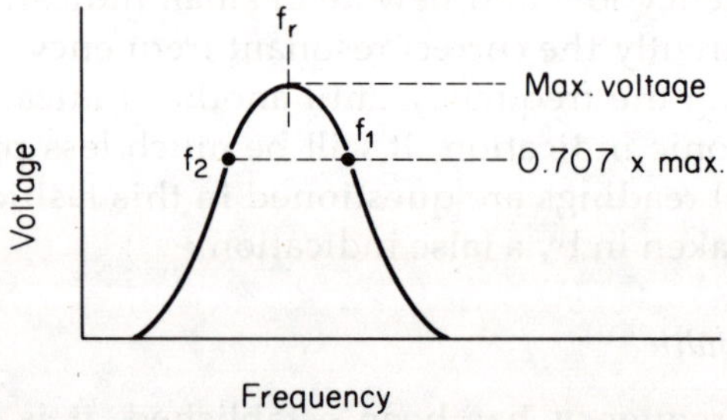

Figure 15–4 Frequency Response

The bandwidth is calculated by subtracting f_2 from f_1. For example, when high half-power frequency is 460 kHz and low half-power frequency is 450 kHz, the bandwidth is

$$\begin{aligned} BW &= f_1 - f_2 \\ &= 460 - 450 = 10 \text{ kHz} \end{aligned}$$

The terms *bandwidth* and *bandpass* are often confused. The previous explanation should have clarified the meaning of bandwidth. Bandpass is simply the actual frequencies contained within the bandwidth. In the previous example, the bandpass is 450 kHz to 460 kHz.

Determining Q of a Circuit

The Q of a circuit is directly related to resonant frequency and bandwidth in the following manner:

$$Q = \frac{f_r}{BW}; \qquad BW = \frac{f_r}{Q}; \qquad f_r = BW \bullet Q$$

For example, a circuit with a resonant frequency of 455 kHz and a bandwidth of 10 kHz has a Q as shown here:

$$Q = \frac{f_r}{\text{BW}}$$
$$= \frac{450 \text{ kHz}}{10 \text{ kHz}}$$
$$= 45$$

MEASURING SWITCHING CHARACTERISTICS

Many transistors, both bipolar and field-effect, are designed for *switching* applications. It is important to match the characteristics of the switching component to the circuit where it will be used. The switching characteristics are listed on the data sheet, and they can be measured by the use of an oscilloscope and a square-wave signal generator. The oscilloscope should be a dual trace with good transient characteristics and a wide-frequency response. The horizontal deflection should be time-calibrated.

Measuring Bipolar Transistors

In order to measure the switching characteristics of a transistor, it is necessary to place it in a properly biased circuit which closely simulates operating conditions. The signal generator then provides input pulses, and the input and output are compared on the oscilloscope. A sample test circuit is illustrated in Figure 15–5.

The transistor is properly biased and the voltages are adjusted to

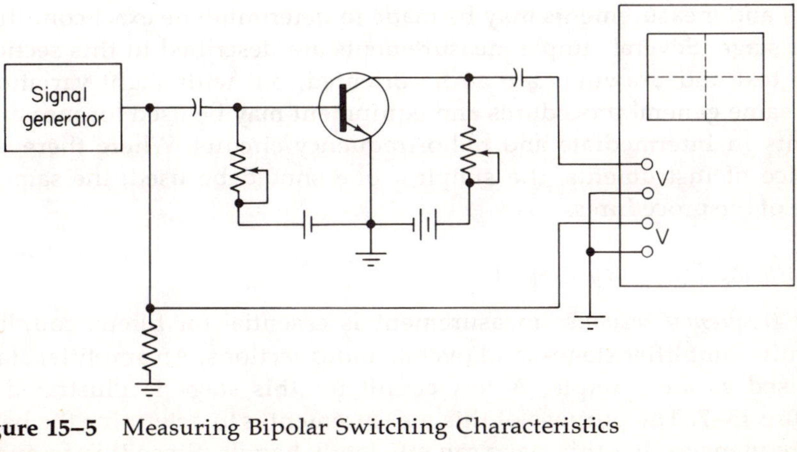

Figure 15–5 Measuring Bipolar Switching Characteristics

the values specified in the data sheet. A pulse from the signal generator is applied directly to one vertical channel of the oscilloscope. The same pulse is applied to the other vertical channel after being processed through the transistor. The amplitude and pulse width of the signal generator output should be set to the values shown on the data sheet.

The signal generator output pulse appears on one trace, and the transistor output appears on the other trace. The critical times such as rise time, delay time, storage time, and fall time can now be measured along the horizontal axis. These times must be compared to times indicated on the data sheets to determine whether or not they are within tolerance.

Measuring FET Switching Characteristics

The switching characteristics of the field-effect transistor can be measured by practically the same procedure just described for standard transistors. The circuit for the FET is slightly different, but the signal generator and oscilloscope are connected in the same fashion. This is illustrated in Figure 15–6.

The same times are critical for FETs as for any switching transistor (rise time, delay time, storage time, and fall time). These times are specified on the data sheets, and measurements are intended to verify these data. These measurements are taken as previously described, and the results are compared to the specified data.

AUDIO CIRCUITS

The signal generator is an indispensable piece of equipment when designing, testing, or troubleshooting audio circuits. A great many tests and measurements may be made to determine the exact condition of a stage. Several simple measurements are described in this section. The text and drawings are audio-oriented, but with slight variations the same general procedures and equipment may be used for measurements in intermediate and radio-frequency circuits. Where there is a choice of instruments, the simplest one should be used; the same is true of the procedures.

Measuring Frequency Response

The *frequency response* measurement is essential for filters, coupling circuits, amplifier stages, and overall audio sections. An amplifier stage is used as an example. A test circuit for this stage is illustrated in Figure 15–7. The purpose of this measurement is to determine the band of frequencies that this stage can efficiently handle. Since it is an audio

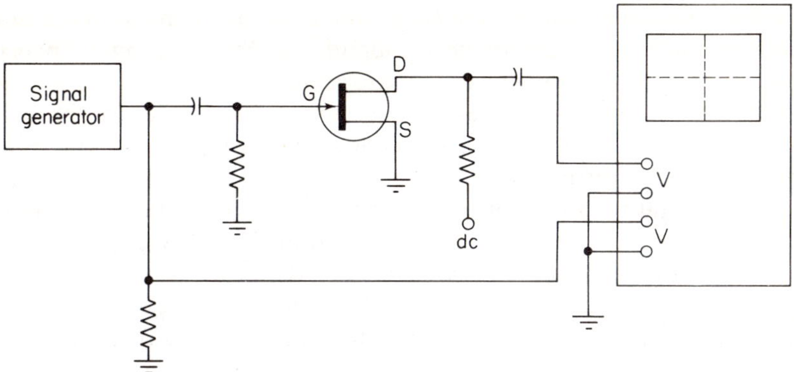

Figure 15–6 Measuring FET Switching Characteristics

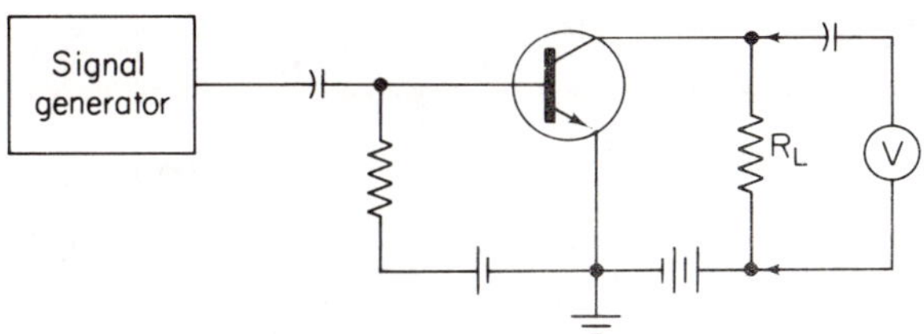

Figure 15–7 Measuring AF Response

stage, it is expected to have a flat response from about 50 Hz to something in excess of 20 kHz. A well-designed stage may have a relatively flat response from 15 Hz to more than 50 kHz.

The signal generator is coupled to the input of the stage, and the electronic voltmeter is connected across the output load resistor. For the best results, the amplifier should be in its own operational circuit. If this is not practical, the test circuit should duplicate the operational circuit as nearly as possible.

The signal generator is set to a low-frequency sine wave output, and the amplitude is set to a value representative of a reasonable audio signal. This amplitude is held constant as the frequency is slowly increased.

The voltmeter is set on ac and has no indication until the signal feeds through the amplifier. As the sine wave frequency approaches the audio band, a small voltage indication will appear on the voltmeter. The voltage will level off to a steady reading as the frequency reaches a value where the amplifier has maximum response. This should happen at about 20 Hz.

The constant voltage reading will hold until the frequency approaches the top of the response band. As the frequency increases above this value, the voltage slowly drops off.

A plot of the voltage and frequency on a simple graph such as that in Figure 15–8 will provide a clear picture of the frequency response of this amplifier stage.

The actual response of the stage is the same as the bandpass, the frequencies being between the low and high half-power points. Since the measurements and graph are in volts, the 0.707 voltage points may be used. The graph in the illustration shows a frequency response from 12 Hz to 75 kHz, and the top of the curve is reasonably flat all the way. A response such as this is obtained from an exceptionally responsive stage.

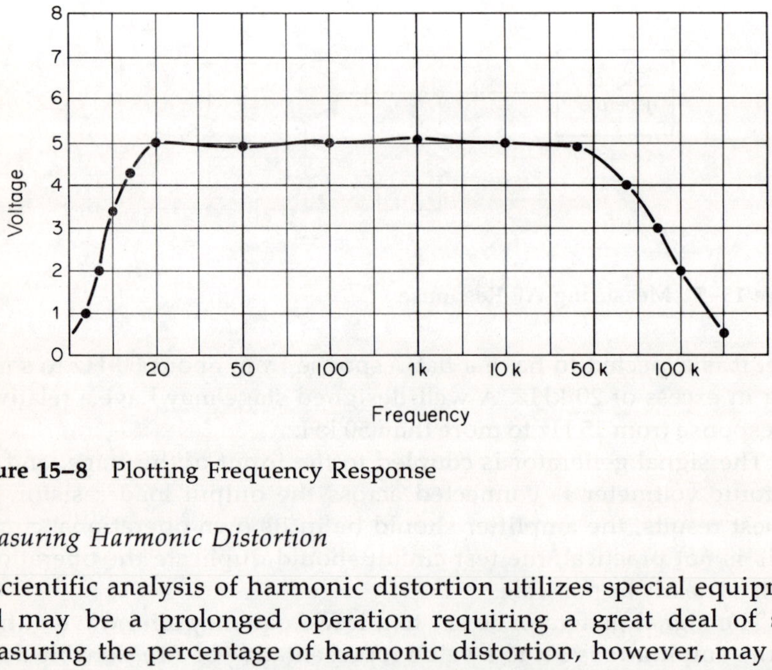

Figure 15–8 Plotting Frequency Response

Measuring Harmonic Distortion

A scientific analysis of harmonic distortion utilizes special equipment and may be a prolonged operation requiring a great deal of skill. Measuring the percentage of harmonic distortion, however, may be a simple process performed by standard shop instruments. One method is illustrated in Figure 15–9.

The procedure is designed to measure the amplifier output under two conditions: one with a filter and one without a filter. Without the filter, the output is the resultant of the fundamental frequency and all the harmonics. With the filter, the output is a result of the harmonics only. The filter eliminates the fundamental frequency.

The signal generator is set for a sine wave output with a reason-

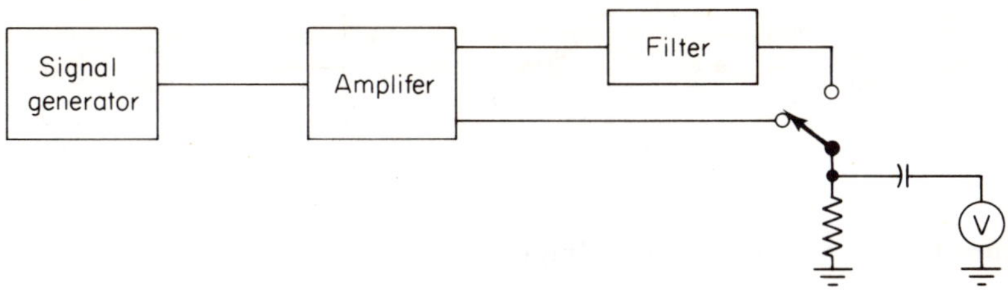

Figure 15–9 Measuring Harmonic Distortion

able voltage value. The filter and the signal generator are then tuned to the same frequency. Some filters can be adjusted and some cannot. But the frequency of the signal generator is adjustable. This may be accomplished by connecting the signal generator directly to the filter and tuning for minimum indication on the voltmeter.

Then with the circuit connected as shown in Figure 15–9, set the signal generator for a strong signal, but be careful not to overdrive the amplifier. Now take one voltage reading without the filter (V_1) and another reading with the filter (V_2). The harmonic distortion percentage is then calculated by

$$\text{Harmonic Distortion Percentage} = \frac{V_2}{V_1} \times 100$$

For example, when the voltage without a filter is 150 mV, and with a filter it measures 5 mV, the percentage of harmonic distortion is

$$\begin{aligned}\text{Harmonic Distortion Percentage} &= \frac{V_2}{V_1} \times 100\\ &= \frac{5\text{ mV}}{150\text{ mV}} \times 100\\ &= 0.033 \times 100\\ &= 3.3\%\end{aligned}$$

Measuring Input Impedance

The signal generator and the electronic voltmeter may be used to measure input impedance by the resistance substitution method. This method is highly accurate when the impedance is mostly resistance and is an approximation when the impedance is mostly reactance. Since impedance must always be matched for maximum transfer of energy, an easy way to measure impedance is a great asset. A test circuit for measuring input impedance is illustrated in Figure 15–10.

The signal generator is set to the type of signal and frequency de-

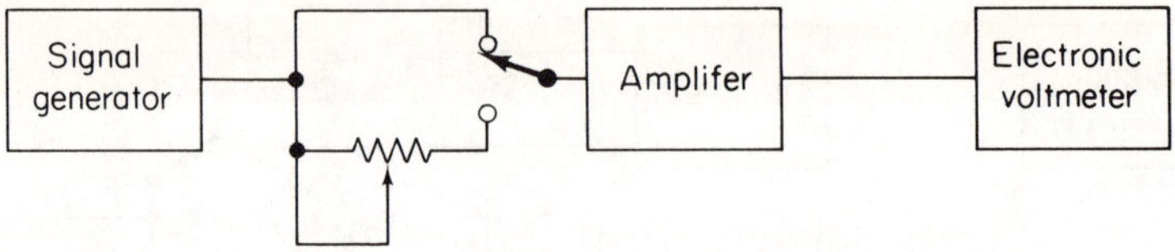

Figure 15–10 Measuring Input Impedance

sired. These should be, as nearly as possible, duplications of what the amplifier will be expected to handle in its own circuit. The test circuit is connected as shown, with the signal generator directly connected to the amplifier input and the electronic voltmeter connected across the amplifier output.

The amplitude of the signal generator output is adjusted to obtain a strong reading on the voltmeter, at least a half-scale deflection. Record this reading. The switch is then changed to place the rheostat in series with the amplifier input. The rheostat is adjusted until the voltmeter indicates exactly half the previous value. Remove the rheostat and measure its resistance; this is equal to the input impedance of the amplifier.

The *rheostat* test circuit loses accuracy when the input impedance is very low (only a few hundred ohms). In this case, a *potentiometer* should be used as illustrated in Figure 15–11.

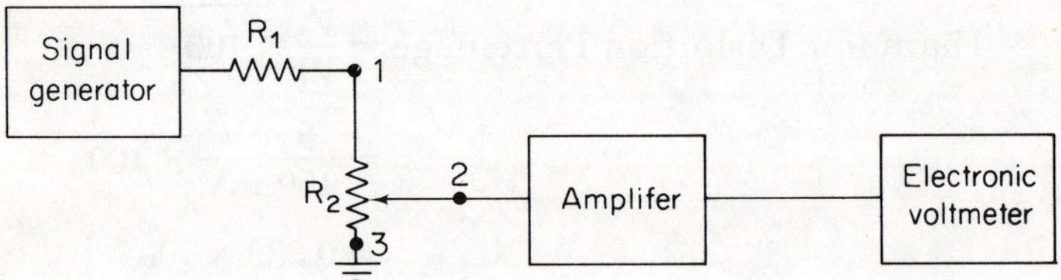

Figure 15–11 Measuring Low Input Impedance

The series resistor (R_1) should be at least *10* times the value of the input impedance; otherwise its value is not critical. The potentiometer (R_2) is set to maximum resistance, with points 1 and 2 shorted directly together. The signal generator output level is adjusted for a convenient reading on the voltmeter, and this reading is recorded. Now R_2 is adjusted until the voltage indication drops to half the recorded level. The resistance of R_2, measured between points 2 and 3, is the same as the input impedance of the amplifier.

Measuring Output Impedance

The output impedance of a circuit or a stage can be measured with a slight variation of the previous procedure. The instruments are connected as shown in Figure 15–12.

The signal generator is connected to the input of the stage being tested, and the electronic voltmeter is connected across the output. The signal generator is set to the desired signal and frequency. The amplitude of the signal-generator output is then increased until a convenient value of ac voltage is indicated on the voltmeter. Record this voltage reading.

Now the rheostat is connected parallel to the amplifier output. The rheostat is adjusted until the voltage drops to half the recorded value. The rheostat is then removed from the circuit and measured. The resistance of the rheostat is equal to the output impedance of the stage under test.

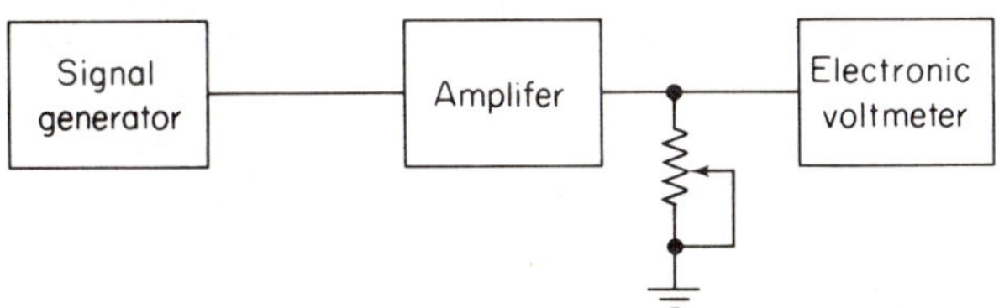

Figure 15–12 Measuring Output Impedance

MEASURING MAGNETIC COMPONENTS

There are certain checks and measurements that can be made on electromagnetic devices such as microphones, speakers, and transformers by using the signal generator. Microphones and speakers need to be checked for distortion and frequency response. Sometimes it is necessary to check transformers for phase relationship, polarity of marking, and percentage of regulation. All of these tests, and many others, can be made with a signal generator, a voltmeter, and an oscilloscope.

Checking Microphones

This procedure assumes that a matching speaker of known quality is available. The frequency response of the *microphone* and the distortion of the signal caused by the microphone can both be checked with one simple arrangement. The equipment and instruments are arranged as indicated in Figure 15–13.

The signal generator is connected to a speaker which has been verified as having good frequency response and being free of distortion. The signal generator is set for a sine wave output and adjusted for a desired amplitude. This output is connected to one vertical channel of a dual-trace oscilloscope as well as to the speaker.

The microphone under test is connected to an amplifier and placed in a position to enable good pickup of the sound from the speaker. The microphone should be shielded from all other sound as nearly as possible. The output of the amplifier is connected to the other vertical channel of the oscilloscope.

The two waveforms can be viewed simultaneously, and distortion is readily apparent. The signal generator can now be tuned through the rated frequency band of the microphone while watching the display for changes in amplitude or evidence of distortion. The amplitude of the signal generator output must, of course, be held constant while changing the frequency. Any change in amplifier output amplitude indicates that the microphone is responding differently to the different frequencies. A decrease in amplitude shows a low response to that frequency.

If there is any doubt that the amplifier is at fault instead of the microphone, a further check is indicated. The amplifier can be checked by removing both microphone and speaker and connecting the signal generator directly to the amplifier input.

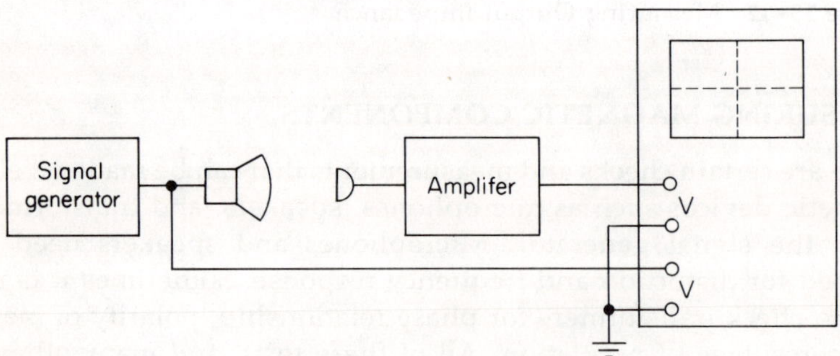

Figure 15–13 Checking Microphone

Checking Speakers

The same procedure described for checking microphones is equally effective for checking speakers. In this case, the microphone and amplifier must be of a known reliable quality. The equipment and instruments are connected as shown in Figure 15–13. Now any distortion or lack of frequency response is caused by the speaker.

Determining Phase Relationships between Windings of a Transformer

A transformer can be a puzzling device when it is disconnected from the circuit and the phase markings are not clear. Many perfectly good transformers have been discarded because the user was uncertain of the phase relation of primary to secondary. This need not happen. A signal generator and a voltmeter can be used to determine the phase relationship, and the check requires only a few moments.

The signal generator is connected to the primary winding and set to the desired frequency and amplitude. The voltmeter is then used to measure the voltage from the secondary. The values of input and output voltages are recorded. Then connections are made as illustrated in Figure 15–14.

Pins 2 and 4 are shorted together, and the voltage is measured between pins 1 and 3. If pins 1 and 3 are in phase, this reading will be equal to the difference between the first two (primary and secondary) measurements. If pins 1 and 4 are in phase, this reading will be the equal to the sum of the first two measurements. For example, with a primary voltage of 10 V and a secondary voltage of 30 V, the following results are obtained:

1. If 1 and 3 are in phase, the voltage between pins 1 and 3 is 20 V.
2. If 1 and 4 are in phase, the voltage between pins 1 and 3 is 40 V.

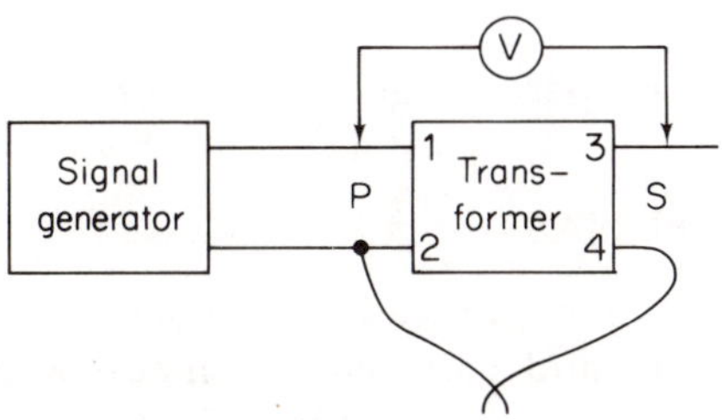

Figure 15–14 Primary-to-Secondary Phase Check

Measuring Transformer Regulation

The *regulation* of a transformer is its ability to maintain a constant output voltage with a changing load. The percentage of regulation can be measured with a test arrangement similar to that in Figure 15–15.

One output voltage reading is taken with the secondary open. If this reading changes only slightly when the switch is closed, regulation is good. The rheostat is then decreased until the ammeter indicates the maximum rated current for the transformer. The voltage indication should remain fairly constant during this change.

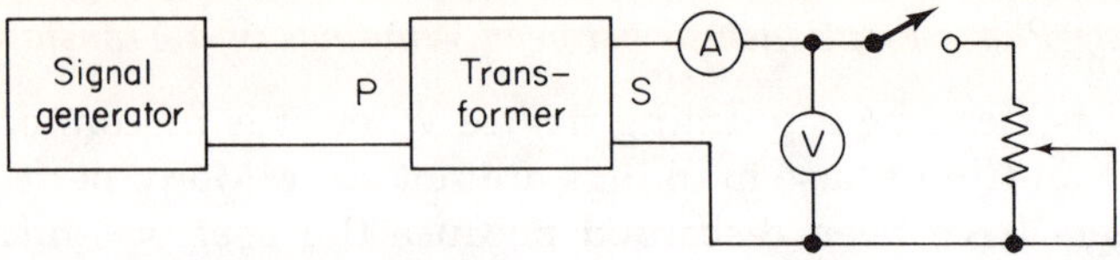

Figure 15–15 Measuring Regulation

If no load voltage is designated as V_1 and the voltage under maximum current conditions (load voltage) is designated as V_2, the percentage of regulation can be calculated as follows:

$$\text{Regulation Percentage} = \frac{V_1 - V_2}{V_1}$$

For example, when no-load voltage is 50 V and full load is 48 V, the percentage of regulation is:

$$\begin{aligned}\text{Regulation Percentage} &= \frac{V_1 - V_2}{V_1} \times 100 \\ &= \frac{50\text{ V} - 48\text{ V}}{50\text{ V}} \times 100 \\ &= \frac{2}{50} \times 100 \\ &= 4\%\end{aligned}$$

The lower the regulation percentage, the better the regulation.

MEASURING ANTENNAS

Several very practical antenna measurements, such as those for antenna resonance, impedance, and gain, can be made with little trouble by utilizing common basic test instruments. These measurements can be performed with a signal generator and an ordinary electronic multimeter.

Measuring Resonant Frequency

The resonant frequency of an antenna is easily measured with a signal generator and an ammeter. The test circuit is illustrated in Figure 15–16. The signal generator is set for the desired output level and a frequency well below the suspected resonant frequency of the antenna. The amplitude of the signal generator output is then held constant as the frequency is gradually increased. The frequency that causes the maximum current indication is the resonant frequency of the antenna.

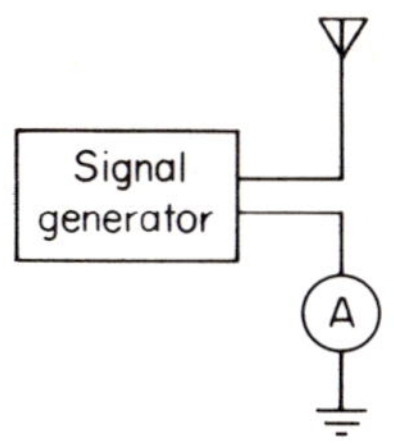

Figure 15–16 Measuring Resonant Frequency

Measuring Antenna Impedance

The impedance of an antenna, along with its associated transmission line, can be easily measured by the resistance substitution method. A test circuit for this measurement is illustrated in Figure 15–17.

The signal generator is adjusted to the resonant frequency of the antenna as previously described, and the current indication is recorded. Now the switch is changed to position 2, and the capacitor is set for maximum current. The rheostat is now adjusted until the previously recorded indication is obtained on the ammeter. At this time, the resistance of the rheostat is equal to the impedance of the antenna and its associated transmission line. This is the radiation impedance.

An assumption was made here that the antenna was tuned with an inductor. Therefore, a tunable capacitor was used in the substitution circuit. If the antenna is tuned with a capacitor, the substitution circuit should contain a tunable inductor. The procedure is the same in either case.

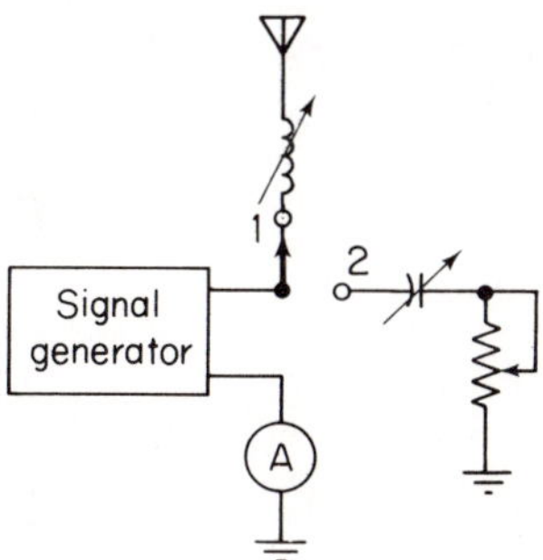

Figure 15–17 Measuring Impedance

Measuring Antenna Gain

The gain of a receiver antenna is a ratio between the voltage developed on the antenna and the voltage developed on a reference dipole of the same size. The reference dipole is constructed of a thin wire, and its gain provides the reference of 0 dB. The antenna gain is generally plotted on a graph in terms of dB above this reference level. When the voltage on the antenna is designated as V_1 and that on the reference dipole as V_2, then

$$\text{Gain} = 10 \log \frac{V_1}{V_2}$$

Figure 15–18 illustrates a test circuit for making this measurement. The signal generator is connected to a transmitter antenna and takes the place of the transmitter during this test. The signal generator is set to the desired frequency and amplitude.

The reference dipole is then placed in a predetermined position, and the voltage induced on it is measured with an electronic voltmeter. The voltage is recorded as V_2.

The reference dipole is then removed, and the test antenna is put in the exact place and position previously held by the dipole. The induced voltage on the antenna is now measured and recorded as V_1. Care must be exercised to put the antenna and reference dipole in exactly the same place and position during the voltage measurements. It is also assumed that all other conditions are identical.

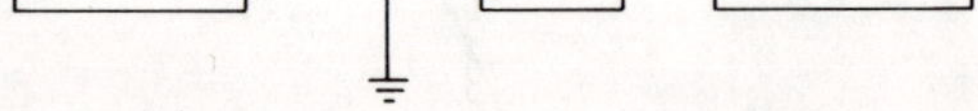

Figure 15–18 Measuring Gain

part three

TECHNIQUES FOR MAINTENANCE, TROUBLESHOOTING, AND MALFUNCTION ANALYSIS

The best instruments in the world will not compensate for a lack of professional techniques. The best instrument available is the mind. It is used as an indispensable part of every test and measurement. The purpose of this part of the book is to outline logical, time-tested techniques that should be practiced. These techniques are not purported to be the ultimate with no room for improvement, but they do work and should be followed until others are proved to be better.

chapter sixteen

ELECTRONIC SYSTEM MAINTENANCE

The definition of an *electronic system* will be used here in its broadest form. Any group of electronic devices integrated to work together for a common purpose is covered by this definition of a system. A *transceiver* is a system. It combines a transmitter and a receiver into a common case, and its purpose is two-way communications. Systems range from the transceiver to huge electronic complexes. Along the way there are broadcasting systems, navigation systems, satellite tracking systems, relay systems, telephone exchange systems, computer systems, and many systems too numerous to mention.

The purpose of such a breakdown is to distinguish an electronic system maintenance environment from that of the laboratory and from the electronic equipment repair facilities. Most engineers and technicians will, sooner or later, be engaged in electronic system maintenance. This chapter is intended to provide an overall picture of this environment and to simplify the many tasks that must be performed in system maintenance.

THE LABORATORY

Electronic personnel engaged in *laboratory* activities normally spend most of their time in creative pursuits, such as research, design, building, and testing of electronic devices. As a rule, the laboratory provides a variety of modern, sophisticated instruments. Many of these instruments and testing practices have been covered in previous chapters. Chapter 18 includes some laboratory projects of a general nature and further use of instruments. Portions of this chapter and the entire Chapter 17, Troubleshooting, are also applicable to the laboratory environment.

EQUIPMENT REPAIR

The average electronic *repair shop* is limited in scope despite the essential service these shops perform in society. The repair shop is generally an operation licensed by a community to engage in specified types of repair and service on electronic devices. The majority of the work in this environment is *troubleshooting*, *repair*, and *alignment* of commercial entertainment equipment. The troubleshooting chapter should be particularly helpful to people associated with the repair environment. In addition, some aspects of system maintenance are applicable to this type of repair work.

SYSTEM MAINTENANCE

In the system maintenance environment, the prime objective is to keep the system functioning efficiently with a bare minimum of downtime. Activities include both preventive and corrective maintenance, but the emphasis is on preventive maintenance. A great deal of skill is required in this environment, and the ability to work under pressure is a decided asset. Proficiency must be developed and maintained at a high level, and this is difficult to accomplish. When the equipment is functioning properly, there is little chance to practice; when it is not functional, the urgency to restore operation builds to near panic in a very short time.

Knowledge of the Equipment

A thorough knowledge of the equipment which composes the system is imperative. This knowledge is available from four sources: (1) technical manuals, (2) other experienced personnel, (3) the equipment record, and (4) personal experience. All of these must be utilized to compensate for the fact that there is little time to practice on the equipment. The proficient maintenance technician will know the capabilities and limitations of his system. He will know the theory of operation and be able to read and interpret all schematics, wiring, and block diagrams. He will know the location of each test point and will be able to use all the available test instruments.

Most systems are equipped with built-in indicators such as meters, counters, and trouble lights. These are intended to reveal when the system is functioning properly and to provide trouble symptoms when it is not. Some built-in indicators help to isolate the problem when a malfunction develops. The technician who views any part of the built-in indicators as a mystery has not completed his education, and he is not as proficient as he should be. He must be thoroughly familiar with all normal indications and be able to evaluate the significance of any abnormal indication.

The location and function of each control are essential knowledge. Manipulation of operating controls can peak up the performance during normal operation. The same controls may intensify the trouble symptoms and furnish valuable clues as to the nature and seriousness of troubles.

More than 50% of the downtime (inoperative time) of a system is directly the fault of the technicians who maintain it. At first glance that appears to be a rash statement, but it is a fact. When the system fails because of improper preventive maintenance, the entire downtime could have been avoided. When more than the minimum required time is used to repair a trouble, a part of that time should have been saved. In either case, the fault rests with the technicians. All conscientious personnel want to be responsible for as little downtime as is absolutely necessary, and the procedures here should help toward that end.

Using the Maintenance Record

Maintenance personnel of most major systems are required to keep a maintenance record. This is sometimes in the form of a log and sometimes a more detailed record. In any case, it should be a complete history of maintenance. It should show all malfunctions, how they were corrected, and the extent of the downtime. Such a record should be kept, even if it is not required. Knowing what the equipment has done in the past furnishes clues as to what it will do in the future.

Knowledge of past malfunctions (symptoms, tests, corrective actions, etc.) can save valuable time when a failure occurs. There probably will be no time for studying the record when the equipment is down, but there is plenty of time between failures for this purpose. This will help the diligent technician to avoid previous mistakes, whether they are his or another's. It will enable him to avoid blind alleys and fruitless efforts that have already been explored.

Preventive Maintenance

Improving the proficiency of the technician results in improved efficiency of the system. Therefore, self-improvement is a form of preventive maintenance. It is a very profitable form of preventive maintenance when you consider the fact that a large portion of the downtime could be avoided by improved skill and better techniques.

Preventive maintenance is the best type of system maintenance. It can be performed with the equipment in operation as part of the daily routine. Preventive maintenance tasks are listed, detailed, and scheduled in the maintenance manuals. These tasks are calculated to achieve maximum efficiency of system performance, insure continuity of service, reduce the number of breakdowns, and lengthen the useful life of the system. These tasks consist principally of cleaning, testing, lubricating, and inspecting the equipment. They are aimed at early detection of conditions which can cause trouble and the correction of these conditions before they can lead to equipment failure.

Preventive maintenance inspections can be grouped into three categories: continuous, daily, and periodic. Monitoring the system's built-in indicators for any deviation from the norm should be a continuous process. This is a running operational check for peak system performance. It should be made at least several times during a work shift. Any tendency from the norm should be recorded, carefully observed, and analyzed.

The daily inspection is still a visual inspection. It includes monitoring for proper operation, but it also includes several other mechanical aspects of the equipment. The daily inspection should detect such items as dirt, corrosion, loose connections, mechanical defects, faulty insulation, and other visual signs of impending trouble. All faults should be logged and scheduled for correction in accordance with the seriousness of their nature. Minor adjustments may be included in these daily inspections to peak up the voltage or current readings.

Periodic inspections are scheduled in the maintenance manual. These are functional checks, tests, measurements, adjustments, and other actions essential to continued trouble-free operation. Bench tests as well as replacement and maintenance of parts are included.

Inspection Routines

Figure 16–1 is a sample of a *preventive maintenance inspection* (PMI) schedule. This is derived from a maintenance manual for a typewriter keyboard.

The PMI routine number identifies the procedure to be followed. The maintenance log or record shows when the PMI was last performed and what maintenance was performed at that time. Each of the routines is spelled out in great detail. Here is a sample routine from the same maintenance manual.

Recommended schedule		
Routine number	Interval in days	
	Continuous operation	Standby operation
1–1	45	90
1–2	45	90
1–3	30	90
1–4	45	90
1–5	90	120
1–6	As required	As required

Figure 16–1 Partial Inspection Schedule

ROUTINE NUMBER 1–4—EQUIPMENT DESIGNATOR TYPEWRITER KEYBOARD

PMI Routine—Typewriter-Ribbon Feed Check

1. General Information
- **(a)** This procedure checks the condition of the typewriter-ribbon feeding mechanism.
- **(b)** Record all maintenance activities executed in the performance of this routine in the equipment maintenance record.
- **(c)** Approximately 15 minutes are required to perform this routine.

2. Test Instruments Required
- **(a)** Electronic Circuit Plug-in Test Set

3. Materials Required
- **(a)** Six-inch scale.
- **(b)** Hand-cycle tool.
- **(c)** Other hand tools as indicated by visual inspection.
- **(d)** Routine 2–1, Ribbon Mechanism.

4. Procedure
- **(a)** Remove power from typewriter.
- **(b)** Disconnect typewriter from associated equipment.
- **(c)** Connect typewriter to test set.
- **(d)** Place paper in typewriter.
- **(e)** Turn on ac power to test set.

CAUTION: Damage may result if typewriter is turned on before test set is energized.

- **(f)** Turn on typewriter.
- **(g)** Unlock typewriter keyboard.
- **(h)** Move ribbon lift lever to high position.
- **(i)** Install a new ribbon.

(j) Type a diagonal (/) about 12 times while checking for ribbon spill or jerk. If either condition is noted, adjust tension on ribbon ratchet spring.
(k) Use six-inch scale and check distance from bottom edge of ribbon to the lowest portion of the diagonal imprint. Adjust ribbon lift control link as necessary to make this distance equal to 1/16 in.
(l) Turn off ac power to test set.
(m) Remove top and bottom typewriter covers.
(n) Depress a key lever while manually unlocking keyboard by pushing lock bail to the rear.
(o) Hand-cycle a character and observe that the ribbon feed pawl feeds two teeth of ratchet plus a slight overthrow. If feed is incorrect, adjust the ribbon feed plate.
(p) If above procedures fail to make corrections or other deviations are noted, perform Routine 2–3, Ribbon Mechanism Adjustment.
(q) Replace covers.
(r) Return typewriter to operational status.

A complex system may have several books filled with these PMI routines. The previous sample was given to provide an idea of the type of detailed guidance that is available.

Corrective Maintenance

Corrective maintenance consists of those activities necessary to restore operation after the system has failed. This includes troubleshooting, repair, and testing. The *troubleshooting* step locates the faulty component. The *repair* step repairs the equipment by adjusting, repairing, or replacing the faulty part. After repair, the system is subjected to a complete operational *check* to make sure the fault has been corrected.

Troubleshooting consists of analyzing the symptoms and locating the trouble. The time required for troubleshooting is generally inversely proportional to the technician's knowledge of the system and his ability to use the available test instruments. In the long run, corrective maintenance is no more important than preventive maintenance. Both types are essential to proper operation over a prolonged period. No matter how conscientiously the preventive maintenance is performed, electronic parts do fail. When this occurs, the skill at corrective maintenance takes on the primary role.

Troubleshooting is common to all aspects of electronic maintenance and must not be identified with system maintenance alone. For this reason, troubleshooting is treated at length in the next chapter.

chapter seventeen

TROUBLESHOOTING

Troubleshooting is a logical procedure that should be applied to the isolation of trouble in all types of electronic equipment. There is little difference in the procedure between troubleshooting for a transistor radio or for a complex electronic system. The technician who approaches all troubleshooting problems in a logical, systematic manner will consistently locate troubles faster than one who tries haphazard shortcuts. Occasionally a hit or miss method of troubleshooting will locate a trouble. A lucky guess may pay off once, but the chances are against it. Many valuable man-hours are lost by individuals who fail to apply the principles of reasoning in their troubleshooting procedures.

No amount of logic can replace a knowledge of the equipment, and no troubleshooting should be attempted without this knowledge. Armed with this knowledge and a logical approach, the technician can functionally divide any electronic equipment and test it in an orderly and professional manner.

THE SIX-STEP PROCEDURE

The six-step procedure in troubleshooting is a standardized approach designed to save time and needless repairs. A great many highly proficient technicians and engineers probably cannot name these six steps, but it is a safe bet that they do apply them either consciously or unconsciously. Here are the six steps:

1. Recognizing symptoms.
2. Enhancing the symptoms.
3. Listing probable faulty functions.
4. Isolating the faulty function.
5. Localizing trouble to a specific circuit.
6. Isolating the faulty component.

Step 1—Recognizing Symptoms

The first step is the recognition of symptoms. A trouble symptom is a sign that the equipment has a disorder or a malfunction. This implies that there is a standard set of normal indications for comparison. The normal symptoms are usually pretty well impressed on the troubleshooter's mind from day-to-day observation. Slight deviations from the norm are probably recorded in areas where a record is maintained. Without knowledge of the equipment and its normal indications, any trouble symptoms short of a complete breakdown are likely to pass unnoticed.

The normal indications for a black and white television receiver are as follows:

1. A clear, properly contrasted picture of an actual scene.

2. The picture is centered within the vertical and horizontal boundaries of the screen.

3. A clearly understandable reproduction of voice and music sounds.

Possible abnormal indications or trouble symptoms of the same television receiver are:

1. Picture rolling vertically.
2. Lack of contrast.
3. Picture tearing.
4. Garbled or distorted sounds.

The symptoms just listed are readily *recognizable* because they are deviations from the normal indications. Symptoms in other equipment may be a trouble light, a low current reading, voltage out of tolerance, or a loss of power.

During the process of performing its assigned job, electronic equipment yields information that can be seen or heard. The experienced technician knows the normal sights and sounds. Thus, the senses of sight and hearing are the first to receive a trouble symptom. Trouble symptoms must be recognized as indications of abnormal performance.

Complete breakdown, of course, is the easiest trouble symptom to recognize. This is probably the one seen most frequently in the repair shop. The equipment simply does not function. Some of these failures are evidenced by: no sound from a speaker, no picture on a CRT, and zero indications on panel meters.

All audible and visual indications may be present, and yet the set may have abnormal performance. When the audible and visual indications do not conform to the manufacturer's specifications, the performance is *degraded.* Degraded performance is a forerunner of equipment breakdown and should be corrected at once. These degraded performance indications range all the way from near-perfect performance to near-total lack of performance. The equipment is still performing its job, but not as well as it should.

Whatever the nature of the trouble symptoms, they must be recognized as a first step in troubleshooting.

Step 2—Enhancing the Symptoms

The second step is intended to extract additional information from the symptoms observed. A particular symptom may be caused by any one of several equipment faults. Jumping from here into schematic analysis can lead to involvement in unrealistic and unnecessary pursuits. The symptom needs to be aggravated and analyzed carefully to extract all the information it is capable of giving. This is a process of symptom elaboration to obtain a more detailed description of the symptom.

Manipulation of operating controls and switches will produce a variety of changes in the circuit conditions. These changes will alter current or voltage readings by changing resistance or impedance of the circuit. This can cause detrimental effects in equipment performance. Observation of all indicators while making changes will, in many cases, reveal some definitive details about the symptoms already observed.

During control manipulation, care must be exercised not to overcontrol. Faulty placement of controls may have caused the present trouble. Additional damage may result if controls are handled carelessly. Electronic circuit components have definite maximum values of voltage and current. If these values are exceeded, burned-out components could be the result. A knowledge of the circuit changes which take place when a control is turned will enable the troubleshooter to think ahead and avoid any damage the control could cause.

In many cases, the trouble symptom will be produced solely by an improper setting of a control or switch. This is not always a careless misadjustment; controls can be moved accidentally. When this happens, a proper adjustment of the controls corrects the trouble and results in a disappearance of the trouble symptom. Many television and radio receivers have been turned in for repair because the power cord was disconnected or some control was set incorrectly.

Even if there is no fault with the controls, the time spent checking them is time well spent. The controls affect a good many circuits and should furnish information for further defining the symptoms. For example, a stereo AM-FM receiver has distorted sound. When the control is moved from AM to FM, the symptom disappears. This eliminates all the FM circuits as possible causes of the trouble, and so hastens the solution of the problem.

Next, the observed symptoms must be fully evaluated in order to completely clarify the symptoms. The indications must be evaluated in relation to one another and in relation to the overall operation of the equipment. It is a good idea to record each bit of data as it is obtained. List the original symptoms; make notes on how the symptoms change as the controls are moved; if moving a control has no effect, make a note of that too. Now sit back for a moment and digest the information before starting the next step.

Step 3—Listing Probable Faulty Functions

Steps 3 and 4 are not always needed; their performance depends on the complexity of the equipment. For instance, a small transistor radio receiver has only one function, so listing the probable faulty functions and isolating the faulty function are meaningless. In this case the troubleshooter should proceed directly to step 5 and start to localize the trouble to a specific circuit.

On the other hand, a transceiver could be considered as having at least six functions. These are: sound pickup, modulation, transmitter, antenna assembly, receiver, and power supply. Any equipment that performs more than one function should be given the benefit of all six troubleshooting steps.

Therefore, we turn again to step 3. The symptoms have been recognized and enhanced by careful analysis. Now, what functions are disabled? Trouble in several places at the same time is not very likely. Is there a single functional area that could cause all the observed symptoms? The symptoms should be studied again—this time along with a functional block diagram of the equipment. Eliminate each functional area if it alone cannot account for all the observed indications, and those that remain comprise a list of suspects. This elimination process has reduced the problem to the realm of possibility. There remain a lot of circuits to be checked, but not the entire system.

Consider the diagram in Figure 17–1. This is a functional block diagram of a standard AM transceiver. Each block performs a specific function, so this is a symbolical representation of the functional units within the equipment. From this diagram and a few carefully analyzed symptoms, two, three, four, or five functions can be eliminated. For example, the original symptoms are the lack of reception and the fact that manipulation of the receiver controls has no effect. The power-on light and the dial lights of the receiver are still lit. Out of the six functional units, only three could be at fault. These are: power supply, antenna, and receiver. That's a good start, but a further reduction of possible faulty functions may be possible.

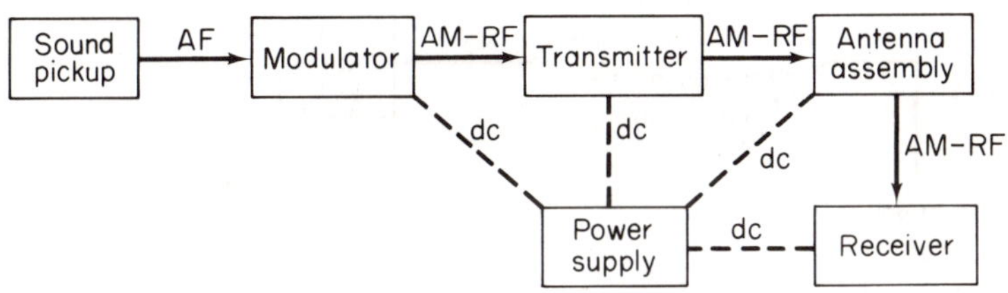

Figure 17–1 Transceiver Functional Diagram

The three suspect units should now be subjected to a test on a faulty unit selector. This selector is constructed as illustrated in Figure 17–2.

Take each of the suspected units in turn and ask, "Can a fault in this unit cause the original symptom?" Knowledge of the equipment and the functional diagram should furnish enough information to answer this question with a definite Yes or No. If not, refer to schematics, wiring diagrams, and any other source of information. The answer must be correct. If the answer is No, eliminate that unit from further consideration. If the answer is Yes, pose the second question, "Can a fault in this unit produce all the associated information?" Again the question is answered Yes or No, after carefully considering all the facts. The answer must be accurate. If it is No, eliminate the unit from further consideration. If it is Yes, keep the unit on the list of suspects.

Since the receiver controls had no effect on the symptom, the second question eliminates the antenna as a possible cause. A faulty antenna could block reception, but it could not prevent the volume control from varying the noise level. This step reduces the possible faulty units to the receiver and the power supply. These units get a positive Yes as an answer to both questions; they remain as suspects.

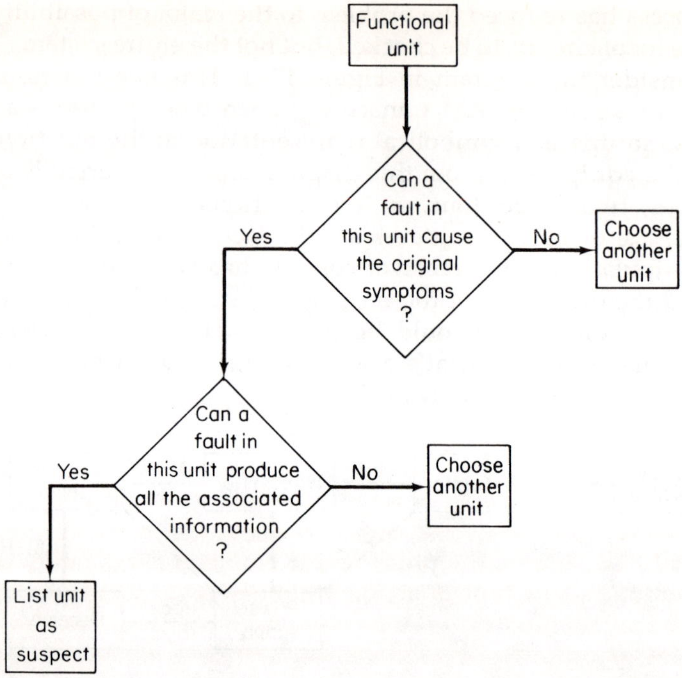

Figure 17–2 Faulty Unit Selector

The trouble is still unknown, but the total area for testing has been reduced by two-thirds. This will undoubtedly save a lot of time that would otherwise have been wasted in testing in areas that could not possibly be at fault. Each of the two remaining units could be at fault, but so far they are only suspect. Each unit has been selected by the analysis of technically valid evidence, but each is only a probable source of trouble. The next step in the procedure will narrow the search area still further.

Step 4—Isolating the Faulty Function

Step 4 is intended to save additional time by logically isolating the functional unit that is most probably faulty. Thus far the procedure has dealt with apparent and obscure performance deficiencies. This led to a logical selection of probable faulty units. No test instruments have been used, except built-in meters and indicators, and no dust covers or drawers have been removed. All actions have been either superficial or rational but the scope of the problem has been drastically reduced. Now, from the list of probables, it is necessary to isolate the functional unit that is responsible for the failure.

Testing will be done during this step, but it should be of a limited nature. These tests combined with the same logical process should lead to a definite identification of the faulty function. Which unit on the list of suspects should be checked first? Consider the block diagram, accessibility of test points, and past experience with troubles of this nature. Consider which unit will yield the most information, no matter whether it proves good or bad. Remember that test points that require disassembly of the equipment for access should be avoided as a first choice. Further, considering past history, check if the past three failures of this type were caused by power-supply problems, and if so, check the power supply.

The results of each test should lead to the next logical test, and the combined tests should isolate the faulty function. This system will succeed if all actions up to this point have been accurate, but mistakes do occur at times. When the suspected faulty units are proved to be good, this shows a previous wrong conclusion some place along the line. Backtrack all the way to step 1. Reevaluate the data; the written notes will help. Locate the point where reasoning was faulty and pick up a new track. This type of action will be necessary only occasionally, provided that the first three steps have been followed carefully. When it is necessary, the troubleshooter should be armed with extra data which enable him to zero in on the faulty area in a very short time. At least he now knows some units that cannot be at fault.

Even when preliminary tests indicate that a particular unit is bad, hasty conclusions should be avoided. The troubleshooter should draw on his equipment knowledge and think back over the ground that he has covered. He should ask, "What trouble symptoms can be produced when this unit fails?" If this leads him back to the original symptoms and accounts for all the associated information, he can be sure that he has the right unit. The matching of the theoretical symptoms with the actual symptoms should leave no room for doubt.

Step 5—Localizing Trouble to a Specific Circuit

In the case of very simple equipment, the troubleshooter may arrive at step 5 directly from step 2. With more complex equipment, he may arrive only after careful consideration of all aspects of the information uncovered through the first four steps. Now he can break out the test instruments and remove the dust covers. But when doing so, he knows that he is not performing a useless task. He knows that the trouble must exist in this functional unit.

If the faulty unit is a radio receiver, the test instruments should consist of a multimeter and an oscilloscope. Some other specialized instruments may be indicated for other units. But the troubleshooter should not begin arbitrary measurements. He should first consult the *servicing block diagrams* in his maintenance manuals. A sample of such a diagram is illustrated in Figure 17–3.

The servicing block diagram separates the units into circuit groups. Each of these circuit groups performs an electronic subfunction that is vital to the task assigned to this functional unit. This diagram is for the receiver section of a transceiver, but there are other diagrams for all other functions. Notice that this diagram is broken into five groups. They are: RF section, frequency converter section, IF section, detector section, and audio section. The task to be performed in troubleshooting in step 5 is to localize the fault to a circuit, but first it must be determined which of these five circuit groups contains the malfunction. This can be determined with no more than four readings on the oscilloscope.

Notice that there are four test points marked with stars. Only these points need be considered while narrowing the problem down to a circuit group. Should the readings be taken in numerical order? Not necessarily. Which measurements will eliminate the greatest number of circuits? Number 4 looks like a good bet.

Connect the oscilloscope to the test point, Star 4, and take a reading. If the audio signal is present, the job is finished. The bad circuit has to be in the audio circuit group. If the signal is not present, the audio group has been eliminated as a cause; it must be a good circuit. Next try Star 2. If the IF signal is present, the trouble is isolated to either the IF section or the detector. Now a check at Star 3 will pin it down. A signal here isolates the trouble to the detector. No signal here isolates the trouble to the IF section. Backing up for a moment, if *no signal* appears at Star 2, it eliminates all circuits from that point forward; the trouble has to be in either the RF section or the frequency converter section. One more check will pin it down; take a reading at Star 1. A signal here labels the frequency converter section as the bad circuit group. No signal at this point fixes the blame on the RF section.

The method used here in isolating the faulty circuit group is sometimes referred to as a *bracketing process*. At first the brackets enclosed the entire receiver. The brackets were then closed in after each measurement until they enclosed only the faulty circuit group. This method is further illustrated in Figure 17–4.

Group 1, shown in part (a), contains a block for each circuit group in the faulty unit. This group has an input but no output. The brackets enclose the entire unit. Take measurements at points calculated to eliminate nearly half of the remaining circuit groups with each measurement. A measurement at point C verifies an input to group 3. This eliminates groups 1 and 2 and moves the brackets as shown in part (b).

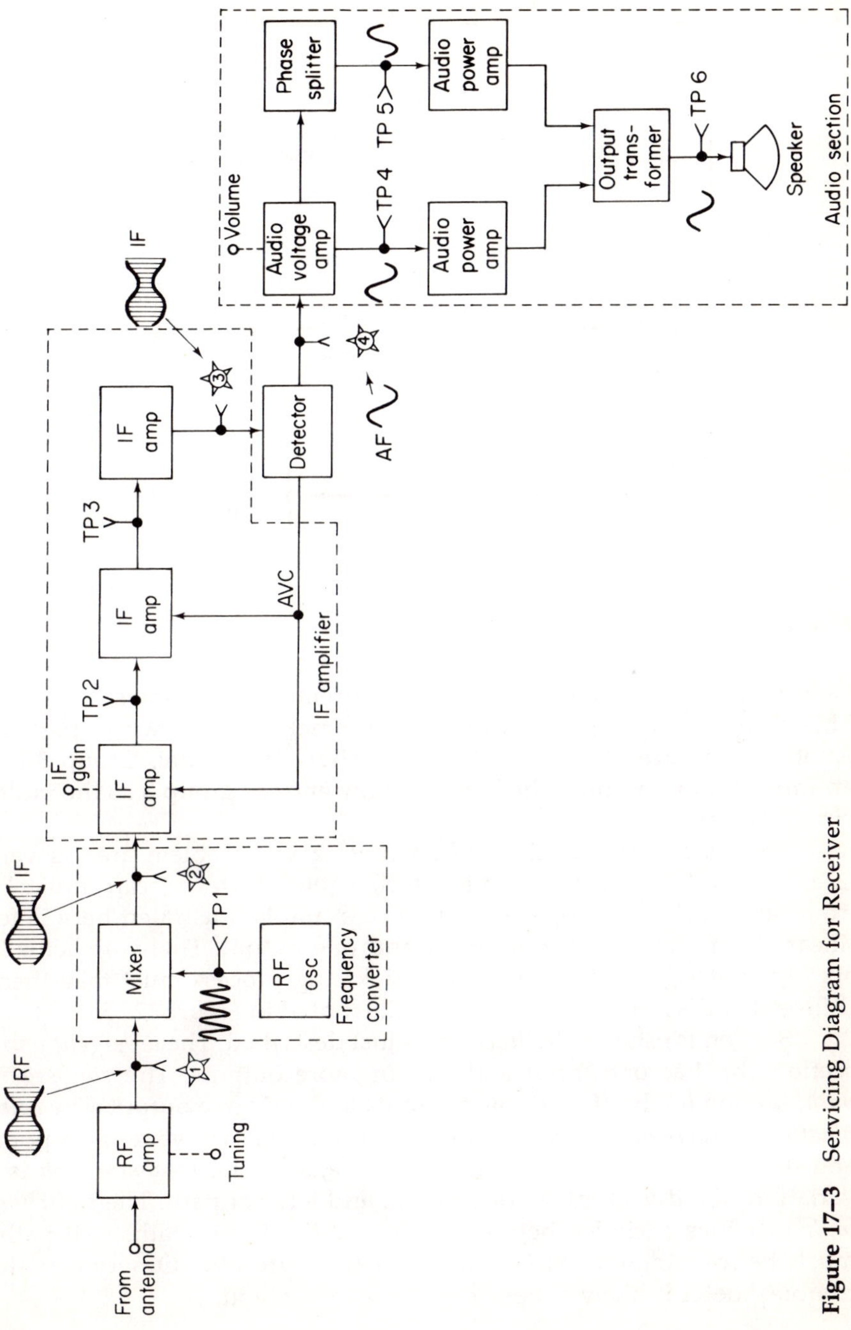

Figure 17–3 Servicing Diagram for Receiver

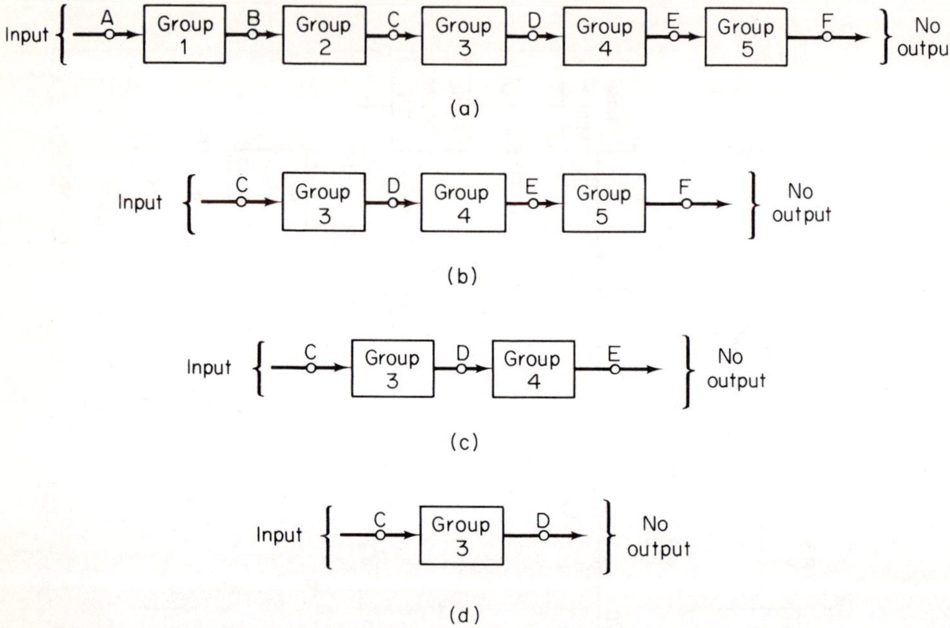

Figure 17–4 The Bracketing Process

Now measure at either D or E. A measurement at E shows no signal. This eliminates group 5 and moves the brackets as shown in part (c). Another measurement, at D this time, shows no signal. Group 3 has an input but no output. The brackets now enclose group 3 as the faulty circuit group.

This bracketing method of narrowing the problem area is very effective, and it is a time-saving technique. However, care must be exercised in constructing the chart. The example illustrated here is for linear signal paths with one input and one output. There are actually six types of signal paths, and the bracketing process must take these different paths into consideration, as illustrated in Figure 17–5.

Section (a) shows the *linear path* just described. The *divergent path*, section (b), has one input and two or more outputs. The *convergent path*, section (c), has two or more inputs and only one output. The *combination convergent-divergent path* has inputs that converge at a point and then spread out into multiple paths again. The *feedback path* is a variation of a divergent, a convergent, and a linear path. The *switching path* provides a choice between two or more linear paths. All paths must be considered during the isolation process. Otherwise, the troubleshooter is likely to be led to a false conclusion.

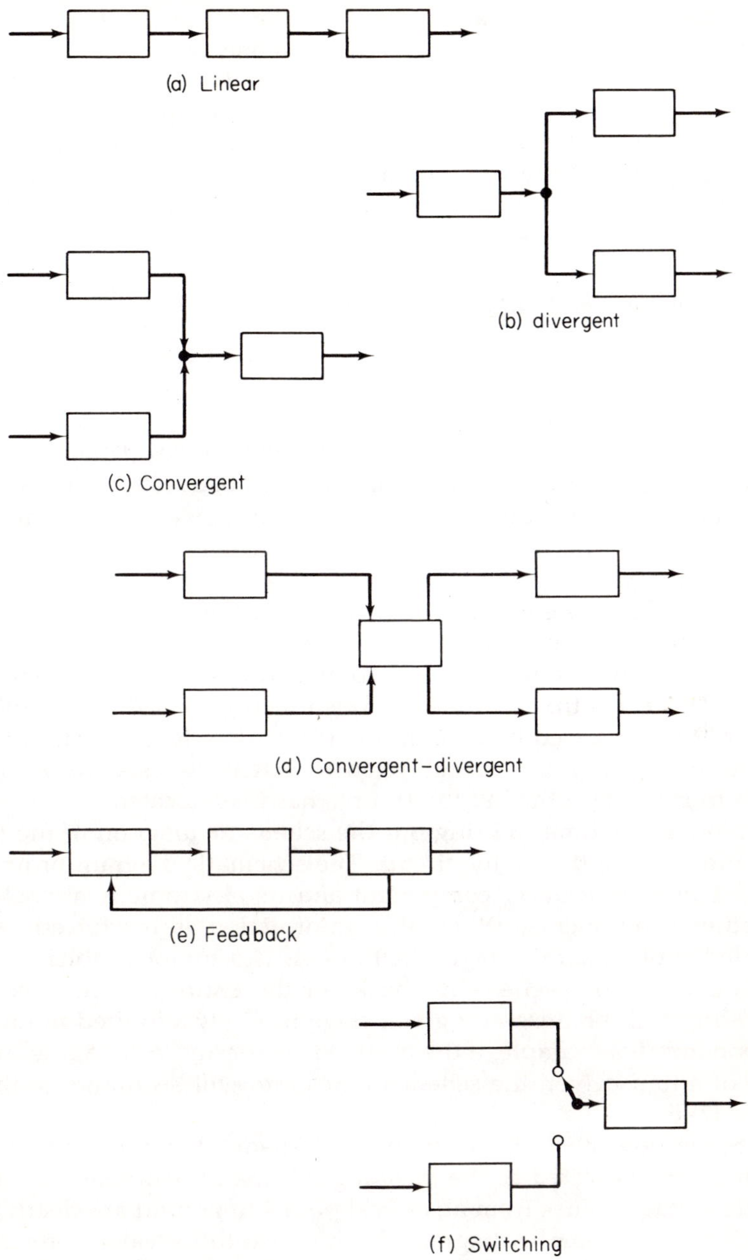

Figure 17–5 Types of Signal Paths

Once the faulty circuit group is determined, a few more readings will serve to isolate the faulty circuit in the group. In the case of the receiver in Figure 17–3, two groups contain only one circuit each. These are the RF section and the detector section. The frequency converter section has only two stages. The audio section is apparently the worst with its six stages. However, if the symptom is a complete loss of sound, the convergent and divergent paths make it very easy. It must be either the transformer or the speaker. A single reading at TP (test point) 6 will isolate the trouble to one or the other.

Step 6—Isolating the Faulty Component

Extensive use of logic and very little actual testing will generally carry the troubleshooter through the first five steps and isolate the faulty circuit. He is now ready for step 6—to determine exactly which of the components in that circuit has failed. Occasionally a quick visual inspection will accomplish this task. A loose connection, a bad solder joint, an open pin, a dark tube, or a cold transistor—all are possible failures that might be spotted by inspection. Such defects as charred resistors and shorted leads are also easy to spot.

More often the entire stage will appear perfectly normal, and additional testing will be required to ferret out the faulty component. Furthermore, even after one bad component is located, it cannot be assumed that it is the only one. An open resistor or a shorted capacitor may have been caused by excessive current. Therefore, the stage where the trouble exists must be subjected to extensive tests and measurements to make sure that all the trouble has been located.

Now is the time to bring out the schematic diagram of the faulty unit and analyze the faulty circuit. The schematic diagram provides a detailed picture of every component and its electronic relationship to the other components. When this information is reinforced with a knowledge of how this stage functions, it is a minor problem to perform detailed, comprehensive checks of the entire circuit. It is not a staggering task because any given stage has only a limited number of components. For example, if the trouble is in the audio voltage amplifier stage of a transceiver, the schematic diagram will be similar to that in Figure 17–6.

Since power is still on from the last measurements, the voltage readings are indicated as the most logical first measurement in step 6. The dc voltage values from all critical points to ground are clearly indicated on the schematic diagram. In this case those values are circled. The values of grid, cathode, plate, and applied voltage are quickly checked by either a voltmeter or an oscilloscope. Most troubleshooters prefer the oscilloscope at this point in the process. It has already been in use and is ready for action. A touch of the probe to tube pins 1 and 2 on each side of R_{47} and the dc values are either verified or deviations are noted.

If one (or more) voltage(s) is wrong, why is it wrong? This question indicates that all resistances should be measured. For that matter, even if all voltages are correct, the resistances should still be measured. After all, this is the faulty stage; something is definitely bad.

Before making a resistance measurement, the troubleshooter may want to verify his previous conclusion that the stage has an input signal but no output signal.

When all signal and voltage measurements have been completed, power is removed and resistances are measured with an ohmmeter. In the sample stage schematic diagram, there are no parallel resistors to interfere with the readings. Therefore, the measurement directly across each resistor should be the value indicated on the schematic ± the tolerance figure. In many cases, resistor measurements are more complicated. The schematic diagram then indicates vital resistance values from specified points to ground or between specified points. This enables a valid resistance measurement without the necessity of disconnecting all resistors from the circuit.

By this time the primary trouble has undoubtedly been found. But there are two more tests that should be made. While the power is off and the ohmmeter is handy, check the capacitors with the ohmmeter. Then check the tube or transistor. It is a small matter in most cases to remove the tube and check it with a tube checker. When the stage is transistorized, the transistor should be subjected to an in-circuit test.

The troubleshooter now knows which component has failed and whether or not there are secondary problems.

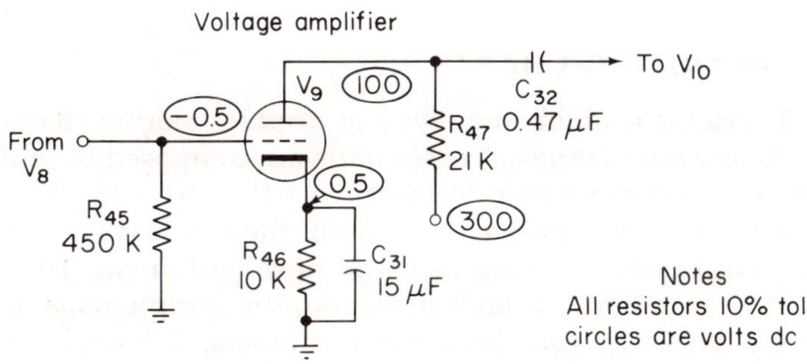

Figure 17–6 Sample Schematic Diagram of a Stage

Repairing and Rechecking

Theoretically the troubleshooter's job is finished; he has found the trouble. And with a little practice, he will consistently locate troubles in less time than it takes to read the six steps of troubleshooting. But practically speaking, his job isn't over because in most cases the troubleshooter is also the technician who does the repair. He now gets out his tools and removes all the faulty and marginal components and replaces them with identical (within tolerance) components.

When the circuit is all back in one piece, each of the measurements must be taken again. Why? Because the repair work quite frequently causes other troubles to develop. After each part of the stage checks out to be all right, the overall performance of the equipment must be checked with the repaired circuit in operation. If there is any deterioration in system performance, it is probably the result of the repaired section. If the performance has been degraded a noticeable amount, several adjustments may be required to bring it up to the specified standard.

VARIATIONS IN TROUBLESHOOTING PROCEDURES

The experienced technician and engineer scarcely need reminding that some parts of nearly all systems operate with dangerously high voltages. Notably among these are cathode-ray tubes and transmitter circuits. The same six-step troubleshooting procedure should be followed in all cases, but the type of measurements taken at any point should be dictated by prudence. This prudence is influenced by knowledge of the equipment. If the measurement is essential, make it; but use the indicated precautions to protect the troubleshooter, the equipment, and the test instruments.

TROUBLESHOOTING CHARTS

Nearly all systems contain some type of troubleshooting charts as part of the maintenance manuals. These charts are composed from theoretical troubles for the most part. In most cases the actual trouble will not appear on the charts. However, a study of these charts will be helpful in developing troubleshooting skills for new equipment. These charts list the primary method of fault detection, the symptoms, and several possible troubles that can cause the symptoms. A sample of such a chart is shown in Figure 17–7. This chart is for an amplifier stage in a transmitter that has built-in meters for monitoring the dc current for both control grid and plate. The symptoms are detected on one of these meters.

There are many aids to troubleshooting, but none are more valuable than knowledge of the equipment and the practice of a logical approach.

Meter	Symptoms	Possible trouble
Grid current	Low reading	1. Weak grid drive 2. Plate of previous stage mistuned 3. Bad tube
	Zero reading	1. No grid drive 2. No RF in plate of previous stage 3. Bad tube
	High reading	1. Open screen resistor 2. No screen voltage 3. No plate voltage
Plate current	Low reading	1. Weak tube 2. Open grid resistor
	Zero reading	1. Bad tube 2. No grid drive 3. Open screen resistor 4. No plate voltage 5. Open filament lead
	High reading	1. No grid drive 2. Mistuned plate tank

Figure 17–7 Troubleshooting Chart

chapter eighteen

MALFUNCTION ANALYSIS

Engineers and technicians involved in equipment design seldom have an opportunity to work with complete systems. Rather, they spend a good deal of time designing, building, and testing individual stages. This is similar to the activity of a troubleshooter in isolating the faulty component in a circuit. This chapter is devoted to trouble analysis of individual stages. It is intended to improve the techniques of the experienced and to aid the beginner in developing sound troubleshooting skills.

A variety of sample circuits are used. These are all common circuits, and they will be recognized by nearly everyone concerned. For this reason the theory of operation and the normal functions are covered only briefly. The emphasis is on the symptoms and the possible troubles that can cause these symptoms.

Since each stage has a limited number of components, there are a limited number of malfunctions that are possible in any given stage. The techniques described in trouble analysis of these common circuits are applicable to all types of circuits, whether newly designed or established parts of a functional system.

AUDIO AMPLIFIERS

Trouble analysis of a few types of audio amplifiers should be helpful to everyone because audio amplifiers are used in so many different kinds of equipment.

Voltage Amplifier

Most radio and television receivers have an audio voltage amplifier located between the detector and the power amplifiers. Its purpose is to accept the low-voltage signal from the detector and amplify it to provide sufficient drive for the power amplifiers. The active component in this circuit, and in most other circuits, may be either an electron tube, a bipolar transistor, or a field-effect transistor. Figure 18–1 shows a circuit of each type to illustrate the similarities.

Notice that the three circuits are identical, with the exception of the amplifying device. In practice, of course, the voltage levels and component values will differ. When an electron tube is used, it is generally a triode as shown here. The bipolar transistor is usually in a common-emitter configuration, although other arrangements are sometimes used. This is an NPN, but the PNP serves just as well. The field-effect transistor here is an N channel in a common-source arrangement, but it is frequently seen as a P channel.

Each of the three circuits is a self-biasing, Class A audio amplifier with degenerative feedback. Several different types of coupling and feedback arrangements may be expected. Circuit failure usually appears in one of three ways: no output, reduced output, or distorted output.

When the symptom is input with no output, a check of the dc voltages will usually pinpoint the defective component. If voltages are normal, the trouble is likely to be faulty coupling capacitors; frequently this is a faulty solder connection. The fault may be an open-load resistor or an open-biasing resistor. In either case, the amplifier would be cut off. The amplifying device itself could also be bad, but this is less likely with this symptom.

The reduced-output symptom probably indicates a bad amplifying device. This is especially true if the output has gradually deteriorated over a period of time. A change in value of the biasing components sometimes causes this symptom. This alters the bias and affects the operational characteristics of the amplifier. Also a reduced-load impedance will reduce the output by lowering the gain of the stage. An open bypass capacitor may be the problem, but this is the fault of a connection more often than the capacitor. An open lead or a cold solder joint to a capacitor produces an effective open capacitor and results in excessive degeneration.

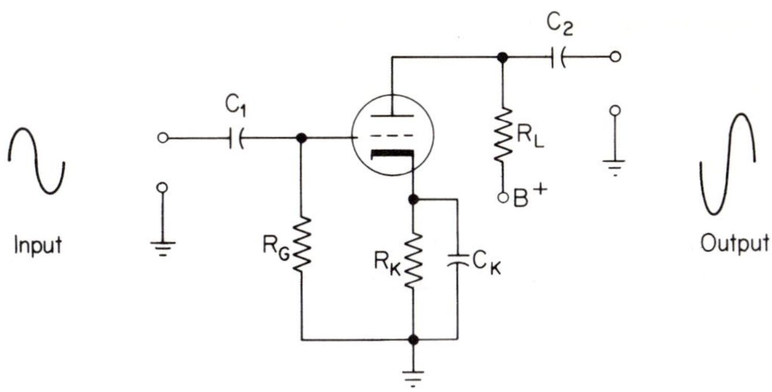

(a) Electron tube

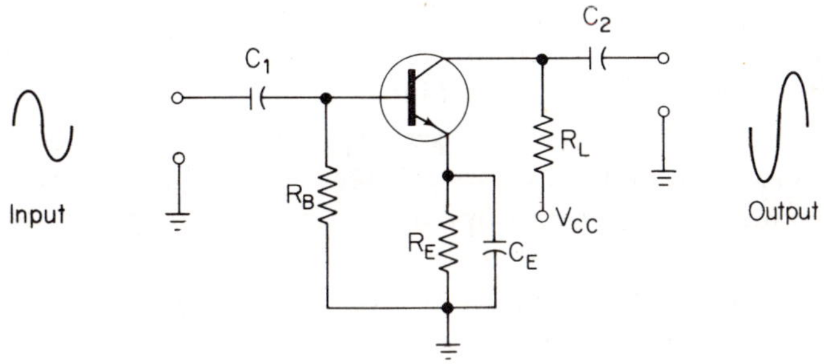

(b) Bipolar transistor

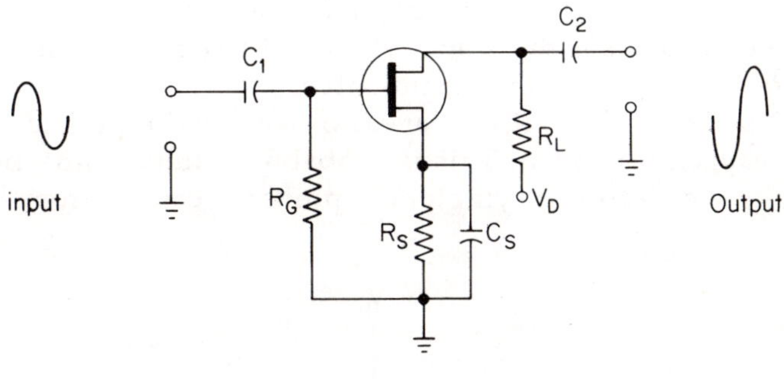

(c) Field effect transistor

Figure 18–1 Audio-Voltage Amplifiers

The distorted output is the usual symptom for too much gain. This can be caused by a number of faults. The amplifying device may have changed its characteristics, but other components are more likely. An open-biasing resistor will cause a loss of bias with resultant excessive gain. Faulty decoupling capacitors may cause too much regenerative feedback. Open components in a degenerative feedback circuit may cause the same effect.

The recommended troubleshooting procedures are as follows:

1. Perform a close visual inspection.
2. Check the dc voltages.
3. Measure point to point resistances.
4. Check capacitors with ohmmeter.
5. Check the amplifying device.
6. Always check for secondary damage, and after repair, perform an operational check.

Single-Stage Phase Splitter

The audio *phase splitter* is an amplifier that produces two outputs equal in amplitude and 180° out of phase. A schematic diagram for a single-stage phase splitter is shown in Figure 18–2.

This common emitter with one input and two outputs is only one of many versions of the phase splitter. This split load arrangement, which produces equal and opposite outputs in respect to ground, is sometimes called a phase inverter. Failure in this circuit can cause one of four symptoms: (1) no output, (2) only one output, (3) reduced outputs, or (4) distorted outputs.

The no-output symptom can only be caused by a trouble that causes the amplifying device to cut off. In the example shown, any open resistor would disable the stage. The amplifying device may be bad, but one of the resistors is more probable.

A loss of only one output leaves no room for doubt. The associated coupling capacitor is open. This is probably a faulty connection.

The reduced-output symptom is probably caused from an ampli-

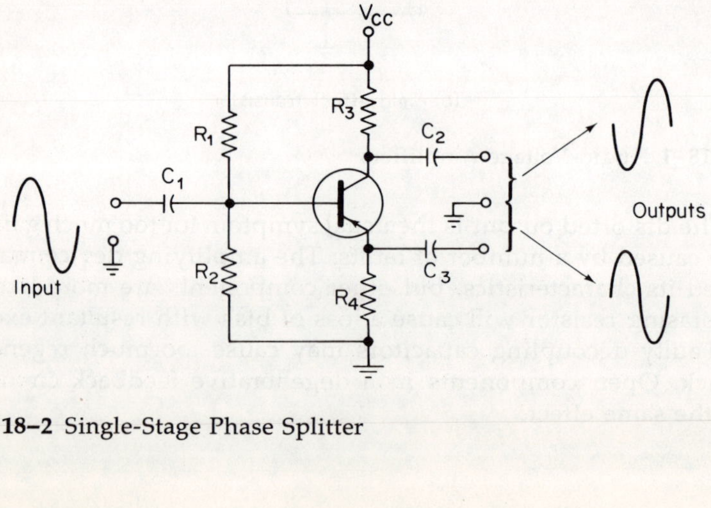

Figure 18–2 Single-Stage Phase Splitter

fying device that has weakened with age. A change of resistance in the input circuit could cause this symptom by changing the bias or reducing the amplitude of the input signal. Cold joints in the solder connections can cause attenuation that results in a low output.

The distorted output is caused from excessive gain. This may be due to a change in resistance in the biasing circuit. Peak clipping may result from a bias that is either too high or too low. A defective amplifying device can cause distortion, but the resistors are more probable.

Paraphase Amplifier

The *paraphase amplifier* is composed of two stages of amplification. Its purpose is to amplify an input signal and produce two outputs that are equal in amplitude and 180° out of phase. It is sometimes called a *phase inverter*, and it is often used instead of a single-stage phase splitter. The outputs are used to drive power amplifiers. A transistorized paraphase amplifier schematic diagram is illustrated in Figure 18–3.

Each of the stages should be considered as a separate amplifier for trouble analysis. The first stage is a common amplifier with a certain amount of degeneration in the emitter circuit. It can produce the regular symptoms of no output, reduced output, or distorted output. The analysis and trouble isolation procedures are the same as described for voltage amplifiers.

The second stage is an amplifier, but the feedback is designed to cancel the gain. The input through C_2 should be approximately the same amplitude as, and 180° out of phase with, the output. The symptoms are no output, reduced output, or distorted output. The analysis of the failure and troubleshooting procedures are the same as for voltage amplifiers.

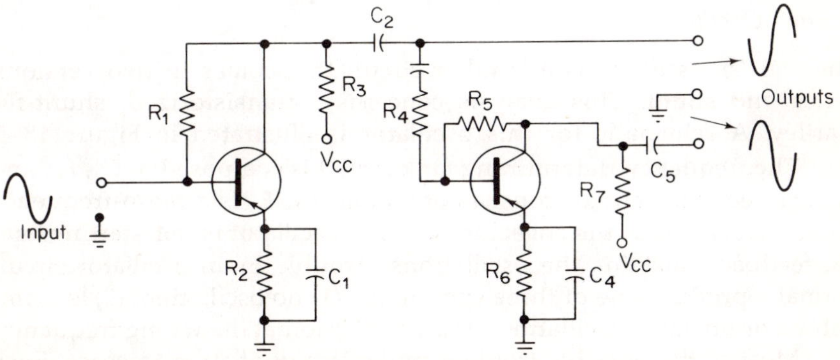

Figure 18–3 Paraphase Amplifier

Push-Pull Amplifier

The *push-pull amplifier* is normally the last stage in a receiver. It is the power amplifier that drives the speaker. This stage requires two inputs equal in amplitude and 180° out of phase. The single output is a high-power, distortion-free signal. The schematic of a transistorized version is shown in Figure 18–4.

This is a Class A amplifier consisting of two transistors connected back-to-back. It is a single-stage amplifier with two amplifying devices. The resistor section composed of R_1, R_2, and R_3 is a balance network to compensate for the fact that no two amplifying devices conduct exactly the same. The rheostat (R_1) can be adjusted for a balanced output.

The symptom of no output from a push-pull amplifier is very definitive. The amplifying devices can be eliminated at once. It is highly improbable that they would both go bad at the same time. This narrows the trouble to a component that will disable both the amplifying devices.

The reduced output normally signals failure of half the stage; that is, one amplifying device has no output. This is either a bad amplifying device or a bad component which disables one-half of the push-pull stage. A biasing resistor could cause this problem.

The distorted-output symptom is usually caused by improper bias resulting from components changing values.

OSCILLATORS

Oscillator circuits are familiar to everyone associated with electronics. They are found in receivers, transmitters, computers, timers, and test instruments; and this is naming only a few of the places where they are used. Since these circuits appear in so many places, trouble analysis on a few common types of oscillators seems to be appropriate.

Hartley Oscillator

The *Hartley oscillator* is a familiar circuit that comes in two versions: series and shunt. This analysis concerns a transistorized, shunt-fed Hartley. A schematic for this oscillator is illustrated in Figure 18–5.

The frequency-determining tank circuit is composed of C_2, L_1, and L_2. The feedback circuit consists of C_4 and L_1. L_3 is a radio-frequency choke that serves as the collector load. The oscillator is self-starting, and the feedback sustains the oscillations. Trouble in an oscillator circuit normally produce one of three symptoms: (1) no oscillation, (2) low amplitude or unstable oscillation, or (3) oscillation at the wrong frequency.

Most of the possible troubles will cause oscillation to cease. Such conditions include an open or shorted transistor; an open or shorted

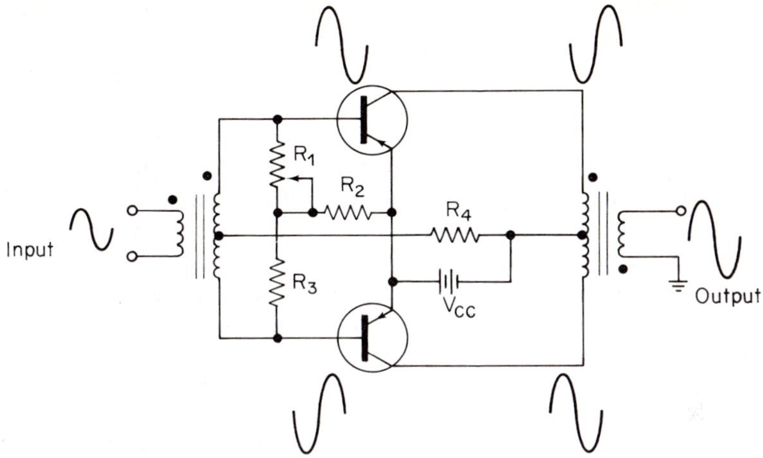

Figure 18–4 Push-Pull Amplifier

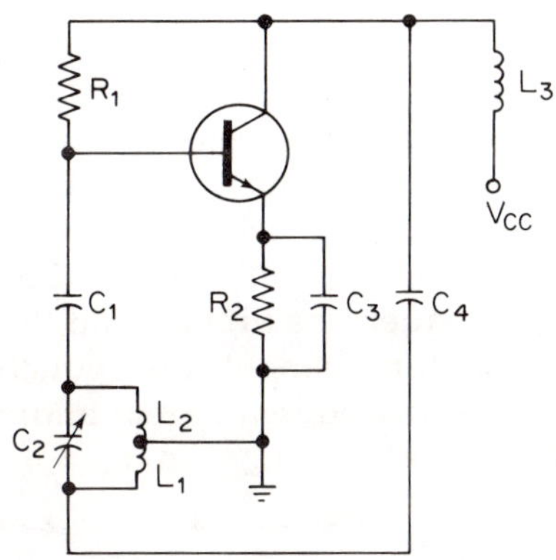

Figure 18–5 Hartley Oscillator

capacitor: C_1, C_2, or C_4; or an open coil: L_1, L_2, L_3, or an open resistor: R_2. The transistor is likely to deteriorate with age, especially under high-temperature operating conditions. If C_4 opens, feedback ceases and stops oscillation. If C_4 shorts, the positive V_{CC} to L_1 stops oscillation; this short-circuits the biasing circuit. If C_1 opens, there is no coupling from the tank to the transistor base, no amplification, no feedback, and no oscillation. If C_1 shorts, negative V_{CC} through L_2 to the base causes the transistor to cut off and stops oscillation.

If C_2 opens or shorts, there is no tank circuit to oscillate, and this capacitor is not easy to check. There is no definite check except to disconnect the capacitor from the circuit. The tank is also destroyed by an opening in either L_1 or L_2. An opening in either R_2 or L_3 will disable the transistor and stop oscillation. In some cases opening R_1 will cause oscillation to cease, but this is improbable.

Instability comes in two forms: frequency and amplitude. Frequency instability is usually a symptom caused by poor connections, poor insulation, or changes in value of coils or capacitors. Amplitude or reduced output are symptoms that can be caused by improper bias, degenerating conditions, or an opened or shorted C_3.

Incorrect frequency is a common symptom, and this can be caused by many deteriorating conditions. Major frequency changes occur with changes in the transistor characteristics or changes in the value of the tank-circuit components.

Care must be exercised when checking oscillator circuits, and test instruments must have a common ground. Meters should be used with high impedance probes, and the operational circuit must not be excessively loaded. Misuse of test instruments will cause faulty indications, and it may cause damage to the circuit.

Colpitts Oscillator

The *Colpitts oscillator* is another familiar circuit. The schematic diagram for one transistorized version is illustrated in Figure 18–6.

In appearance this Colpitts oscillator is very similar to the shunt-Hartley, but it is a more stable oscillator. The Colpitts is also capable of reliable oscillation at frequencies higher than the Hartley is capable of. Since the Colpitts oscillator is capable of producing a sine wave output of constant amplitude and frequency from audio through the high RF

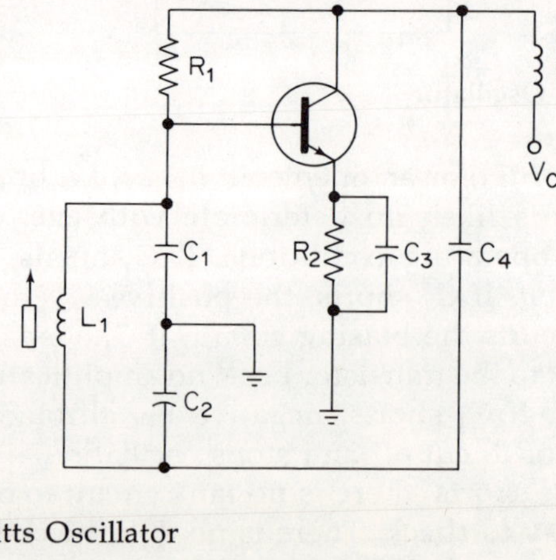

Figure 18–6 Colpitts Oscillator

range, it is a very popular oscillator for receiver local oscillators, signal sources in signal generators, and for many applications requiring variable-frequency oscillators.

The frequency is determined by a tank circuit composed of a tunable coil and two capacitors. The feedback to sustain oscillation is supplied through a capacitor voltage divider. The voltage across C_1 is the input to the transistor, and the quantity of feedback is determined by the ratio between C_1 and C_2.

The symptom, no oscillation, will appear with any failure that disables either the transistor or the tank circuit. An open or a short of the transistor or in C_1, C_2, or C_4 will stop oscillation. A bad transistor stops all current. Opening C_4 stops feedback to the tank circuit. Shorting C_4 applies dc to both sides of the tank circuit. Opening or shorting in C_1 or C_2 disables the tank circuit. An open in either coil or in R_2 will stop oscillation. Opening L_1 disables the tank circuit. Opening L_2 or R_2 disables the transistor circuit.

Instability in amplitude or reduced output is likely to occur as a result of improper biasing. This is usually caused by characteristic changes due to extreme temperatures. Reduced amplitude means too much bias or excessive degeneration, and this can be caused by increased resistance in the bias circuit (R_1 or R_2) or an open in C_3. A leaky coupling capacitor, C_4, will reflect a resistance through the inductance of the tank that appears between the base and collector of the transistor; this also changes bias.

Frequency instability indicates problems with the tank circuit. This can be caused by poor mechanical connections or deteriorated insulation. Partial shorts between windings of L_1 can also cause frequency instability. Improper conduction of a faulty transistor may cause frequency shifts, but this will more than likely appear as amplitude instability.

Incorrect output frequency is not as common as with the Hartley oscillator. When it does happen, it is a result of deteriorating conditions. The deterioration may be in the tank-circuit components, transistor, or coupling. A change in parameters causes a more pronounced frequency shift when operating at higher frequencies.

Clapp Oscillator

The *Clapp oscillator* is considered to be a variation of the Colpitts. It is used to produce a sine wave within the RF range. This oscillator is frequently employed in signal generators and as a variable frequency oscillator for general use over the high and very high frequency ranges. Figure 18–7 is a schematic diagram of a transistorized Clapp oscillator.

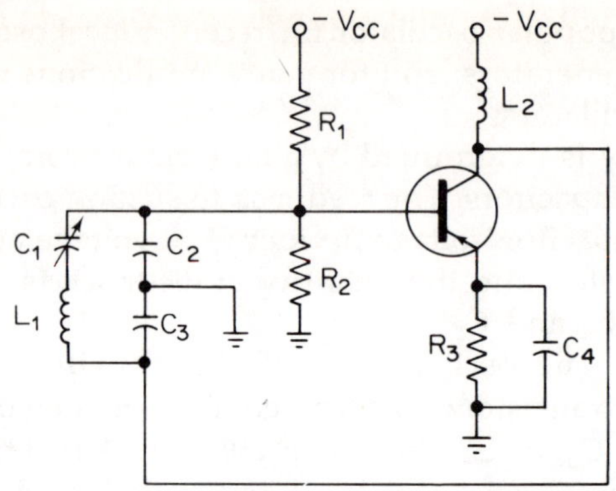

Figure 18–7 Clapp Oscillator

The frequency determining device is a series resonant circuit composed of L_1 and C_1. C_2 and C_3 form a voltage divider that determines the amplitude of the feedback voltage. The transistor has fixed bias developed by R_1 and R_2. R_3 and C_4 are emitter components to prevent instability and degeneration. The tuned circuit is designed for a high Q under load, which gives the Clapp oscillator greater stability than the basic Colpitts.

The no-oscillation symptom is an indication of an opened or shorted condition that disables either the transistor or the tuned circuit. Any opened component, with the exception of C_4, can cause this symptom; this includes the transistor. Shorting of capacitor C_1, C_2, or C_3 will disable the tuned circuit and stop oscillation. Poor connections may effectively produce one or more of these conditions and result in no oscillation.

Reduced or unstable amplitude of oscillation is generally caused by a change in bias resistance or a defective bypass capacitor in the emitter circuit. Loss of gain due to transistor deterioration may also cause a reduced or unstable output.

Frequency instability is generally caused by poor connections that affect the tank-circuit components.

Incorrect frequency can be caused by changes in parameters that vary the distributive capacitance. Variation in the inductance of the tuned circuit will cause a frequency shift. Shorting of a few coils on the inductor will change the inductance. A change in the value of the tuning capacitor would cause frequency shifts also, but this probably would not be noticed until the capacitor failed completely.

Electron-Coupled Oscillator

The *electron-coupled oscillator* produces an RF signal that is constant in both amplitude and frequency. It is widely used in equipment that requires good stability, provided the waveform is not a critical factor. Figure 18–8 is a schematic diagram of a pentode electron-coupled oscillator.

The electron-coupled oscillator can be constructed by using any of several types of oscillator circuits. The amplifying device may be a triode, a tetrode, a pentode, or a beam power tube. In this example, the circuit is a modified, shunt-fed Hartley.

The pentode amplifier operates Class C. The bias is of the grid-leak variety. The oscillator is self-starting and extremely stable.

The no-oscillation condition can be caused from a short or open that disables either the electron tube or the tank circuit. However, in this case, it is possible to have oscillation in the tank circuit and still have no output. This is possible because the plate of the tube is the output circuit, and the screen grid serves as a plate for the feedback circuit. This latter condition is indicative of a bad tube or an open in the plate load, L_4.

Oscillations will be stopped when the tank circuit is disabled; this can be caused by a number of faults. An open or short in the tank will prevent oscillation; lack of feedback will prevent oscillation; and/or failure of the amplifier to conduct will prevent oscillation. R_1 or C_3 opened or C_3 shorted will destroy the biasing circuit and stop conduction. C_2 opened will stop feedback. C_2 shorted will complete a path from the tank circuit to *B+*. C_4 shorted will stop the screen grid current and destroy feedback; the same is true with an open in R_2 or L_3.

Reduced or unstable oscillation is evidence of bad connections or some deteriorating condition such as a weak tube or a change in component values. This type of symptom is clearly a result of improper energy coupling; therefore shorts, connections, and changed values need to be carefully checked.

Changes that affect distributive capacitance can result in a frequency shift. This is also true for shorted components in the screen grid circuit or for shorted windings in either L_1 or L_2.

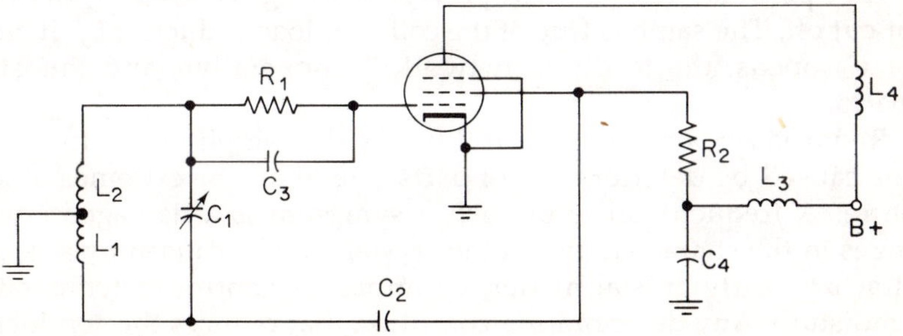

Figure 18–8 Electron-Coupled Oscillator

Crystal Oscillator

There are several types of *crystal oscillators*. The *Colpitts*, the *Miller*, and the *Pierce* are among the most popular. Crystal oscillators are likely to be used in any circuit that requires a radio-frequency signal of extreme stability. The example used here is a Colpitts crystal oscillator. It is generally employed at the higher radio frequencies. Figure 18–9 is a schematic of a transistorized version of this oscillator.

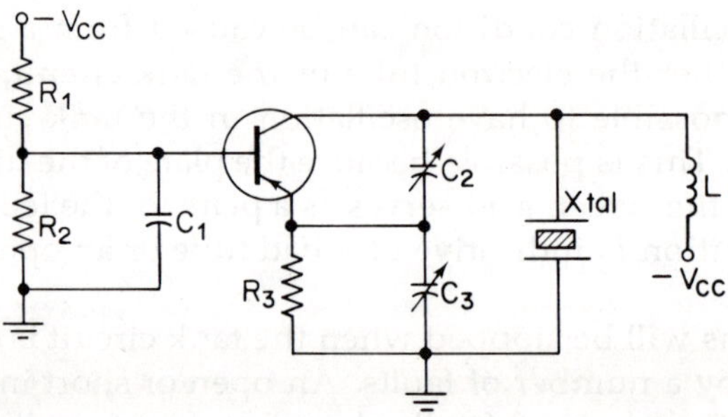

Figure 18–9 Colpitts Crystal Oscillator

The Colpitts crystal oscillator uses external capacitive divider feedback to sustain oscillation of the crystal. C_1 and C_2 provide this feedback. Another voltage divider composed of R_1 and R_2 supplies a fixed bias for easy starting. The quartz crystal, operating on the piezoelectric effect, produces oscillation at a stable frequency. The frequency of the oscillator is not limited to the fundamental crystal frequency; it may be equivalent to any one of several harmonics of this basic frequency.

A symptom of no oscillation is an indication of a trouble that stops the transistor from conducting, stops feedback, or disables the crystal itself. The easiest way to check a suspected crystal is to substitute one that is known to be good. This substitution can be made without breaking the circuit.

An open-bias resistor will prevent starting by keeping the transistor cut off. The same is true of the collector load inductor, L_1. If either C_2 or C_3 opens, the feedback network is inoperative, and this stops oscillation.

Reduced amplitude or instability of either amplitude or frequency can be caused by deterioration of parts due to age or extremes of temperature. A frequent cause of such a symptom is a damaged crystal. Changes in the characteristics of the crystal may be caused by age, partial fracture, dirty crystal holder, or improper control of temperature and moisture. Any deteriorating condition that reduces the conduction of the transistor can also cause some form of reduced or unstable output.

Oscillation at an incorrect frequency is more than likely due to a faulty crystal. Changes in the value of the feedback capacitors, C_2 and C_3, could cause this problem, but since they are tunable, the result would probably go unnoticed until one of them failed completely. Simple measurements such as those for voltage, resistance, and continuity will generally isolate the faulty component.

BIBLIOGRAPHY

Cooper, William David, *Electronic Instrumentation and Measurement Techniques*, Prentice-Hall, Englewood Cliffs, N.J. 1970.

Farkas, L. L., *Electronic Testing*, McGraw-Hill, N.Y. 1966.

Leach, Donald P., *Transistor Circuit Measurements*, McGraw-Hill, N.Y. 1968.

Lenk, John D., *Data Book for Electronic Technicians and Engineers*, Prentice-Hall, Englewood Cliffs, N.J. 1968.

Lenk, John D., *Handbook of Electronic Test Equipment*, Prentice-Hall, Englewood Cliffs, N.J. 1969.

Lenk, John D., *Handbook of Practical Electronic Tests and Measurements*, Prentice-Hall, Englewood Cliffs, N.J. 1971.

Littauer, Raphael, *Pulse Electronics*, McGraw-Hill, N.Y. 1968.

PHILCO, *Electronic Troubleshooting*, Prentice-Hall, Englewood Cliffs, N.J. 1966.

Prensky, Sol D., *Electronic Instrumentation*, Prentice-Hall, Englewood Cliffs, N.J. 1971.

Soisson, Harold E., *Electronic Measuring Instruments*, McGraw-Hill, N.Y. 1961.

Thomas, Harry E., and Carole A. Clarke, *Handbook of Electronic Instruments and Measurement Techniques*, Prentice-Hall, Englewood Cliffs, N.J. 1967.

Wedlock, Bruce D., and James K. Roberge, *Electronic Components and Measurements,* Prentice-Hall, Englewood Cliffs, N.J. 1969.

Zbar, Paul B., *Electronic Instruments and Measurements*, McGraw-Hill, N.Y. 1965.

APPENDICES

A

Unit Conversions

From	To	Multiply by
pico	micro	10^{-6}
pico	milli	10^{-9}
pico	unit	10^{-12}
micro	pico	10^{6}
micro	milli	10^{-3}
micro	unit	10^{-6}
milli	pico	10^{9}
milli	micro	10^{3}
milli	unit	10^{-3}
unit	pico	10^{12}
unit	micro	10^{6}
unit	milli	10^{3}
unit	kilo	10^{-3}
unit	mega	10^{-6}
kilo	mega	10^{-3}
kilo	unit	10^{3}
mega	kilo	10^{3}
mega	unit	10^{6}

Converting from ↓	Converting to →						
	mega	kilo	unit	milli	micro	nano	pico
mega	–	3 →	6 →	9 →	12 →	15 →	18 →
kilo	← 3	–	3 →	6 →	9 →	12 →	15 →
unit	← 6	← 3	–	3 →	6 →	9 →	12 →
milli	← 9	← 6	← 3	–	3 →	6 →	9 →
micro	← 12	← 9	← 6	← 3	–	3 →	6 →
nano	← 15	← 12	← 9	← 6	← 3	–	3 →
pico	← 18	← 15	← 12	← 9	← 6	← 3	–

Resistor Color Code

EIA STANDARD RS-172

MILITARY STANDARD MIL-R-11E

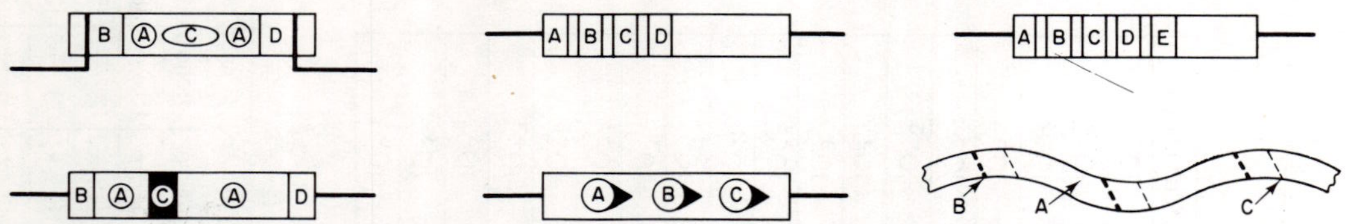

Color	1st Digit A	2nd Digit B	Multiplier C	Tolerance D	Failure Rate* E
Black	0	0	1	—	—
Brown	1	1	10	± 1%	1.0
Red	2	2	100	± 2%	0.1
Orange	3	3	1,000	± 3%	0.01
Yellow	4	4	10,000	± 4%	0.001
Green	5	5	100,000	—	—
Blue	6	6	1,000,000	—	—
Violet	7	7	10,000,000	—	—
Gray	8	8	100,000,000	—	—
White	9	9	—	—	Solderable*
Gold	—	—	0.1	± 5%	—
Silver	—	—	0.01	±15%	—
No color	—	—	—	±20%	—

*Band E, when used on composition resistors, indicates percent failure per 1,000 hours. On film resistors, a white band E indicates solderable terminal.

Insulation coding

EIA: Insulated resistors with axial leads are designated by a background of any color except black. The usual color is natural tan. Noninsulated resistors with axial leads are designated by a black background color.

MILITARY (MIL): Same as EIA with the addition of: Noninsulated resistors with radial leads designated by a black background color or by a background the same color as the first significant figure of the resistance value.

C

Decibel conversion chart

Power ratio	Voltage and current ratio	Decibels ← (+) (−) →	Voltage and current ratio	Power ratio
1.000	1.000	0.0	1.000	1.000
1.023	1.012	0.1	0.9886	0.9772
1.047	1.023	0.2	0.9772	0.9550
1.072	1.035	0.3	0.9661	0.9333
1.096	1.047	0.4	0.9550	0.9120
1.122	1.059	0.5	0.9441	0.8913
1.148	1.072	0.6	0.9333	0.8710
1.175	1.084	0.7	0.9226	0.8511
1.202	1.096	0.8	0.9120	0.8318
1.230	1.109	0.9	0.9016	0.8128
1.259	1.122	1.0	0.8913	0.7943
1.585	1.259	2.0	0.7943	0.6310
1.995	1.413	3.0	0.7079	0.5012
2.512	1.585	4.0	0.6310	0.3981
3.162	1.778	5.0	0.5623	0.3162
3.981	1.995	6.0	0.5012	0.2512
5.012	2.239	7.0	0.4467	0.1995
6.310	2.512	8.0	0.3981	0.1585
7.943	2.818	9.0	0.3548	0.1259
10.00	3.162	10.0	0.3162	0.10000
12.59	3.548	11.0	0.2818	0.07943
15.85	3.981	12.0	0.2515	0.06310
19.95	4.467	13.0	0.2293	0.05012
25.12	5.012	14.0	0.1995	0.03981
31.62	5.632	15.0	0.1778	0.03162
39.81	6.310	16.0	0.1585	0.02512
50.12	7.079	17.0	0.1413	0.01995
63.10	7.943	18.0	0.1259	0.01585
79.43	8.913	19.0	0.1122	0.01259
100.00	10.000	20.0	0.1000	0.01000

D

Mica Capacitor Color Code

EIA STANDARD RS-153A

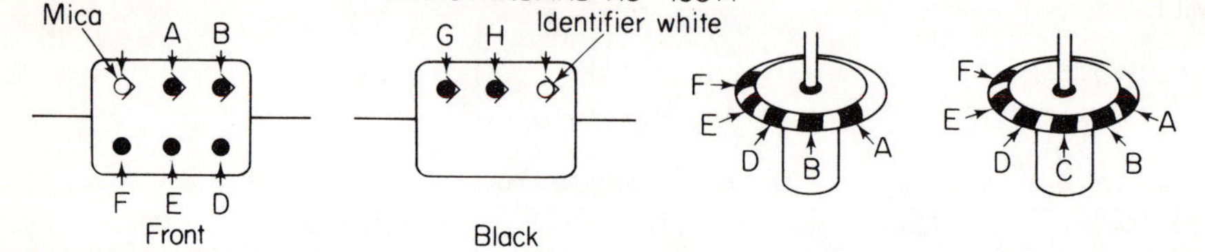

Color	Digits of capacitance (pF)			Multiplier	Tolerance %	Characteristic—see table below	Working voltage	Operating temperature
	A	B	C	D	E*	F	G	H
Black	0	0	0	1	±20	A	—	—
Brown	1	1	1	10	± 1	B	100 V. dc	—
Red	2	2	2	100	± 2	C	—	-55 to +85°C
Orange	3	3	3	1,000	± 3	D	300 V. dc	—
Yellow	4	4	4	10,000	—	E	—	-55 to +125°C
Green	5	5	5	—	± 5	—	500 V. dc	—
Blue	6	6	6	—	—	—	—	—
Violet	7	7	7	—	—	—	—	—
Gray	8	8	8	—	—	—	—	—
White	9	9	9	—	—	—	—	—
Gold	—	—	—	0.1	—	—	1,000 V. dc	—
Silver	—	—	—	0.01	±10	—	—	—

* or ± 1 pF whichever is greater

DESCRIPTION OF CHARACTERISTIC

Characteristic	Temperature coefficient (parts per million per °C)	Maximum capacitance drift	Minimum insulation resistance (megohms)
A	±1000	±(5% + 1 pF)	3000
B	±500	±(3% + 1 pF)	6000
C	±200	±(0.5% + 0.5 pF)	6000
D	±100	±(0.3% + 0.1 pF)	6000
E	-20 to +100	±(0.1% + 0.1 pF)	6000
—	—	—	—
—	—	—	—

VOLTAGE RATING

(Indicated by dimensions rather than color coding)

Maximum inches			Style	Capacitance (pF)	Rating (v d-c)
Long	Wide	Thick			
$\frac{51}{64}$	$\frac{15}{32}$	$\frac{7}{32}$	20	5-510 560-1000	500 300
$1\frac{7}{64}$	$\frac{15}{32}$	$\frac{7}{32}$	25	5-1000 1100-1500	500 300
$\frac{53}{64}$	$\frac{53}{64}$	$\frac{9}{32}$	30	470-6200 Over 6200	500 300
$\frac{53}{64}$	$\frac{53}{64}$	$\frac{3}{8}$	35	3300-6200 Over 6200	500 300
$1\frac{1}{32}$	$\frac{41}{64}$	$\frac{11}{32}$	40	100-2400 2700-7500 Over 7500	1000 500 300

Mica Capacitor Color Code

Obsolete Style

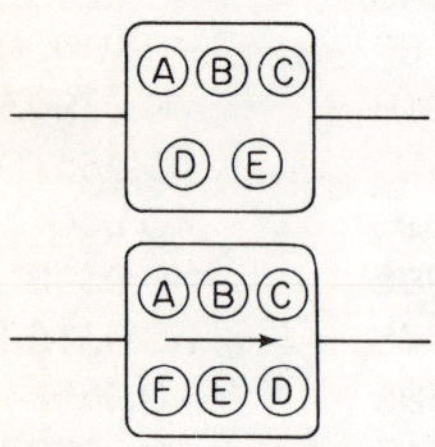

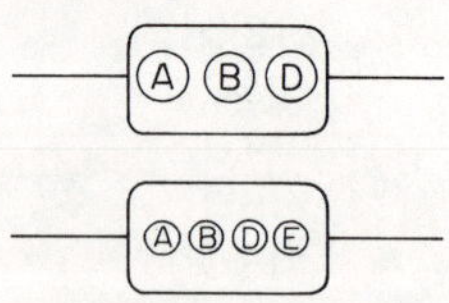

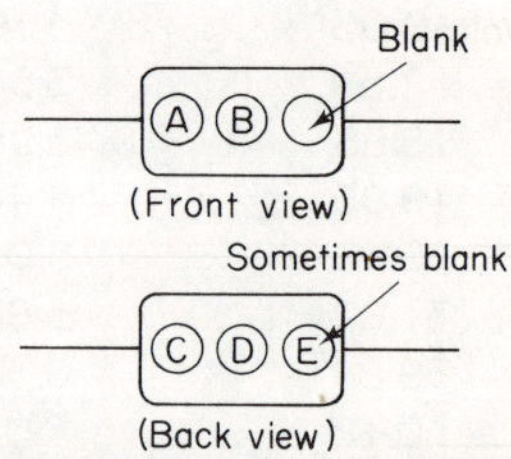

Dot color	Digits of capacitance (pF)			Multiplier	Tolerance %	Voltage rating (v d-c)
	A	B	C	D	E	F
Black	0	0	0	1	±20	—
Brown	1	1	1	10	± 1	100
Red	2	2	2	100	± 2	200
Orange	3	3	3	1,000	± 3	300
Yellow	4	4	4	10,000	± 4	400
Green	5	5	5	100,000	± 5	500
Blue	6	6	6	1,000,000	± 6	600
Violet	7	7	7	10,000,000	± 7	700
Gray	8	8	8	100,000,000	± 8	800
White	9	9	9	1000,000,000	± 9	900
Gold	—	—	—	0.1	± 5	1,000
Silver	—	—	—	0.01	±10	2,000
No color	—	—	—	—	±20	500

Ceramic Capacitor Color Code

EIA STANDARD RS-198

MILITARY STANDARD MIL-C-20D

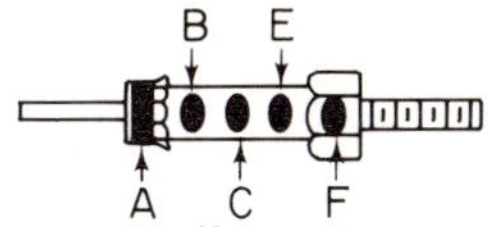

Stand-off capacitors
(EIA ONLY)

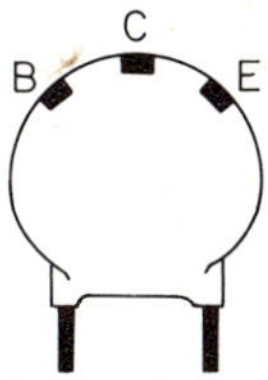

3-dot disc capacitors
(RETMA ONLY)
(Voltage rating is always 500 V.,
tolerance is always −0.)

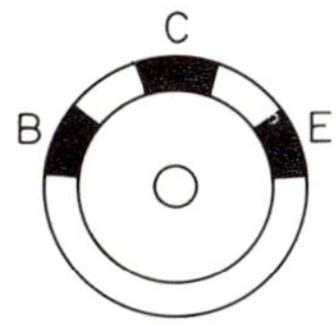

3-dot button capacitors
(EIA ONLY)

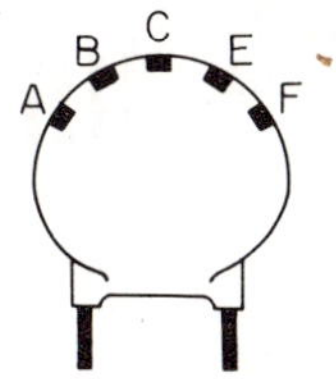

5-dot disc capacitors
(EIA ONLY)
(Voltage rating is always 500 V;)

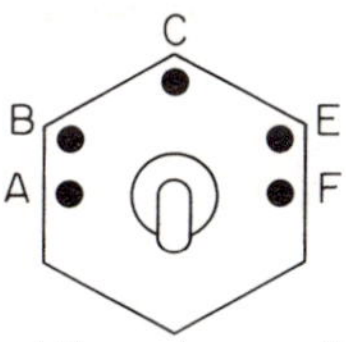

Feed through capacitors
(EIA ONLY)

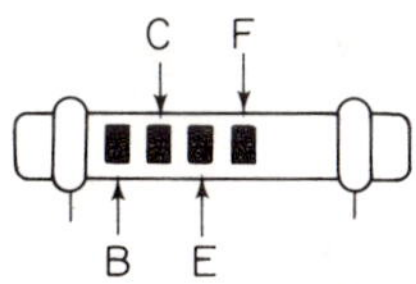

High capacity tubulars
(Insulated or non-insulated)

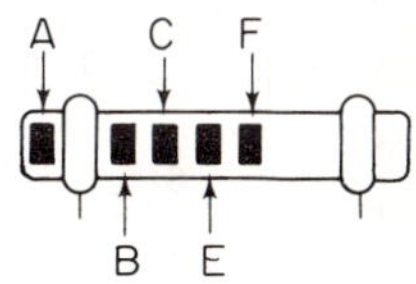

Temperature compensating
tubulars

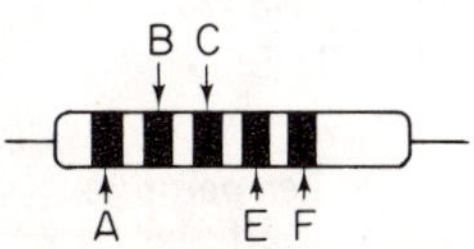

Tubular capacitors
(Voltage rating is always 500 V.)

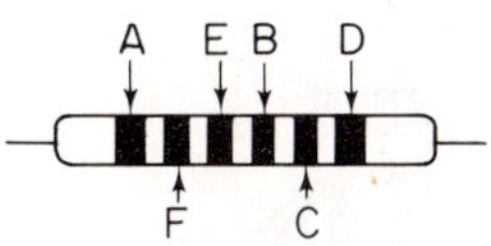

Tubular capacitors
(Old RMA)

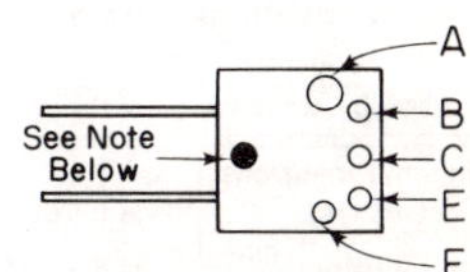

MIL style CC
rectangular

Note: Styles CC-60 through CC-71 will be color coded here with Green = 500 and Brown = 150, working volts dc.

Color	Digits of Capacitance (pF)			Multiplier E	Tolerance F		Temp. Coef. A (Parts per million per °C.)	
	B	C	D		10 pF or less (pF)	Over 10 pF (%)	EIA	MILITARY†
Black	0	0	0	1	± 2.0	± 20*	± 0	± 0
Brown	1	1	1	10	± 0.1*	± 1	− 33	− 30
Red	2	2	2	100	± 0.25	± 2	− 75	− 80
Orange	3	3	3	1,000	—	± 2.5*	−150	−150
Yellow	4	4	4	10,000*	—	—	−220	−220
Green	5	5	5	—	± 0.5	± 5	−330	−330
Blue	6	6	6	—	—	—	−470	−470
Violet	7	7	7	—	—	—	−750	−750
Gray	8	8	8	0.01	0.25*	—	+150 to −1500	—
White	9	9	9	0.1	1.0	± 10	+100 to −750	—
Gold	—	—	—	—	—	—	—	+100

*EIA only † Per charts in MIL-C-20D

Paper Capacitor Color Code

MILITARY STANDARD MIL-C-91A

(Commercial codes are same except as noted)

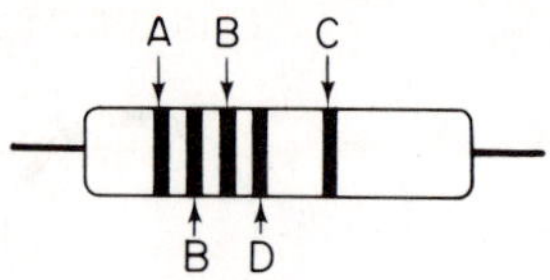

Tubular capacitors
(commercial only)

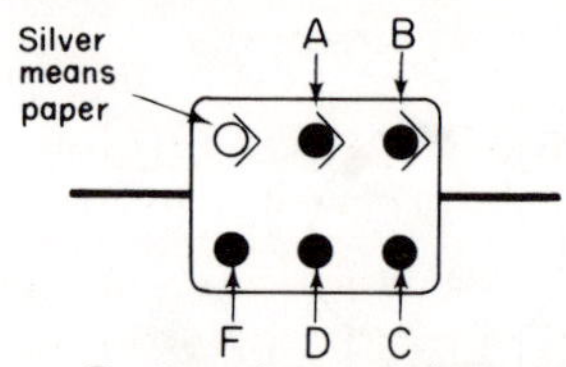

Rectangular capacitors

Color	Digits of capacitance (pF) A	Digits of capacitance (pF) B	Multiplier C	Tolerance % D	Tubular voltage rating (v dc) E	Temp. rating °C and characteristic F
Black	0	0	1	± 20	—	85-A
Brown	1	1	10	—	100	85-E
Red	2	2	100	—	200	—
Orange	3	3	1,000	± 30	300	—
Yellow	4	4	10,000	—	400	—
Green	5	5	—	—	500	—
Blue	6	6	—	—	600	—
Violet	7	7	—	—	700	—
Gray	8	8	—	—	800	—
White	9	9	—	—	900	—
Gold	—	—	—	—	1,000	—
Silver	—	—	—	± 10	—	—

Voltage rating for rectangular capacitors

(Indicated by dimensions rather than color coding)

Maximum dimensions (inches) Length	Width	Thick-ness	Style CN	Capacitance (pF)	Voltage rating (v dc)
$\frac{51}{64}$	$\frac{15}{32}$	$\frac{7}{32}$	20	1000 2000-6000 10,000	400 200 120
$\frac{57}{64}$	$\frac{37}{64}$	$\frac{17}{64}$	22	2000-3000 6000-10,000 20,000	400 300 120
$\frac{53}{64}$	$\frac{53}{64}$	$\frac{9}{32}$	30	1000-2000 3000 6000-10,000 20,000	800 600 400 120
$\frac{53}{64}$	$\frac{53}{64}$	$\frac{11}{32}$	35	3000 6000-10,000 20,000	800 600 300
$1\frac{1}{4}$	$\frac{41}{64}$	$\frac{9}{32}$	41	3000 - 6000 10,000 20,000 30,000	600 400 300 120
$1\frac{15}{32}$	$\frac{49}{64}$	$\frac{11}{32}$	42	1000-6000 10,000-20,000 30,000 50,000 100,000	1000 600 400 300 120
$1\frac{15}{32}$	$\frac{49}{64}$	$\frac{13}{32}$	43	10,000 20,000-30,000 50,000-100,000 200,000	1000 600 400 120

Typographically marked tubular ceramics

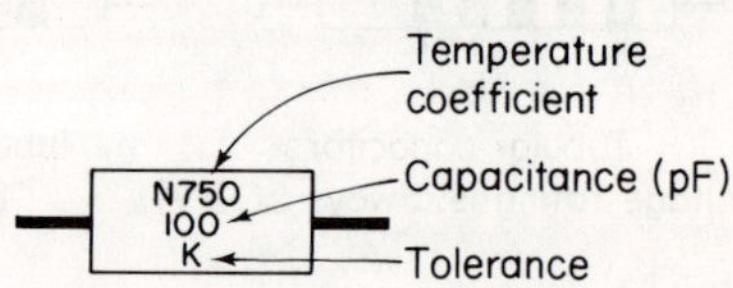

Jan letter	Tolerance 10 pF or less	Tolerance Over 10 pF
C	± 0.25 pF	....
D	± 0.5 pF	....
F	± 1.0 pF	± 1%
G	± 2.0 pF	± 2%
J	....	± 5%
K	....	± 10%
M	....	±20%

E

Frequency Allocations

Frequency	Classification	Abbreviation
3-30 kHz	Very low frequencies	VLF
30-300 kHz	Low frequencies	LF
300-3000 kHz	Medium frequencies	MF
3-30 MHz	High frequencies	HF
30-300 MHz	Very high frequencies	VHF
300-3000 MHz	Ultra high frequencies	UHF
3-30 GHz	Super high frequencies	SHF
30-300 GHz	Extremely high freq.	EHF
300-3000 GHz	---	---

Modified Modulation Index Table

Modulation Index	Number of Effective Side Bands		Band Width
	Above carrier	Below carrier	(f = Mod. Freq.)
0.1	1	1	2f
0.2	1	1	2f
0.3	1	1	2f
0.4	1	1	2f
0.5	2	2	4f
1	3	3	6f
2	4	4	8f
3	6	6	12f
4	7	7	14f
5	8	8	16f
6	9	9	18f
7	10	10	20f
8	12	12	24f
9	13	13	26f
10	14	14	28f
11	16	16	32f
12	17	17	34f
13	18	18	36f
14	19	19	38f
15	20	20	40f

F

Short- and Open-circuit Parameters

With input current, i_1 and output current, i_2, output voltage, V_2, and input voltage, V_1

Generalized:

"Y" closed-circuit and "Z" open-circuit parameters

$y_{11} = 1/h_{11}$, short-circuit input admittance
$\mu_{12} = h_{12}$, reverse open-circuit voltage gain
$a_{21} = h_{21}$, forward short-circuit current gain
$Z_{22} = 1/h_{22}$, open-circuit output impedance using the following:

Hybrid parameters

h_{11} = input resistance
h_{12} = voltage feedback ratio
h_{21} = forward current ratio
h_{22} = output admittance

Common-emitter hybrid parameters	Hybrid analogy
h_{ie} = input resistance	h_{11}
h_{re} = voltage feedback ratio	h_{12}
h_{fe} = forward current gain	h_{21}
h_{oe} = output admittance	h_{22}

Hybrid Conversion Formulas using Common-emitter Parameters

Network to IRE parameters

Hybrid	Admittance	Impedance	Conductance
$h_{11} = h_i$	$y_{11} = y_i$	$z_{11} = z_i$	$g_{11} = g_i$
$h_{21} = h_f$	$y_{21} = y_f$	$z_{21} = z_f$	$g_{21} = g_f$
$h_{12} = h_r$	$y_{12} = y_r$	$z_{12} = z_r$	$g_{12} = g_r$
$h_{22} = h_o$	$y_{22} = y_o$	$z_{22} = z_o$	$g_{22} = g_o$

$\alpha = h_{fb}$ $\qquad \beta = h_{fe}$

Common base

$$h_{ib} = \frac{h_{ie}}{1 + h_{fe}} \qquad h_{fb} = \frac{h_{fe}}{1 + h_{fe}}$$

$$h_{rb} = \frac{(h_{ie} \times h_{oe})}{1 + h_{fe}} \qquad h_{ob} = \frac{h_{oe}}{1 + h_{fe}}$$

Common collector

$$h_{ic} = h_{ie} \qquad h_{fc} = -(1 + h_{fe})$$

$$h_{rc} = 1 - h_{re} \qquad h_{oc} = h_{oe}$$

$$\alpha = \frac{h_{fe}}{1 + h_{fe}} \qquad r_e = \frac{h_{re}}{h_{oe}}$$

$$r_c = \frac{1 + h_{fe}}{h_{oe}} \qquad r_b = h_{ie} - \left[\frac{h_{re}}{h_{oe}} \times (1 + h_{fe})\right]$$

dc Operating Parameters

Symbol	Definition	Symbol	Definition
V_{EB1}, V_{CB1}, V_{CE}	Terminal voltages—emitter to base, collector to base, collector to emitter	BV_{CBO1}, BV_{EBO}	Respective breakdown voltages from collector and emitter to base at specified current
I_{B1}, I_{E1}, I_C	dc currents in base, emitter, and collector	BV_{CEO}, BV_{CER}, BV_{CES}, BV_{CEX}, V_{RT}	Breakdown voltage from collector to emitter under the following conditions: BV_{CEO} = with base open-circuited; BV_{CER} = with base connected to emitter through a resistor; BV_{CES} = with base shorted to emitter; BV_{CEX} = with base reverse biased; V_{RT} = "Reach-through" or "punch-through" voltage
$I_{CO} = I_{CBO}$	Collector leakage current (reverse voltage with emitter open)—positive for NPN, negative for PNP types		
$I_{EO} = I_{EBO}$	Emitter leakage current (reverse voltage with collector open)-positive for NPN, negative for PNP types		
R_{B1}, R_{E1}, R_C	Ohmic resistance internal to transistor in series respectively with base, emitter, and collector		
α	Normal alpha, small-signal common-base forward-current transfer ratio from emitter to collector with output ac short-circuited	dc beta (β)	Collector current gain with emitter grounded
		(H_{FE})	Forward current transfer

INDEX